Innovative Energy Strategies for CO$_2$ Stabilization

The vast majority of the world's climate scientists believe that the build-up in the atmosphere of the heat-trapping gas carbon dioxide will lead to global warming in the next century unless we burn less coal, oil and natural gas. At the same time, it is clear that energy must be supplied in increasing amounts if the developed world is to avoid economic collapse and if developing countries are to attain wealth. *Innovative Energy Strategies for CO$_2$ Stabilization* discusses the feasibility of increasingly efficient energy use for limiting energy requirements as well as the potential for supplying energy from sources that do not introduce carbon dioxide into the atmosphere.

The book begins with a discussion of concerns about global warming and the relationship between the growing need to supply energy to the globe's population and the importance of adaptive decision making strategies for future policy decisions. The book goes on to analyze the prospects for Earth-based renewables: solar, wind, biomass, hydroelectricity, geothermal and ocean energy. The problems of transmission and storage that are related to many renewable energy options are discussed. The option of energy from nuclear fission is considered in light of its total possible contribution to world energy needs and also of the four cardinal issues on its acceptance by the public: safety, waste disposal, proliferation of nuclear weapons, and cost. A separate chapter reviews the potential of fusion reactors for providing a nearly limitless energy supply. The relatively new idea of harvesting solar energy on satellites or lunar bases and beaming it to Earth using microwaves is then explored in detail. Finally, the possibility of geo-engineering is discussed.

Innovative Energy Strategies for CO$_2$ Stabilization will be essential reading for all those interested in the development of "clean" energy technologies, including engineers and physicists of all kinds (electrical, mechanical, chemical, industrial, environmental, nuclear), and industrial leaders and politicians dealing with the energy issue. It will also be used as a supplementary textbook on advanced courses on energy.

Robert G. Watts is a Professor of Mechanical Engineering at Tulane University in Louisiana. His current research interests are in climate modeling, the socio-economic and political aspects of energy policy, and the physics of sea ice. His publications on these and other topics have appeared in *Climate Change, Journal of Geophysical Research* and *Nature* as well as the mechanical engineering literature. Professor Watts is the author of *Keep Your Eye on the Ball: Curveballs, Knuckleballs, and Fallacies of Baseball* (with A. Terry Bahill; W. H. Freeman publishers, 1991, 2000) and is editor of *Engineering Response to Global Climate Change* (Lewis Publishers, 1997). He is a member of the American Society of Mechanical Engineers, and has been an ASME Distinguished Lecturer. Recently, he gave the prestigious George Hawkins Memorial Lecture at Purdue University.

T0183004

INNOVATIVE ENERGY STRATEGIES FOR CO₂ STABILIZATION

edited by

ROBERT G. WATTS

Department of Mechanical Engineering Tulane University

CAMBRIDGE
UNIVERSITY PRESS

CAMBRIDGE UNIVERSITY PRESS
Cambridge, New York, Melbourne, Madrid, Cape Town, Singapore, São Paulo, Delhi

Cambridge University Press
The Edinburgh Building, Cambridge CB2 8RU, UK

Published in the United States of America by Cambridge University Press, New York

www.cambridge.org
Information on this title: www.cambridge.org/9780521807258

First published 2002
This digitally printed version 2008

A catalogue record for this publication is available from the British Library

Library of Congress Cataloguing in Publication data

Innovative energy strategies for CO_2 stabilization / edited by Robert G. Watts.
p.   cm.
Includes bibliographical references and index.
ISBN 0 521 80725 5
1. Renewable energy sources. 2. Power resources–Environmental aspects.
3. Atmospheric carbon dioxide–Environmental aspects. I. Watts, Robert G.
TJ808 .I56 2002
333.79′14–dc21   2001037357

ISBN 978-0-521-80725-8 hardback
ISBN 978-0-521-08782-7 paperback

Acknowledgments

This book is an outgrowth of a workshop that was held at the Aspen Global Change Institute during the summer of 1998. Some of the participants in the workshop did not contribute directly to the authorships of the chapters in this book. Yet their contributions to the central ideas of the book were of considerable importance, and I gratefully acknowledge their participation in the lively discussions at the workshop. Their input is reflected in the chapters that appear in this book. Summaries of their presentations at the workshop appear in "Elements of Change" edited by Susan Joy Hassol and John Katzenberger and published by the Aspen Global Change Institute. In particular, two of the conveners whose innovative ideas were directly responsible for the creation of the workshop deserve special thanks. The leadership and creativity of Drs Martin I. Hoffert of New York University and Ken Caldeira of Lawrence Livermore Laboratory were instrumental in the development of the ideas expressed in this book. I am greatly indebted to them.

John Katzenberger and the staff of the Aspen Global Change Institute deserve thanks for hosting this and many other workshops that deal with varied aspects of global change.

I, along with all of the authors, am deeply indebted to Frances Nex, our copy editor, whose skill in finding glitches both large and small have much improved the book.

Finally, I thank Dr Matt Lloyd of Cambridge University Press, my editor, for his patience and persistence during the evolution of this book.

Robert G. Watts
Tulane University
New Orleans
September, 2000

Contents

Contributors

Gene D. Berry	Energy Analysis, Policy and Planning, Lawrence Livermore National Laboratory
David R. Criswell	Institute of Space Systems Operations, University of Houston
Hadi Dowlatabadi	Center for Integrated Study of the Human Dimensions of Climate Change, Carnegie Mellon University
Susan J. Hassol	Aspen Global Change Institute
Atul K. Jain	Department of Atmospheric Sciences, University of Illinois
Patrick Keegan	National Renewable Energy Laboratory
David W. Keith	Department of Engineering and Public Policy, Carnegie Mellon University
Robert Krakowski	Systems Engineering & Integration Group, Los Alamos National Laboratory
Alan D. Lamont	Energy Analysis, Policy and Planning, Lawrence Livermore National Laboratory
Robert J. Lempert	RAND
Arthur W. Molvik	Fusion Energy Division, Lawrence Livermore National Laboratory
John L. Perkins	Fusion Energy Division, Lawrence Livermore National Laboratory
Michael E. Schlesinger	Department of Atmospheric Sciences, University of Illinois
Walter Short	National Renewable Energy Laboratory
Neil D. Strachan	Center for Integrated Study of the Human Dimensions of Climate Change, Carnegie Mellon University

Robert G. Watts   Department of Mechanical Engineering, Tulane
                      University
Richard Wilson   Department of Physics, Harvard University
Donald J. Wuebbles   Department of Atmospheric Sciences, University of
                      Illinois

1

Concerns about Climate Change and Global Warming

1.1 Introduction

Climate is defined as the typical behavior of the atmosphere, the aggregation of the weather, and is generally expressed in terms of averages and variances of temperature, precipitation and other physical properties. The greenhouse effect, the ability of certain gases like carbon dioxide and water vapor to effectively trap some of the reemission of solar energy by the planet, is a necessary component to life on Earth; without the greenhouse effect the planet would be too cold to support life. However, human activities are increasing the concentration of carbon dioxide and several other greenhouse gases, resulting in concerns about warming of the Earth by 1–5 °C over the next century (IPCC, 1996a). Recent increases in global averaged temperature over the last decade already appear to be outside the normal variability of temperature changes for the last thousand years. A number of different analyses strongly suggest that this temperature increase is resulting from the increasing atmospheric concentrations of greenhouse gases, thus lending credence to the concerns about much larger changes in climate being predicted for the coming decades. It is this evidence that led the international scientific community through the Intergovernmental Panel on Climate Change (IPCC, 1996a) to conclude, after a discussion of remaining uncertainties, "Nonetheless, the balance of the evidence suggests a human influence on global climate". More recent findings have further strengthened this conclusion. Computer-based models of the complex processes affecting the carbon cycle have implicated the burning of fossil fuels by an ever-increasing world population as a major factor in the past increase in concentrations of carbon dioxide. These models also suggest that, without major policy or technology changes, future concentrations of CO_2 will continue to increase, largely as a result of fossil fuel burning. This chapter briefly reviews the state of the science of the concerns

1

Table 1.1 *Summary of trends in observed climatic variables (WMO, 1998; Harvey, 2000). Note that NH implies Northern Hemisphere and SH implies Southern Hemisphere*

Variable	Analysis Period	Trend or Change
Surface air temperature and sea surface temperature (SST)	1851–1995	$0.65 \pm 0.15\,°C$
Alpine glaciers	Last century	Implies warming of 0.6–1.0 °C in alpine regions
Extent of snowcover in the NH	1972–1992	10% decrease in annual mean
Extent of sea ice in the NH	1973–1994	Downward since 1977
Extent of sea ice in the SH	1973–1994	No change, possible decrease between mid 1950s and early 1970s
Length of the NH growing season	1981–1991	12 ± 4 days longer
Precipitation	1900–1994	Generally increasing outside tropics, decreasing in Sahel
Heavy precipitation	1910–1990	Growing in importance
Antarctic snowfall	Recent decades	5–20% increase
Global mean sea level	Last century	1.8 ± 0.7 mm/year

about climate change that could result from fossil fuels and other human related emissions.

1.2 The Changing Climate

There is an extensive amount of evidence indicating that the Earth's climate has warmed during the past century (see Table 1.1). Foremost among this evidence are compilations of the variation in global mean sea surface temperature and in surface air temperature over land and sea. Supplementing these indicators of surface temperature change is a global network of balloon-based measurements of atmospheric temperature since 1958. As well, there are several indirect or *proxy* indications of temperature change, including satellite observations (since 1979) of microwave emissions from the atmosphere, and records of the width and density of tree rings. The combination of surface-, balloon-, and satellite-based indicators provides a more complete picture than could be obtained from any given indicator alone, while proxy records from tree rings and other indicators allow the temperature record at selected locations to be extended back for a thousand years. Apart from temperature, changes in the extent of alpine glaciers, sea ice, seasonal snow cover, and the length of the growing season have been documented that are consistent with the evidence that the climate is warming (e.g., IPCC, 1996a; Vaughn and Doake, 1996; Johannessen *et al.*, 1999). Less certain, but also consistent, changes appear to have occurred in precipitation, cloudiness, and interannual temperature and rainfall variability.

As a starting point, paleoclimatic records of past climate changes should be a useful guide as to what one might expect if the climate is warming. During warmer climates in the past, high latitudes have warmed more than lower latitudes (Hoffert and Covey, 1992). Mountain glaciers should retreat. Sea level should rise. The current climate change is showing all of these features (Haeberli, 1990; Diaz and Graham, 1996).

Thermometer-based measurements of air temperature have been systematically recorded at a number of sites in Europe and North America as far back as 1760. However, the set of observing sites did not attain sufficient geographic coverage to permit a rough computation of the global average land temperature until the mid-nineteenth century. Land-based, marine air, and sea surface temperature datasets all require rather involved corrections to account for changing conditions and measurement techniques. Analyses of these records indicates a global mean warming from 1851 to 1995 of about $0.65 \pm 0.05\,°C$ (Jones *et al.*, 1997a, b).

As shown in Figure 1.1, the increase in temperature has occurred in two distinct periods. The first occurred from roughly 1910–1945, while the second is since 1976. Recent warming has been about $0.2\,°C$ per decade. Very large changes have occurred in the last decade, with 1998 being the warmest year in the global temperature record. The highest ten years in global surface temperature have been since 1980, with eight of them occurring in the last eleven years.

In addition to limited sampling of temperature with altitude through balloon-borne instruments, satellite-based sensors, known as microwave sounding units (MSUs), are being used to examine global temperature changes in the middle troposphere (mainly the 850–300 HPa layer), and in the lower stratosphere (~50–100 Hpa). None of the channels sample at the ground. The MSU measurements have been controversial because some earlier versions of the satellite dataset had indicated a cooling in the lower troposphere in contrast to the warming from the ground-based instruments. However, several errors and problems (e.g., due to decay in the orbit of the satellite) with the MSU data have been found, and the latest analyses of MSU corrected for these problems show a warming (about $0.1\,°C$ per decade), albeit somewhat smaller than that found at the ground (NRC, 2000). These analyses also suggest that the cooling effect of decreasing ozone in the lower stratosphere (as a result of chlorine and

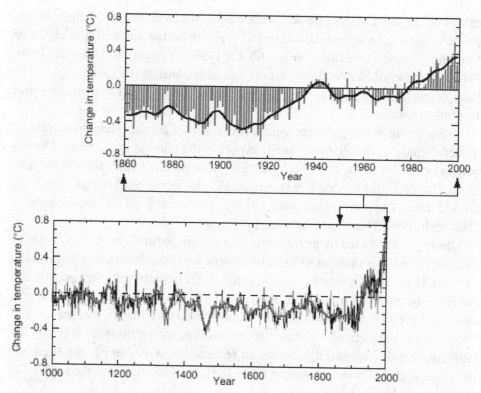

Figure 1.1   Variations of the Earth's surface temperature for the last 1000 years. The top panel shows the combined annual land-surface and sea-surface temperature anomalies for 1861 to 1999, relative to the average of the 1961 to 1990 period. This figure is an update by P. D. Jones of the analysis previously done for IPCC (1996a). The bottom panel shows the Northern Hemispheric temperature reconstruction over the last 1000 years from proxy data in combination with the instrumental data record (Mann *et al.*, 1999).

bromine from human-related emissions of chlorofluorocarbons and other halocarbons) may have led to the difference in upper tropospheric and ground-level temperature trends.

The 1910–1945 warming primarily occurred in the Northern Atlantic. In contrast, the most recent warming has primarily occurred at middle and high latitudes of the Northern Hemisphere continents in winter and spring, while the northwest portion of the Northern Atlantic and the central North Pacific Oceans have shown year-around cooling. Significant regional cooling occurred in the Northern Hemisphere during the period from 1946 to 1975.

Proxy temperature indicators, such as tree ring width and density, the chemical composition and annual growth rate in corals, and characteristics of annual layers in ice cores, are being used at a number of locations to extend temperature records back as much as a thousand years (Jones *et al.*, 1998; Mann *et al.*, 1999; Bradley, 2000). As seen in Figure 1.1, the reconstruction indicates the decade of the 1990s has been warmer than any time during this millennium and that 1998 was the warmest year in the 1000-year record (Mann *et al.*, 1999). Using a different approach, based on underground temperature measurements from boreholes, Huang *et al.* (2000) found temperature changes over the last 500 years that are very similar to the trend in Mann *et al.* (1999). The basic conclusion is the same, that the late-twentieth century warming is unprecedented in the last 500 to 1000 years.

Recent studies (for example, Boer *et al.*, 2000; Delworth and Knutson, 2000; Wigley, 1999) with state-of-the-art numerical models of the climate system have been able to match the observed temperature record well, but only if they include the effects of greenhouse gases and aerosols. These studies indicate that natural variability of the climate system and solar variations are not sufficient to explain the increasing temperatures in the 1980s and 1990s. However, natural variability and variations in the solar flux are important in fully explaining the increase in temperature in the 1910–1945 period. Emissions from large volcanic eruptions resulting in sulfate aerosols and other aerosols in the lower stratosphere are also important in explaining some of the short-term variations in the climate record.

Levitus *et al.* (2000) have used more than five million measurements of the temperature of the world ocean at various depths and locations to show that the temperatures have increased in the middle depths by an average of about 0.06 °C between the 1950s and the mid-1990s. Watts and Morantine (1991) had previously suggested, based on data of mid-depth Atlantic Ocean temperature changes reported by Roemmich and Wunsch (1984), that much of the global warming temperature signature lay in the deep ocean.

Any changes in climate associated with increasing levels of carbon dioxide would also be expected to result in cooling stratospheric temperatures. The stratosphere indeed is cooling (Angell, 1999). While part of the cooling in the lower stratosphere can be explained by the observed decrease in stratospheric ozone, such changes in ozone can only explain part of the observed temperature change. The increase in CO_2 is also necessary to explain the changes in lower stratospheric temperatures (Miller *et al.*, 1992).

1.3 The Changing Atmospheric Composition

Without human intervention, concentrations of many atmospheric gases would be expected to change slowly. Ice core measurements of the gases trapped in ancient ice bubbles indicate this was the case before the last century.

However, since the beginning of the industrial age, emissions associated with human activities have risen rapidly. Agriculture, industry, waste disposal, deforestation, and especially fossil fuel use have been producing increasing amounts of carbon dioxide (CO_2), methane (CH_4), nitrous oxide (N_2O), chlorofluorocarbons (CFCs) and other important gases. Due to increasing emissions, atmospheric levels of these greenhouse gases have been building at an unprecedented rate, raising concerns regarding the impact of these gases on climate. Some of the gases, such as CFCs, are also responsible for large observed depletions in the natural levels of another gas important to climate, ozone. Of these gases, two, carbon dioxide and methane, are of special concern to climate change. These two gases are discussed in some detail in the sections below. Under the international Montreal Protocol and its amendments, emissions of CFCs and other halocarbons are being controlled and their atmospheric concentrations will gradually decline over the next century. Emissions leading to atmospheric concentrations of sulfate and other aerosol particles are also important to climate change and are further discussed below. Unless stated otherwise, most of the discussion below is based on the most recent IPCC and WMO international assessments (IPCC, 1996a; WMO, 1998) of global change, with concentrations and trends updated as much as possible, such as data available from NOAA CMDL (National Oceanic and Atmospheric Administration's Climate Monitoring and Diagnostics Laboratory).

1.3.1 Carbon dioxide

Carbon dioxide has the largest changing concentration of the greenhouse gases. It is also the gas of most concern to analyses of potential human effects on climate. Accurate measurements of atmospheric CO_2 concentration began in 1958. The annually averaged concentration of CO_2 in the atmosphere has risen from 316 ppm (parts per million, molar) in 1959 to 364 ppm in 1997 (Keeling and Whorf, 1998), as shown in Figure 1.2. The CO_2 measurements exhibit a seasonal cycle, which is mainly caused by the seasonal uptake and release of atmospheric CO_2 by terrestrial ecosystems. The average annual rate of increase over the whole time period is about 1.2 ppm or 0.4% per year, with the rate of increase over the last decade being about 1.6 ppm/yr. Measurements of CO_2 concentration in air trapped in ice cores indicate that the pre-industrial concentration of CO_2 was approximately 280 ppm. This data indicates that carbon dioxide concentrations fluctuated by ± 10 ppm around 280 ppm for over a thousand years until the recent increase to the current 360+ ppm, an increase of over 30%.

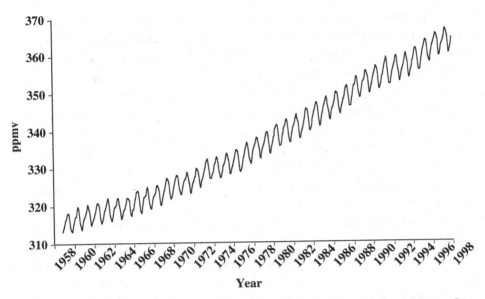

Figure 1.2   Observed monthly average CO_2 concentration (ppmv) from Mauna Loa, Hawaii (Keeling and Whorf, 1998). Seasonal variations are primarily due to the uptake and production of CO_2 by the terrestrial biosphere.

Why has the atmospheric concentration of CO_2 increased so dramatically? Analyses with models of the atmosphere–ocean–biosphere system of the carbon cycle, in coordination with observational analyses of the isotopes of carbon in CO_2, indicate that human activities are primarily responsible for the increase in CO_2. Two types of human activities are primarily responsible for emissions of CO_2: fossil fuel use, which released about 6.0 GtC into the atmosphere in 1990, and land use, including deforestation and biomass burning, which may have contributed about an additional 1.6 ± 1.0 GtC. Evaluations of carbon releases from vegetation and soils based on changes in land use indicate that land use decreased carbon storage in vegetation and soil by about 170 Gt since 1800. The added atmospheric CO_2 resulting from human activities is redistributed within the atmospheric, oceanic, and terrestrial biospheric parts of the global carbon cycle, with the dynamics of this redistribution determining the corresponding rise in atmospheric CO_2 concentration. In the future, as the amount of CO_2 increases in the atmosphere and in the ocean, it is expected that the oceans will take up a smaller percentage of the new emissions. Analyses of the carbon budget previously had implied that a mismatch existed between observed levels of CO_2 and known loss processes. This discrepancy suggested that a missing carbon sink has existed during recent decades. This sink now appears to be largely explained through increased net carbon storage by the terrestrial biomass stimulated by the CO_2 fertilization effect (increased growth in a higher CO_2 concentration atmosphere) (Kheshgi *et al.*, 1996).

Carbon dioxide is emitted when carbon-containing fossil fuels are oxidized by combustion. Carbon dioxide emissions depend on energy and carbon content, which ranges from 13.6 to 14.0 MtC/EJ for natural gas, 19.0 to 20.3 for oil, and 23.9 to 24.5 for coal. Other energy sources such as hydro, nuclear, wind, and solar have no direct carbon emissions. Biomass energy, however, is a special case. When biomass is used as a fuel, it releases carbon with a carbon-to-energy ratio similar to that of coal. However, the biomass has already absorbed an equal amount of carbon from the atmosphere prior to its emission, so that net emissions of carbon from biomass fuels are zero over its life cycle.

Human-related emissions from fossil fuel use have been estimated as far back as 1751. Before 1863, emissions did not exceed 0.1 GtC/yr. However, by 1995 they had reached 6.5 GtC/yr, giving an average emission growth rate slightly greater than 3 percent per year over the last two and a half centuries. Recent growth rates have been significantly lower, at 1.8 percent per year between 1970 and 1995. Emissions were initially dominated by coal. Since 1985, liquids have been the main source of emissions despite their lower carbon intensity. The regional pattern of emissions has also changed. Once dominated by Europe and North America, developing nations are providing an increasing share of emissions. In 1995, non-Annex I (developing countries; includes China and India) nations accounted for 48 percent of global emissions.

Future CO_2 levels in the atmosphere depend not only on the assumed emission scenarios, but also on the transfer processes between the major carbon reservoirs, such as the oceans (with marine biota and sediments) and the terrestrial ecosystems (with land use changes, soil and forest destruction). Recent work for the new IPCC assessment shows, based on projections of fossil-fuel use and land use changes, that the concentration of CO_2 is expected to increase well above current levels by 2100 (75 to 220% over pre-industrial concentrations). As discussed later, none of these scenarios lead to stabilization of the CO_2 concentration before 2100.

1.3.2 Methane

Although its atmospheric abundance is less than 0.5 percent that of CO_2, on a molecule by molecule basis, a molecule of CH_4 is approximately 50 times more effective as a greenhouse gas in the current atmosphere than CO_2. When this is combined with the large increase in its atmospheric concentration, methane becomes the second most important greenhouse gas of concern to climate change. Based on analyses of ice cores, the concentration of methane has more than doubled since pre-industrial times. In the year 1997, the globally averaged atmospheric concentration of methane was about 1.73 ppmv (Dlugokencky *et al.*, 1998).

Continuous monitoring of methane trends in ambient air from 1979 to 1989 indicates that concentrations had been increasing at an average of about 16 ppb (~1 percent per year). During much of the 1990s, the rate of increase in methane appeared to be declining. Although the cause of the longer-term global decline in methane growth is still not well understood, it may be that much of the earlier rapid increase in methane emissions from agricultural sources is now slowing down. However, in 1998 the CH_4 growth rate increased to about 10 ppb per year (Figure 1.3b). There are some indications that this increase in the growth rate may be due to a response of emissions from wetlands in the Northern Hemisphere responding to warm temperatures. In 1999, the growth rate decreased to about 5 ppb per year (Dlugokencky, NOAA CMDL, private communication, 2000).

Methane emissions come from a number of different sources, both natural and anthropogenic. One type of human related emission arises from biogenic sources from agriculture and waste disposal, including enteric fermentation, animal and human wastes, rice paddies, biomass burning, and landfills. Emissions also result from fossil fuel-related methane sources such as natural gas loss, coal mining, and the petroleum industry. Methane is emitted naturally by wetlands, termites, other wild ruminants, oceans, and by hydrates. Based on recent estimates, current human-related biogenic and fossil fuel-related sources of methane are approximately 275 and 100 $TgCH_4$/yr while total natural sources are around 160 $TgCH_4$/yr.

1.3.3 Sulfuric and other aerosols

Emissions of sulfur dioxide and other gases can result in the formation of aerosols that can affect climate. Aerosols affect climate directly by absorption and scattering of solar radiation and indirectly by acting as cloud condensation nuclei (CCN). A variety of analyses indicate that human-related emissions of sulfur, and the resulting increased sulfuric acid concentrations in the troposphere, may be cooling the Northern Hemisphere sufficiently to compensate for a sizable fraction of the warming expected from greenhouse gases. As the lifetime in the lower atmosphere of these aerosols is typically only about one week, the large continual emissions of the aerosol precursors largely determines the impact of the aerosols on climate. Large volcanic explosions can influence climate for periods of one to three years through emissions of sulfur dioxide, and the resulting sulfate aerosols, into the lower stratosphere.

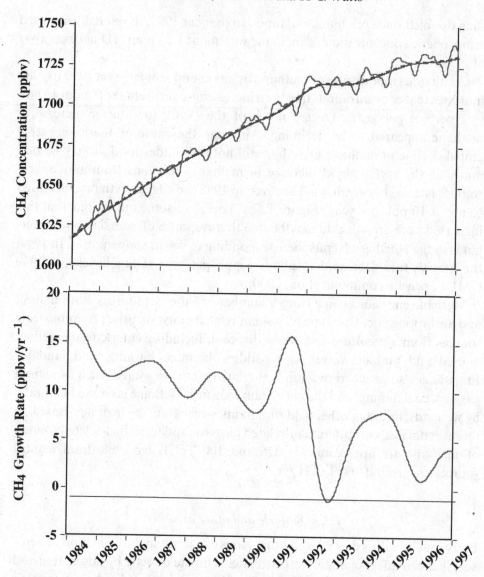

Figure 1.3   Globally averaged atmospheric CH_4 concentrations (ppbv) derived from NOAA Climate Monitoring Diagnostic Laboratory air sampling sites (Dlugokencky *et al.*, 1998). The solid line is a deseasonalized trend curve fitted to the data. The dashed line is a model (that accounts for CH_4 emissions and loss in the atmosphere) estimated calculated trend that is fitted to the globally average values. (b) Atmospheric CH_4 instantaneous growth rate (ppbv/year) which is the derivative with respect to the trend curve shown in (a).

Over half of the sulfur dioxide, SO_2, emitted into the atmosphere comes from human-related sources, mainly from the combustion of coal and other fossil fuels. Most of these emissions occur in the Northern Hemisphere. Analyses indicate that anthropogenic emissions have grown dramatically during this century. Other SO_2 sources come from biomass burning, from volcanic eruptions, and from the oxidation of di-methyl sulfide (DMS) and hydrogen sulfide (H_2S) in the atmosphere. DMS and H_2S are primarily produced in the oceans. Atmospheric SO_2 has a lifetime of less than a week, leading to formation of sulfuric acid and eventually to sulfate aerosol particles. Gas-to-particle conversion can also occur in cloud droplets; when precipitation doesn't soon occur, the evaporation of such droplets can then leave sulfate aerosols in the atmosphere.

Other aerosols are also important to climate. Of particular interest are the carbonaceous aerosols or black carbon (soot) aerosols that are absorbers of solar and infrared radiation, and can thus add to the concerns about warming.

1.4 Radiative Forcing and Climate Change

A perturbation to the atmospheric concentration of an important greenhouse gas, or the distribution of aerosols, induces a radiative forcing that can affect climate. Radiative forcing of the surface–troposphere system is defined as the change in net radiative flux at the tropopause due to a change in either solar or infrared radiation (IPCC, 1996a). Generally, this net flux is calculated after allowing for stratospheric temperatures to re-adjust to radiative equilibrium. A positive radiative forcing tends on average to warm the surface; a negative radiative forcing tends to cool the surface. This definition is based on earlier climate modeling studies, which indicated an approximately linear relationship between the global mean radiative forcing at the tropopause and the resulting global mean surface temperature change at equilibrium. However, recent studies of greenhouse gases (e.g., Hansen *et al.*, 1997) indicate that the climatic response can be sensitive to the altitude, latitude, and nature of the forcing.

The resulting change in radiative forcing can then drive changes in the climate. A positive radiative forcing tends on average to warm the Earth's surface; a negative radiative forcing tends to cool the surface. Changes in radiative forcing can occur either as a result of natural phenomena or due to human activities. Natural causes for significant changes in radiative forcing include those due to changes in solar luminosity or due to concentrations of sulfate aerosols following a major volcanic eruption. Human related causes include the changes in atmospheric concentrations of greenhouse gases and in aerosol loading discussed earlier.

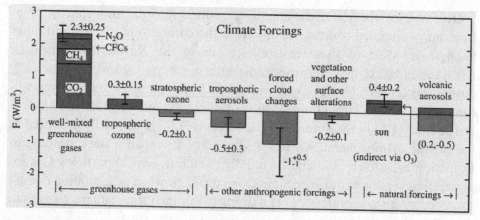

Figure 1.4   Estimated change in globally and annually averaged anthropogenic radiative forcing (Wm^{-2}) resulting from changes in concentrations of greenhouse gases and aerosols from the pre-industrial time to 1998 and to natural changes in solar output from 1850 to 1998. The error bars show an estimate of the uncertainty range. Based on Hansen *et al.* (1998).

1.4.1 Explaining the past record

As shown in Figure 1.4, analyses of the direct radiative forcing due to the changes in greenhouse gas concentrations since the beginning of the Industrial Era (roughly about 1800) give an increase of about 2.3 Wm^{-2} (Hansen *et al.*, 1998). To put this into perspective, a doubling of CO_2 from pre-industrial levels would correspond to about 4 Wm^{-2}; climate models studies indicate this would give 1.5 to 4.5°C increase in global temperature. Approximately 0.5 Wm^{-2} of the increase has occurred within the last decade. By far the largest effect on radiative forcing has been the increasing concentration of carbon dioxide, accounting for about 64 percent of the total change in forcing. Methane has produced the second largest change in radiative forcing of the greenhouse gases.

Changes in the amounts of sulfate, nitrate, and carbonaceous aerosols induced by natural and human activities have all contributed to changes in radiative forcing over the last century. The direct effect on climate from sulfate aerosols occurs primarily through the scattering of solar radiation. This scattering produces a negative radiative forcing, and has resulted in a cooling tendency on the Earth's surface that counteracts some of the warming effect from the greenhouse gases. In the global average, increases in amounts of carbonaceous aerosols, which absorb solar and infrared radiation, have likely counteracted some of the effect of the sulfate aerosols. Aerosols can also produce an indirect radiative forcing by acting as condensation nuclei for cloud formation.

There is large uncertainty in determining the extent of radiative forcing that has resulted from this indirect effect, as indicated in Figure 1.4.

Changes in tropospheric and stratospheric ozone also affect climate, but the increase in tropospheric ozone over the last century and the decrease in stratospheric ozone over recent decades have had a relatively small combined effect on radiative forcing compared to CO_2. The radiative forcing from the changes in amount of stratospheric ozone, which has primarily occurred over the last few decades, mostly as a result of human-related emissions of halogenated compounds containing chlorine and bromine, is generally well understood. However, the changes in concentration of tropospheric ozone over the last century, and the resulting radiative forcing, are much less well understood.

Change in the solar energy output reaching the Earth is also an important external forcing on the climate system. The Sun's output of energy is known to vary by small amounts over the 11-year cycle associated with sunspots, and there are indications that the solar output may vary by larger amounts over longer time periods. Slow variations in the Earth's orbit, over time scales of multiple decades to thousands of years, have varied the solar radiation reaching the Earth, and have affected the past climate. As shown in Figure 1.4, solar variations over the last century are thought to have had a small but important effect on the climate, but are not important in explaining the large increase in temperatures over the last few decades.

Evaluation of the radiative forcing from all of the different sources since pre-industrial times indicates that globally-averaged radiative forcing on climate has increased. Because of the hemispheric and other inhomogeneous variations in concentrations of aerosols, the overall change in radiative forcing is much greater or much smaller at specific locations over the globe.

Any changes induced in climate as a result of human activities, or from natural forcings like variations in the solar flux, will be superimposed on a background of natural climatic variations that occur on a broad range of temporal and spatial scales. Analyses to detect the possible influence of human activities have had to take such natural variations into consideration. As mentioned earlier, however, recent studies suggest that the warmer global temperatures over the last decade appear to be outside the range of natural variability found in the climate record for the last four hundred to one thousand years.

1.4.2 Projecting the future changes

In order to study the potential implications on climate from further changes in human-related emissions and atmospheric composition, a range of scenarios for future emissions of greenhouse gases and aerosol precursors has been produced by the IPCC Special Report on Emission Scenarios (SRES), for use in modeling studies to assess potential changes in climate over the next century for the current IPCC international assessment of climate change. None of these scenarios should be considered as a prediction of the future, but they do illustrate the effects of various assumptions about economics, demography, and policy on future emissions. In this study, four SRES "marker" scenarios, labeled A1, A2, B1, and B2, are investigated as examples of the possible effect of greenhouse gases on climate. Each scenario is based on a narrative storyline, describing alternative future developments in economics, technical, environmental and social dimensions. Details of these storylines and the SRES process can be found elsewhere (Nakicenovic *et al.*, 1998; Wigley, 1999). These scenarios are generally thought to represent the possible range for a business-as-usual situation where there have been no significant efforts to reduce emissions to slow down or prevent climate changes.

Figure 1.5 shows the anthropogenic emissions for four of the most important gases to concerns about climate change, CO_2, N_2O, CH_4, and SO_2. Carbon dioxide emissions span a wide range, from nearly five times the 1990 value by 2100 to emissions that rise and then fall to near their 1990 value. N_2O and CH_4 emission scenarios reflect these variations and have similar trends. However, global sulfur dioxide emissions in 2100 have declined to below their 1990 levels in all scenarios, because rising affluence increases the demand for emissions reductions. Note that sulfur emissions, particularly to mid-century, differ fairly substantially between the scenarios. Also, the new scenarios for sulfur emissions are much smaller than earlier analyses (e.g., IPCC, 1996a), largely as a result of increased recognition worldwide of the importance of reducing sulfate aerosol effects on human health, on agriculture, and on the biosphere.

In this study the global climate change consequences of SRES scenarios were calculated with the reduced form version of our Integrated Science Assessment Model (ISAM) (Jain *et al.*, 1996; Kheshgi *et al.*, 1999). The model consists of several gas cycle sub-models converting emissions of major greenhouse gases to concentrations, an energy balance climate model for the atmosphere and ocean, and a sea level rise model. In this study, updated radiative forcing analyses (Jain *et al.*, 2000) for various greenhouse gases have been used. Based on results from the carbon cycle submodel within ISAM, Figure 1.6 shows the derived changes in concentrations of carbon dioxide for the four SRES scenarios. Over the next century, CO_2 concentrations continue to increase in each scenario, reaching concentrations from 548 ppm to 806 ppm. Even though emissions decline in some of the scenarios, the long atmospheric lifetime of CO_2 results in continued increases in concentration over the century.

The upper panel of Figure 1.7 shows the derived globally averaged radiative

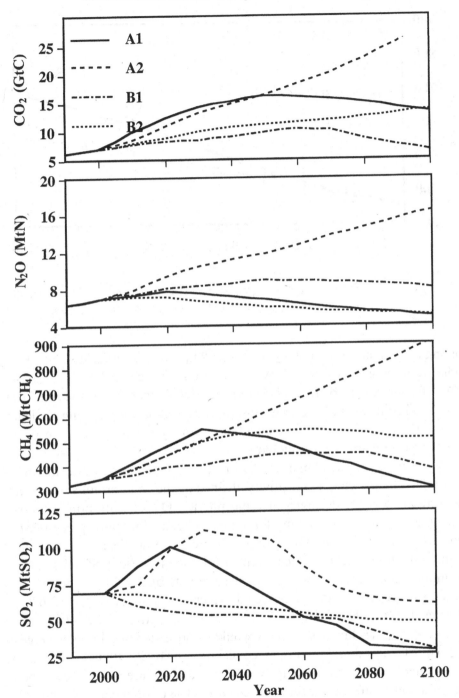

Figure 1.5   Anthropogenic emissions in the SRES marker scenarios for CO_2, CH_4, N_2O, and SO_2. Note that the SRES emission values are standardized such that emissions from 1990 to 2000 are identical in all scenarios.

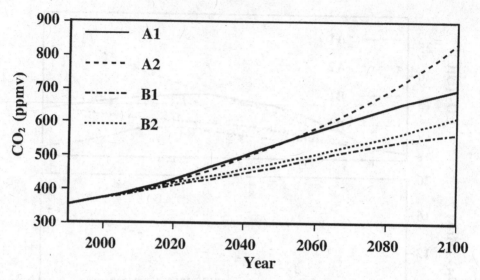

Figure 1.6   Integrated Science Assessment Model (ISAM) estimated CO_2 concentrations for the SRES marker scenarios.

forcing as a function of time for various SRES marker scenarios. Calculated radiative forcing increases to 7 Wm^{-2} by 2100 for high scenario A2 and 4.7 Wm^{-2} for low scenarios B1 and B2 relative to the beginning of the Industrial Era. As a result, each of the scenarios implies a significant warming tendency. Direct effects of aerosols are included in this analysis but indirect effects and effects on ozone are not considered.

The response of the climate system to the changes in radiative forcing is determined by the climate sensitivity, defined as the equilibrium surface temperature increase for doubling of atmospheric CO_2 concentration. This parameter is intended to account for all the climate feedback processes not modeled explicitly. The middle panel of Figure 1.7 shows the model-calculated changes in global mean surface temperature for the various SRES scenarios assuming a central value for the climate sensitivity of 2.5°C. For the four scenarios, surface temperature is projected to increase by about 1.8°C to 2.6°C by 2100 relative to 1990 for this assumed climate sensitivity. The full range of uncertainty in the climate sensitivity would be presented by a broader range of 1.5 to 4.5°C for a doubling of CO_2 concentration (IPCC, 1996a). Accounting for this uncertainty, the scenarios would give an increase in surface temperature of about 1.3 to 5°C for the four scenarios. The bottom panel of Figure 1.7 shows the derived sea level rise for the four scenarios. The difference in future sea level scenarios is much smaller as compared to the temperature scenarios. This is because the sea level rise has a much stronger memory effect due mainly

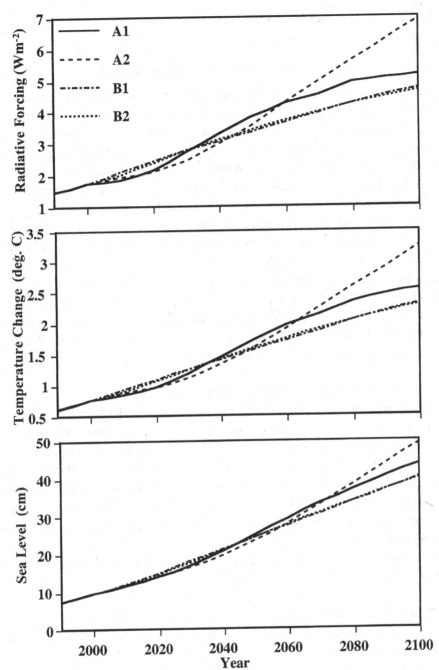

Figure 1.7   Estimated change in radiative forcing, change in globally-averaged surface temperature, and sea-level rise for the SRES marker scenarios as calculated with the ISAM model. The temperature change is estimated assuming a climate sensitivity of 2.5°C for a doubling of CO_2; this sensitivity appears to best represent the climate sensitivity of current climate models (IPCC, 1996a).

to the large thermal inertia of the ocean and hence long time scale of the ocean response.

1.5 Potential Impacts of Climate Change

In the previous sections, we briefly discussed projected changes in climate as a result of current and potential human activities. There are many uncertainties in our predictions, particularly with regard to the timing, magnitude, and regional patterns of climate change. At this point, potential changes in climate globally are better understood than the changes that could occur locally or regionally. However, the impacts of interest from climate change are primarily local to regional in scale. Nevertheless, scientific studies have shown that human health, ecological systems and socioeconomic sectors (e.g., hydrology and water resources, food and fiber production, and coastal systems, all of which are vital to sustainable development) are sensitive to changes in climate as well as to changes in climate variability. Recently, a great deal of work has been undertaken to assess the potential consequences of climate change (e.g., IPCC, 1998). The IPCC (1998) study has assessed current knowledge on how systems would respond to future projections of climate change. Here we restrict our discussion to only a brief overview.

Ecosystems. Ecosystems both affect and are affected by climate. As carbon dioxide levels increase, the productivity and efficiency of water use by vegetation may also increase. As temperature warms, the composition and geographical distribution of many ecosystems will shift as individual species respond to changes in climate. As vegetation shifts, this will in turn affect climate. Vegetation and other land cover determine the amount of radiation absorbed and emitted by the Earth's surface. As the Earth's radiation balance changes, the temperature of the atmosphere will be affected, resulting in further climate change. Other likely climate change impacts from ecosystems include reductions in biological diversity and in the goods and services that ecosystems provide society.

Water resources. Climate change may lead to an intensification of the global hydrological cycle and can have major impacts on regional water resources. Reduced rainfall and increased evaporation in a warmer world could dramatically reduce runoff in some areas, significantly decreasing the availability of water resources for crop irrigation, hydroelectric power production, and industrial/commercial and transport uses. Other regions may see increased rainfall. In light of the increase in artificial fertilizers, pesticides, feedlots excrement, and hazardous waste dumps, the provision of good quality drinking water is anticipated to be difficult.

Agriculture. Crop yields and productivity are projected to increase in some areas, at least during the next few decades, and decrease in others. The most significant decreases are expected in the tropics and subtropics, which contain the majority of the world's population. The decrease could be so severe as to result in increased risk of hunger and famine in some locations that already contain many of the world's poorest people. These regions are particularly vulnerable, as industrialized countries may be able to counteract climate change impacts by technological developments, genetic diversity, and maintaining food reserves.

Livestock production may also be affected by changes in grain prices due to pasture productivity. Supplies of forest products such as wood during the next century may also become increasingly inadequate to meet projected consumption due to both climatic and non-climatic factors. Boreal forests are likely to undergo irregular and large-scale losses of living trees because of the impact of projected climate change. Marine fisheries production are also expected to be affected by climate change. The principal impacts will be felt at the national and local levels.

Sea level rise. In a warmer climate, sea level will rise due to two primary factors: (i) the thermal expansion of ocean water as it warms, and (ii) the melting of snow and ice from mountain glaciers and polar ice caps. Over the last century, the global-mean sea level has risen about 10 to 25 cm. Over the next century, current models project a further increase of 25 to 100 cm in global-mean sea level for typical scenarios of greenhouse gas emissions and resulting climate effects (IPCC, 1996a; Neumann *et al.*, 2000). Figure 1.7 shows the sea level rise calculated with the ISAM model for the four SRES scenarios assuming a climate sensitivity of 2.5 °C for a doubling of the CO_2 concentration. A sea level rise in the upper part of the range could have very detrimental effects on low-lying coastal areas. In addition to direct flooding and property damage or loss, other impacts may include coastal erosion, increased frequency of storm surge flooding, salt water infiltration and hence pollution of irrigation and drinking water, destruction of estuarine habitats, damage to coral reefs, etc. Little change is expected to occur in the Antarctic over the next century, but if there were to be any major melting, it would potentially increase sea level by a large amount.

Health and human infrastructure. Climate change can impact human health through changes in weather, sea level and water supplies, and through changes in ecosystems that affect food security or the geography of vector-borne diseases. The section of IPCC (1996b) dealing with human health issues found that most of the possible impacts of global warming would be adverse.

In terms of direct effects on human health, increased frequency of heat waves would increase rates of cardio-respiratory illness and death. High temperatures would also exacerbate the health effects of primary pollutants generated in fossil fuel combustion processes and increase the formation of secondary pollutants such as tropospheric ozone. Changes in the geographical distribution of disease vectors such as mosquitoes (malaria) and snails (schistosomiasis) and changes in life-cycle dynamics of both vectors and infective parasites would increase the potential for transmission of disease. Non-vector-borne diseases such as cholera might increase in the tropics and sub-tropics because of climatic change effects on water supplies, temperature and microorganism proliferation. Concern for climate change effects on human health are legitimate. However, impacts research on this subject is sparse and the conclusions reached by the IPCC are still highly speculative.

Indirect effects that would result from climatic changes that decrease food production would reduce overall global food security and lead to malnutrition and hunger. Shortages of food and fresh water and the disruptive effects of sea level rise may lead to psychological stresses and the disruption of economic and social systems.

1.6 Policy Considerations

Worldwide concern over climate change and its potential consequences has led to consideration of international actions to address this issue. These actions fall into two broad categories: an adaptive approach, in which people change their lifestyle to adapt to the new climate conditions; and a preventive or "mitigation" approach, in which attempts are made to minimize human-induced global climate change by removing its causes. While it is not our intention here to consider or examine the range of possible policy options, it is important to discuss recent international activities that have resulted in a number of recommendations for emission reductions.

In Rio de Janeiro in 1992, the United Nation Framework Convention on Climate Change (UNFCC) agreed to call for the "stabilization of greenhouse gas concentrations in the atmosphere at a level that would prevent dangerous anthropogenic interference with the climate system." (UN, 1992). While specific concentration levels and time paths to reach stabilization for greenhouse gases were not stated, analyses of illustrative scenarios for future CO_2 concentrations have given some guidance as to what is required to reach CO_2 stabilization at various levels (Enting *et al.*, 1994; Wigley *et al.*, 1996). Figure 1.8 shows the calculated allowable emission levels over time which ultimately stabilize atmospheric CO_2 at levels ranging from 350 to 750 parts per million

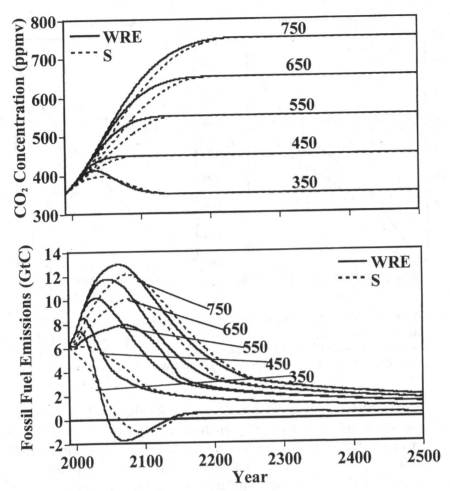

Figure 1.8 CO_2 concentration stabilization profiles and associated fossil CO_2 emissions. The "S" and "WRE" pathways are defined in Enting *et al.* (1994) and Wigley *et al.* (1996).

(ppmv). These calculations were made with the carbon cycle component of the Integrated Science Assessment Model (updated version of the model in Jain *et al.*, 1996). From this figure it is clear that, regardless of the stabilization target, global CO_2 emissions initially can continue to increase, would have to reach a maximum some time in the next century, and eventually must begin a long-term decline that continues through the remainder of the analysis period.

While the reductions in emissions in the stabilization scenarios are projected to lead to measurable decreases in the rate of increase in CO_2 concentrations, no specific commitments to achieve this goal were made until the December 1997 meeting of the Conference of Parties to the FCCC in Kyoto, Japan (UN,

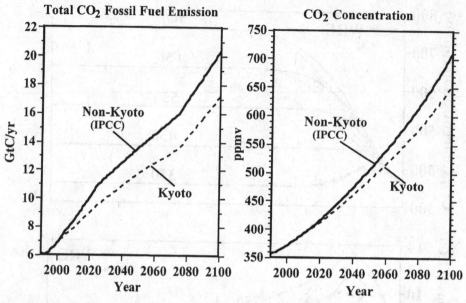

Figure 1.9   Global fossil CO_2 emissions and concentrations where countries follow the various emissions limitations proposed under the Kyoto Protocol. Global emissions and concentrations under no-limitations in a business-as-usual scenario are also given for comparison. Calculations were made with the ISAM model.

1997). At that meeting, developed nations agreed for the first time to reduce their emissions of greenhouse gases by an average of 5.2 percent below 1990 levels. Emission targets range from a return to baseline year emissions for most Eastern European countries up to an 8% reduction for the European Union. Emission limits for the United States under the Kyoto Protocol consist of a 7% reduction below baseline year emission levels. The baseline year relative to which emission reductions are determined is 1990 for CO_2, CH_4 and N_2O, and the choice of either 1990 or 1995 for HFCs, PFCs and SF_6. Mitigation actions can include reductions in any of six greenhouse gases: CO_2, CH_4, N_2O, halocarbons (HFCs), perfluorocarbons (PFCs) and sulfur hexafluoride (SF_6).

   However, should this protocol enter into force in the US (which is currently responsible for 25% of the world's greenhouse gas emissions), and even if its terms were renewed throughout the remainder of the 21st century, it would not achieve the goal of the UNFCC. As Figure 1.9 clearly shows, the long-term effect of the Kyoto Protocol is small. This is due to the fact that Kyoto only legislates emission controls for developed or industrialized nations. In the past, a move towards industrialization has been accompanied by an enormous increase in greenhouse gas emissions. Although emissions from the developed countries listed in the Kyoto Protocol currently account for the majority of global greenhouse gas emissions, most developing nations are already moving towards industrialization. If their relationship between greenhouse gas emissions, fossil fuel use and industrialization follow the paths of other developed nations, emissions from currently developing nations are projected to equal emissions from currently developed nations by 2020 and far surpass them by the end of the century. Thus, emissions from developed nations will make up a smaller and smaller part of the climate change problem as we proceed further into the coming century. For this reason, Kyoto controls on currently developed countries are not enough if we want to prevent dangerous climate change impacts. At the same time, countries in the process of industrialization have the right to be allowed to develop into industrialized nations with higher standards of living and greater wealth. The challenge facing the world community today is how to allow nations the right of development while successfully preventing "dangerous anthropogenic interference with the climate system".

Kyoto is important as the first concrete step in international cooperation. However, stabilizing radiative forcing will require much larger reductions that can only be fully supplied by CO_2 emissions (Hoffert *et al.*, 1998) from all nations. The future emphasis on CO_2 emission reductions from developed and developing countries highlights the importance of energy and transportation technologies that do not emit CO_2 and technologies such as efficiency improvements or carbon capture and sequestration that provide mechanisms by which fossil fuels can continue to play an important role in future global energy systems without concurrent emissions growth.

1.7 Conclusions

Human activities already appear to be having an impact on climate. The latest evaluation for future global warming by 2100, relative to 1990, for a business-as-usual set of scenarios based on varying assumptions about population and economic growth, is by a factor of 1.3 to almost 5°C. Potential economic, social and environmental impacts on ecosystems, food production, water resources, and human health could be quite important, but require much more study. A certain degree of future climatic change is inevitable due to human activities no matter what policy actions are taken. Some adaptation to a changing climate will be necessary. However, the extent of impacts and the amount of adaptation will depend on our willingness to take appropriate policy actions. The consensus grows that we must follow a two-pronged strategy to conduct research to narrow down uncertainties in our knowledge, and, at the same time, take precautionary measures to reduce emissions of greenhouse gases.

Acknowledgments

This study was supported in part by grants from the US Department of Energy (DFEG02–99ER62741), from the US Environmental Protection Agency (CX825749-01) and from the National Science Foundation (DMF9711624).

References

Angell, J. K., 1999: Global, hemispheric, and zonal temperature deviations derived from radiosonds records. In *Trends: A Compendium of Data on Global Change.* Carbon Dioxide Information Analysis Center, Oak Ridge National Laboratory, U.S. Department of Energy, Oak Ridge, Tenn., U.S.A.

Boer, G. J., G. Flato, M. C. Reader, and D. Ramsden, 2000: A transient climate change simulation with greenhouse gas and aerosol forcing: experimental design and comparison with the instrumental record for the 20th century. *Clim. Dyn.*, in press.

Bradley, R., 2000: 1000 years of climate change. *Science*, **288**, 1353–5.

Delworth, T. L. and T. R. Knutson, 2000: Simulation of early 20th century global warming. *Science*, **287**, 2246–2250.

Diaz, H. and N. Graham, 1996: Recent changes in tropical freezing heights and the role of sea surface temperature, *Nature*, **383**, 152–5.

Dlugokencky, E. J., K. A. Masarie, P. M. Lang, and P. P. Tans, 1998: Continuing decline in the growth rate of the atmospheric CH_4 burden. *Nature*, **393**, 447–50.

Enting, I. G., T. M. L. Wigley, and M. Heimann, 1994: Future emissions and concentrations of carbon dioxide: key ocean/atmosphere/land analyses. *CSIRO Division of Atmospheric Research Technical Paper* No. 31, CSIRO Australia 1994.

Haeberli, W., 1990: Glacier and permafrost signals of 20th century warming. *Ann. Glaciol.*, **14**, 99–101.

Hansen, J., M. Sato, and R. Ruedy, 1997: Radiative forcing and climate response. *J. Geophys. Res.*, **102**, 6831–64.

Hansen, J., M. Sato, A. Lacis, R. Ruedy, I. Tegen, and E. Matthews, 1998: Climate forcing in the industrial era. *Proc. Nat. Acad. Sci.*, **95**, 12 753–8.

Harvey, L. D. D., 2000: *Global Warming: The Hard Science*, Addison-Wesley-Longman.

Hoffert, M. I. and C. Covey, 1992: Deriving climate sensitivity from paleoclimate reconstructons, *Nature*, **360**, 573–6.

Hoffert, M. I., K. Caldeira, A. K. Jain, L. D. Harvey, E. F. Haites, S. D. Potter, S. H. Schneider, R. G. Watts, T. M. L. Wigley, and D. J. Wuebbles, 1998: Energy implications of future stabilization of atmospheric CO_2 content. *Nature*, **395**, 881–4.

Huang, S., H. N. Pollack, and P.-Y. Shen, 2000: Temperature trends over the past five centuries reconstructed from borehole measurements. *Nature*, **403**, 756–8.

IPCC (Intergovernmental Panel on Climate Change), 1996a: *Climate Change 1995: The Science of Climate Change. Contribution of Working Group I to the Second Assessment Report of the Intergovernmental Panel on Climate Change*, J. T. Houghton, L. G. Meira Filho, B. A. Callander, N. Harriss, A. Kattenberg, and K. Maskell, eds., Cambridge University Press, Cambridge, U.K., 572 pp.

IPCC (Intergovernmental Panel on Climate Change), 1996b: *Climate Change 1995:*

Impacts, Adaptation, and Mitigation of Climate Change: Scientific-Technical Analysis. The Contribution of Working Group II to the Second Assessment Report of the Intergovernmental Panel on Climate Change. R. T. Watson, M. C. Zinyowera, and R. H. Moss, eds., Cambridge University Press, Cambridge, U.K., 878 pp.

IPCC (Intergovernmental Panel on Climate Change), 1998: *The Regional Impacts of Climate Change: An Assessment of Vulnerability.* R. T. Watson, M. C. Zinyowera, and R. H. Moss, eds., Cambridge University Press, Cambridge, U.K., 517 pp.

Jain, A. K., H. S. Kheshgi, and D. J. Wuebbles, 1996: A globally aggregated reconstruction of cycles of carbon and its isotopes. *Tellus,* **48B,** 583–600.

Jain, A. K., B. P. Briegleb, K. Minschwaner, and D. J. Wuebbles, 2000: Radiative forcings and global warming potentials of thirty-nine greenhouse gases. *J. Geophys. Res.,* **105,** 20773–90.

Johannessen, O. M., E. V. Shalina, and M. W. Miles, 1999: Satellite evidence for an arctic sea ice cover in transformation. *Science,* **286,** 1937–9.

Jones, P. D., T. J. Osborn, and K. R. Briffa, 1997a: Estimating sampling errors in large-scale temperature averages. *J. Climate,* **10,** 2548–68.

Jones, P. D., T. J. Osborn, T. M. L. Wigley, P. M. Kelly, and B. D. Santer, 1997b: Comparisons between the microwave sounding unit temperature record and the surface temperature record from 1979 to 1996: real differences or potential discontinuities? *J. Geophys. Res.,* **102,** 30135–45.

Jones, P. D., K. R. Briffa, T. P. Barnett, and S. F. B. Tett, 1998: High-resolution paleoclimatic records for the last millennium: interpretation, integration and comparison with general circulation model control run temperatures. *Holocene,* **8,** 477–83.

Keeling, C. D. and T. P. Whorf, 1998: *Atmospheric CO_2 concentrations – Mauna Loa Observatory, Hawaii, 1958–1997.* Available at their website: cdiac.esd.ornl.gov/ftp/ndp001/maunaloa.txt.

Kheshgi, H. S., A. K. Jain, and D. J. Wuebbles, 1996: Accounting for the missing carbon sink with the CO_2 fertilization effect. *Climatic Change,* **33,** 31–62.

Kheshgi, H. S., A. K. Jain, and D. J. Wuebbles, 1999: The global carbon budget and its uncertainty derived from carbon dioxide and carbon isotope records. *J. Geophys. Res.,* **104,** 31127–43.

Levitus, S., J. I. Antonov, T. P. Boyer, and C. Stephens, 2000: Warming of the world ocean, *Science,* **287,** 2225–9.

Mann, M. E., R. S. Bradley, and M. K. Hughes, 1999: Northern hemispheric temperatures during the past millennium: inferences, uncertainties, and limitations. *Geophys. Res. Lett.,* **26,** 759–62.

Miller, A. J., R. M. Nagatani, G. C. Tiao, X. F. Niu, G. C. Reinsel, D. Wuebbles, and K. Grant, 1992: Comparisons of observed ozone and temperature trends in the lower stratosphere. *Geophys. Res. Lett.,* **19,** 929–32.

Nakicenovic, N., N. Victor, and T. Morita, 1998: Emissions scenarios database and review of scenarios. *Mitigation and Adaptation Strategies for Global Change,* **3,** 95–120.

Neumann, J. E., G. Yohe, R. Nicholls, and M. Manion, 2000: *Sea-level rise & global climate change.* Pew Center on Climate Change, Arlington, VA.

NRC (National Research Council), 2000: Reconciling observations of global temperature change, National Academy Press, Washington, D.C., 84 pp.

Roemmich, D. and K. Wunsch, 1984: Apparent changes in the climate state of the deep North Atlantic Ocean. *Nature,* **307,** 447–50.

UN (United Nations), 1992: *Framework Convention on Climate Change*. United Nations, New York, 33 pp.

UN (United Nations), 1997: *Kyoto Protocol to the United Nations Framework Convention on Climate Change*. United Nations, New York, 24 pp.

Vaughn, D. G. and C. S. M. Doake, 1996: Recent atmospheric warming and retreat of ice shelves on the Antarctic peninsula. *Nature*, **379**, 328–31.

Watts, R. G. and M. C. Morantine, 1991: Is the greenhouse gas-climate signal hiding in the deep ocean? *Climatic Change*, **18**, iii–vi.

Wigley, T. M. L., 1999: The science of climate change: global and US perspectives. Pew Center on Climate Change, Arlington, VA, 48 pp.

Wigley, T. M. L., R. Richels, and J. Edmonds, 1996: Economic and environmental choices in the stabilization of atmospheric CO_2 concentrations. *Nature*, **379**, 240–3.

WMO (World Meteorological Organization), 1998: *Scientific Assessment of Ozone Layer: 1998*. WMO Global Ozone Research and Monitoring Project, Report No. 44, Geneva, 668 pp.

2

Posing the Problem

2.1 Scenarios

Scenarios are unfolding sequences of interrelated events stemming from prior and ongoing decisions. With regard to climate change and carbon emissions, the outputs of climate and carbon cycle models under different scenarios of population growth, economic development, and fossil and other fuel use have proved to be useful in exploring possible policy options for mitigation of climate change due to greenhouse gas emissions. Owing to the contingent nature of historical evolution, economic and technological development may in fact be unknowable in advance. Nevertheless, the scenario approach is useful because it allows asking "what if?" questions. A variety of scenarios have been published by several research groups. There is, of course, an infinite variety of scenarios that will lead, say, to keeping the future level of CO_2 below a certain level. One of the earliest sets is that presented in the first of the scientific reports from the Intergovernmental Panel on Climate Change (IPCC). Subsequent scenarios have been suggested by Wigley *et al.* (1996) and by Nakicenovic *et al.* (1998).

2.1.1 The Intergovernmental Panel on Climate Change (IPCC) scenarios

The authors of the IPCC report point out that scenarios are not predictions of the future, but merely illustrate the effects of a range of economic, demographic, and policy assumptions. In Climate Change 1992 the IPCC examined six scenarios, including the one called IS92a, which has become known as the Business as Usual scenario (BAU). This scenario assumes a world population growing to 11.3 billions by 2100, economic growth of 2.9% per year until 2025 and 2.3% thereafter, and a set of assumptions about the availability and cost of various fuels. Other scenarios are either more or less optimistic. The reader should refer to the IPCC report for details (Houghton *et al.*, 1992). The IPCC scenarios were constructed under the following constraints: (1) prescribed initial concentrations of CO_2 and the rate of change of concentration set at 1990 values, (2) a range of prescribed stabilization levels and attainment dates, and (3) a requirement that the implied emissions should not change too abruptly. An inverse calculation using a carbon cycle model was then used to determine emission rates required to attain stabilization via the set pathways. These are shown in Figure 1.8 in Chapter 1. We return to the BAU scenario later.

2.1.2 *The Wigley, Richels, Edmonds (WRE) scenarios*

Wigley *et al.* (1996) added an initial constraint. They required that the initial emissions trajectories initially track the BAU scenario but that the higher the concentration target for stabilizing CO_2, the longer the adherence to BAU. Concentration targets can be met when departures from BAU are delayed by 10 to 30 years. Of course, the atmospheric loading of CO_2 increases more rapidly initially than in the IPCC scenarios, but stabilization is ultimately achieved at a prescribed level in either the IPCC scenarios or the WRE scenarios. *Cumulative* emissions are notably higher when emissions follow the BAU scenario longer. Reducing emissions later allows greater total CO_2 production, and therefore total use of fossil fuels. Important economic considerations are implied by this conclusion. Several studies (Nordhaus, 1979; Manne and Richels, 1995; Richels and Edmonds, 1994) have shown that emission reduction pathways that are modest in the early stages and sharper later on are less expensive than those requiring early sharp reductions. This is because (1) future environmental impacts have low present values in economic analyses, (2) capital stock for energy is long-lived and one pays a penalty for retiring it prematurely, and (3) future energy technologies will presumably be more efficient and less costly than present ones. The carbon emissions pathways from WRE are shown in Figure 1.8 in Chapter 1, along with the IPCC pathways for comparison.

Wigley *et al.* (1996) are careful to point out that their results do not suggest a "wait and see" policy. "No regrets" options should be adopted immediately where appropriate. Since capital stocks do have a long lifetime and since new supply options typically take many years to enter the marketplace a sustained effort in research, development and deployment should begin now.

2.1.3 The IIASA (Nakicenovic et al., 1998) scenarios

Nakicenovich *et al.* (1998) discuss in great detail six scenarios ranging from a "high growth" case (A1) that is somewhat similar to the IPCC BAU scenario to cases C1 and C2, which the authors characterize as "ecologically driven". Case C incorporates a variety of different possible ways to reduce CO_2 emissions to one third of current values by 2100. The difference between cases C1 and C2 involves the role of nuclear energy. In case C1 nuclear energy is phased out, while in case C2 a new generation of safe and acceptable nuclear energy is assumed to come on line.

Case B incorporates more modest estimates of economic growth and technological development, particularly in the developing countries. It contains features that are less attractive for developing coutries, though perhaps more realistic, than characteristics in the other cases.

Scenario A is devolved into three cases. Case A1 is a future dominated by natural gas and oil, while in A2 it is assumed that the greenhouse debate is resolved in favor of coal, which is assumed to be extensively used. In case A3 there is increasing reliance on both nuclear energy and renewables.

In all cases the population is assumed to reach 10.1 billions by 2050 and 11.7 billions by 2100. The primary energy intensity, defined as the primary energy needed to produce a unit of economic output, in watts per $US (1990), decreases by 0.8% per year (case B) to 1.4% per year (case C). Gross world product increases dramatically in all scenarios (but less so in C1 and C2), so that the globally averaged per capita income increases even in developing nations to somewhat near the value in developed countries today. However, there will still be quite large disparities between developed and developing countries. According to the IIASA scenarios the GDP per capita in North America will be $US 60000 or more by 2100, while in Latin America and the Caribbean it would be about $16000 to $20000 and in sub-Saharan Africa only $10000, even assuming stable governance, no major regional conflicts, and high social capital (education, the empowerment of women and other conditions conducive to reducing fertility). In centrally planned Asia in only the high growth scenario (scenario A) does the per capita income exceed the OECD 1990 income of $19000 by 2100.

Unlike in the WRE scenarios, all of the IIASA scenarios depart from current trajectories beginning in 1990. In cases A1 and A2, the CO_2 loading of the atmosphere reaches 650 and 750 ppmv respectively and continues to increase thereafter. Even so, of the 33 TWt (terawatts thermal) of primary energy use in 2050, 11.3 TWt comes from nuclear and renewables in case A1 and 9 TWt comes from this combination in case A2. In case C2 world primary energy consumption in 2050 is 18.6 TWt, of which 9.1 TWt comes from nuclear and renewables. Other scenarios produce similar results. These are remarkable numbers, given that current world primary energy consumption stands at approximately 14 TWt . They are in substantial agreement with values that we have derived from a detailed analysis of the IPCC BAU scenario, which we discuss in some detail in the next section.

2.2 Energy Implications of the Scenarios

Carbonaceous fossil fuels (coal, oil and natural gas) represent approximately 80% of the primary global energy supply today (Nakicenovic *et al.*, 1998). The carbon free component of the global energy supply is composed of traditional renewables 10% firewood (presumed sustainably burned), 5% hydropower and 4% nuclear fission. Collectively, the "high tech" renewables (solar electric, solar thermal, wind, biomass, and geothermal) produce less than 1%. In this section we examine what the IIASA scenarios, the IPCC BAU scenario and the WRE scenarios (which lead to specific future equilibrium levels of atmospheric CO_2) imply about the need for carbon free global power supply over the 21st century.

First of all, let us examine some results of the IIASA scenarios. Table 2.1 is a compilation of the total world energy consumption associated with the six scenarios. Values for total primary energy for 1990 and 2050 are read from Table 5.1 of Nakicenovic *et al.* (1998). Values for the year 2100 are those read from graphs in that reference. Since Nakicenovic *et al.* give values in Gtoe (gigatons oil equivalent) a conversion factor $1 Gtoe \times 1.33 = 1 TWt$ is used.

According to the six IIASA scenarios as little as 8 TWt or as much as 13.6 TWt of energy must be supplied by renewables plus nuclear power by 50 years hence. In 100 years the range is from 20 to 42 TWt, that is, from nearly two to about four times global energy consumption today.

Most global energy projections include a reference scenario supposed to represent a "business as usual" or "no policy" projection (Nakicenovic *et al.*, 1998; Raskin and Margolis, 1995). In the IIASA scenarios it is probably best represented by case A1. To characterize "business as usual" energy demand and carbon emissions over the 21st century, the IPCC developed the IS92a (BAU) scenario, incorporating widely accepted population projections and the current consensus on economic development of the World Bank and the United Nations (Leggett *et al.*, 1992). Note that the BAU scenario was not meant to represent particularly high or low rates of energy use, but is a nominal reference case in the absence of intervening policy actions.

To understand the implications of the BAU scenario and the WRE scenarios

Table 2.1 *Primary energy consumption from six IIASA scenarios. Renewables plus nuclear means energy from renewables such as solar, wind, hydroelectricity, biomass and nuclear sources and is shown in bold face. Units are TWt (terawatts thermal). Note that the required energy production from renewables and nuclear by 2050 is in all cases comparable to total world energy production in 1990.*

Scenario	Energy (TWt)	1990	2050	2100
A1	Total	12	33.3	60
	Renewable plus nuclear	2.7	**11.3**	**30**
A2	Total	—	33.3	60
	Renewables plus nuclear	—	**9**	**30**
A3	Total	—	33.3	60
	Renewables plus nuclear	—	**13.6**	**42**
B	Total	—	26.6	46.6
	Renewables plus nuclear	—	**9.6**	**26**
C1	Total	—	18.6	28
	Renewables plus nuclear	—	**8**	**20**
C2	Total	—	18.6	28
	Renewables plus nuclear	—	**9.1**	**23**

for future energy needs Hoffert *et al.* (1998) examined the individual terms in the Kaya identity (Kaya, 1989). In general, the rate at which carbon is emitted (as CO_2) by energy production is given by the identity

$$M_C = N(GDP/N)(E'/GDP)(C/E)$$

expressing the population (N), per capita gross domestic product (GDP/N), primary energy intensity (E'/GDP), E' being the rate of energy use, and carbon intensity (C/E). Here we express the primary energy consumption from all fuel sources in watts and the gross domestic product in (1990 US)$ per year so their ratio, the energy intensity, has units of watt years per 1990 US$. Carbon intensity, the weighted average of the carbon to energy emission factors of all energy sources, has the units of kgC/watt year. For example, from the Kaya identity, fossil fuel emissions in 1990 were $M_C = 5.3 \times 10^9$ persons times $4100 per person per year times 0.49 watt year per $US (1990) times 0.56 kgC per watt year, or about 6.0 GtC/yr. This corresponds to approximately 1.1 metric tons of carbon per person emitted to the atmosphere per year as CO_2 (in 1990). At this time per capita emissions were significantly higher in the developed than in the developing nations (Hammond *et al.*, 1994), but the projected growth of emissions is far greater in developing nations. The rate of population growth is also projected to be much larger in developing nations (Alcamo *et al.*, 1995)

To illustrate the relative contributions of the factors in the Kaya identity, Hoffert *et al.*(1998) evaluated each of them globally over the 210 year period from 1890 to 2100, from historical data before 1990 (Nakicenovic, 1996), and from documents defining IPCC scenarios after 1990 (see Hoffert *et al.*, 1998 for details).

One way to reduce the rate of increase of carbon emissions is to increase the economic productivity of energy, which we define as the inverse of the primary energy intensity, E'/GDP. Data from individual nations indicate that the E'/GDP generally increases during economic development as countries establish heavy infrastructure, and declines only after some lag as economic productivity rises and the economy shifts structurally to less energy intensive activities (for example, to services). Energy intensities of China and India are two to five times the global mean, but they are decreasing. To focus on energy supply issues Hoffert *et al.* (1998) provisionally accepted the IS1992a projections of 1% per year improvements. Note that the IIASA projections range from 0.8% to 1.4%. Achieving this will depend crucially on technology and structural changes adopted by individual nations. We will see later how larger or smaller values change our estimates of the nature of future energy supplies.

Another opportunity for emission reductions of CO_2 is the continuation of the "decarbonization" of the past 100 years that is reflected in the decreasing carbon intensity of the global energy mix (Nakicenovic, 1996). The carbon intensity is the mass of carbon produced from a unit of energy for any given carbon fuel, in kgC/watt year. For example, for direct combustion of natural gas the factor is 0.46 kgC/Watt year (Nakicenovic, 1996), while that for oil is 0.60 and for coal, 0.77. Emission factors decrease as the carbon to hydrogen ratio of the fuel decreases. For wood the factor is 0.89 kgC/watt year. To compute the emission factor for the fuel mix,

$$C/E = \left(\sum_i C_i \times E_i\right)/\sum_i E_i,$$

where E_i is the *i*th source of energy characterized by an emission factor C_i.

In 1990 the value of C/E was 0.56, slightly below whzat it would be if global energy came completely from oil. At that time global energy sources were 2.54 TWt from natural gas, 3.90 TWt from oil and 3.14 from coal, with 1.3 TWt from nuclear and renewables, which are assumed to emit no CO_2.

Both of these factors are decreasing, and they are assumed (by the IPCC and other scenarios) to continue their decline throughout the 21st century. The other two factors in the Kaya identity will continue to increase, however.

Table 2.2 *Global energy production in TWt, world* GDP *in 10^{12} \$US (1990)
and world population in billions as projected by the IPCC IS92a (BAU)
scenario*

Year	Gas	Oil	Coal	Nuclear	Renew.	Total	*GDP*	*N*
1990	2.54	3.90	3.14	0.54	0.76	10.87	21.8	5.3
2000	3.11	4.44	3.71	0.73	0.95	12.92	20.0	6.24
2025	4.44	5.23	6.97	1.78	4.03	22.44	59.3	8.40
2050	4.09	4.56	10.84	2.60	7.51	29.61	97.8	9.95
2075	2.71	3.76	16.34	4.07	10.95	37.83	161.3	10.83
2100	1.33	2.95	21.84	5.55	14.39	46.06	265.9	11.30

To project into the future according to the IPCC 1992a (BAU) scenario we use E' from Pepper *et al.* (1992), N from Leggett *et al.* (1992, IPCC) and *GDP* from the IPCC projections. The population in 1990 was 5.3 billion, and projections are for 8.4 billion in 2025 and 11.3 billion in 2100. The IPCC assumes increases in economic growth of 2.9% per year from 1990 until 2025 and 2.3% from 1990 until 2100. The energy mix and its evolution along with world *GDP* are shown in Table 2.2.

From these numbers one can calculate each of the terms in the Kaya identity. We note that in the BAU scenario 10.11 TWt of renewable plus nuclear (i.e., non-fossil) energy is required globally by 2050 and 19.94 TWt by 2100, even though carbon emissions reach approximately 20 GtC/year, and the CO_2 content of the atmosphere nearly triples. These results are within the range of values in the IIASA scenarios.

We now ask the question "How much renewable energy will be required to follow the WRE scenarios that maintain atmospheric CO_2 levels below specified targets?" First assume that the technology *mix* implied by the BAU scenario, that is, the value of *C/E* calculated from Table 2.2, is maintained. In that case the differences between the primary power available for the BAU and the primary power available for the CO_2 stabilization scenarios represent shortfalls that would presumably have to be made up entirely by carbon free energy sources to achieve the global economic targets of the BAU scenarios. To find the global primary power available for CO_2 stabilization scenarios the carbon emission rates are computed by running the carbon cycle model of Jain *et al.* (1995) in an inverse mode in which the atmospheric concentration paths are constrained to follow those in the WRE stabilization scenarios. Under the assumptions of the BAU scenario economic targets it is now possible to calculate the primary power consumption (from the fossil fuel mix) that results in these emission rates from the Kaya identity.

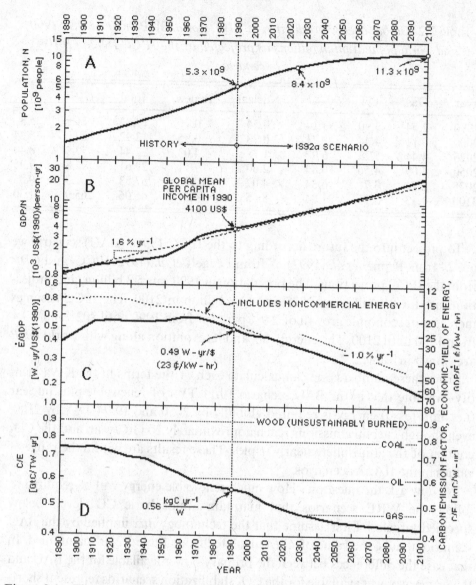

Figure 2.1   The evolution of factors governing the rate of global fossil-fuel carbon emissions in the Kaya identity. Historical curves and future projections computed for the IPCC IS92a (BAU) scenario. Reprinted with permission from *Nature* (29 October, 1998), © 1998 Macmillan Magazines Limited.

The top curve in Figure 2.1 shows the evolution of primary power required to meet the economic goals of the BAU scenario. Also shown are the contributions from the fossil fuels: coal, oil, and natural gas. The dashed lines shown on that figure represent the primary power from fossil fuel that is allowed following the WRE scenarios for stabilization of CO_2 at various levels. Carbon emissions from the various scenarios are shown in Figure 2.2.

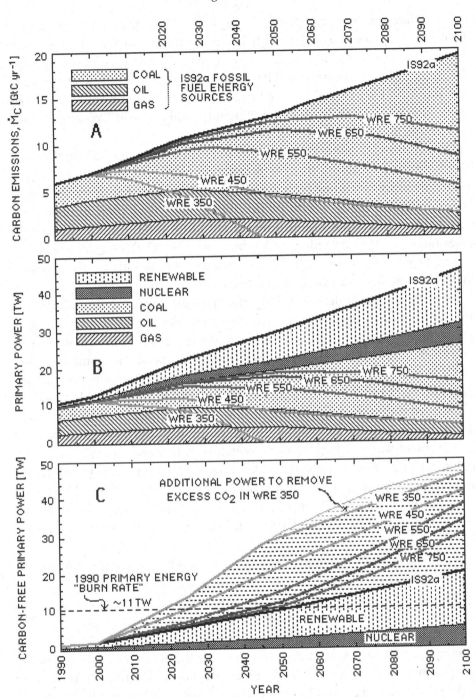

Figure 2.2 Fossil-fuel emissions and primary power in the 21st century for the IPCC IS92a (BAU) and the WRE CO$_2$ stabilization scenarios. Reprinted with permission from *Nature* (29 October, 1998), © 1998 Macmillan Magazines Limited.

The differences between the energy requirements of the BAU scenario and the WRE scenarios must presumably be made up from nuclear and renewable sources. These are shown in Figure 2.2C. Even in the BAU scenario, approximately 10 TWt must be produced from renewables and nuclear energy by 2050, and about 20 TWt by 2100, more or less in agreement with the IIASA scenarios. To maintain CO_2 levels at twice preindustrial levels 30 TWt must be produced from non-fossil sources by 2100.

The numbers are rather daunting. Let us begin to examine their implications by noting the probability that the population will be at least as large as that used in most scenarios and with the hopeful assumption that we will be able to increase the per capita wealth of developing nations at some nominal rate, reasonably close to that assumed by the IIASA or the IPCC scenarios. What other factors in the Kaya identity can be manipulated? The candidates are the primary energy intensity (E'/GDP) and the carbon intensity (C/E). Reducing the carbon intensity implies a switch to carbon free energy.

As we have seen, in the opinions of the many experts involved in the IIASA and IPCC studies, improvements (decreases) in the primary energy intensity will be between 0.8 and 1.4% per year. Among those who believe that it can be improved much more rapidly is Art Rosenfeld (personal communication, 2000). Rosenfeld points out that historically the US primary energy intensity has improved from 2.26 watt years per \$US (1990) in 1850 to 0.49 in 1995, or a sustained decrease of about 1% per year. However, there were periods when the decrease reached as high as 5% per year. Between 1855 and 1880 the improvement was nearly 50% as we switched from wood to coal; during the 1920s there was another rapid decrease as we switched from coal to natural gas and oil. There has also been a recent decrease greater than 1% per year (about 2.2% per year since 1975), mostly due to energy price increases during the OPEC years (about 1974 until 1985) and the recent rapid growth in the US economy.

There is some evidence that the primary energy intensity is improving rapidly in China (Rosenfeld, personal communication), having declined by 5% per year for the past 25 years.

Figure 2.3 shows how improvements in the primary energy intensity affect our conclusions about the need for more non-fossil energy sources in the future. If improvements can be sustained at 2% per year the need for carbon free power remains modest even until 2100. However, periods of very rapid improvement in the past have been driven by changes in technology (changing from wood to coal and from coal to oil and gas) or by the scarcity of a resource (during the OPEC years). A sustained decrease of more than 1% per year in the future will be very difficult to manage without substantial policy intervention in the form

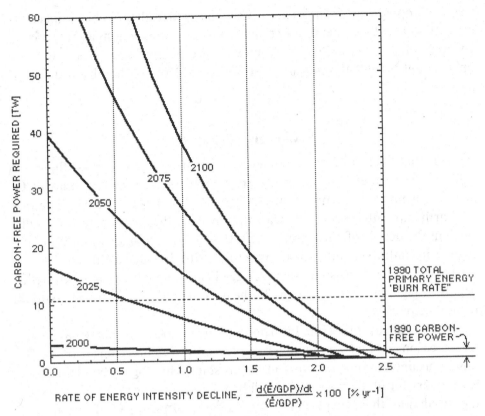

Figure 2.3 Twenty-first century trade-offs between carbon free power required and energy intensity for CO_2 stabilization at twice the preindustrial atmospheric concentration. Reprinted with permission from *Nature* (29 October, 1998), © 1998 Macmillan Magazines Limited.

of carbon taxes or other incentives. Whether this can be implemented globally, as it would have to be, remains seriously in question. It will require a supranational authority having the power to somehow impose its will on sovereign nations. Whether this is possible or even desirable is problematic (Pendergraft, 1999).

The factors in the Kaya identity are not likely to be independent of each other. The rate of population growth, for example, depends on the *GDP* per capita and on education (particularly of women). Per capita wealth is not likely to increase very rapidly without abundant, inexpensive energy. The most abundant and inexpensive source of energy in China is coal, and China is rapidly developing an energy infrastructure based on coal. It is difficult to see how they can reduce *C/E* very far below that of coal, 0.77 kgC/watt year. Once this infrastructure is established, they will not want to replace it soon. If an international treaty is proposed to tax emissions of CO_2, they will in all probability not cooperate. It is likely that "mandatory reductions" of CO_2 emissions as proposed in Kyoto would in any case only be supported by developed countries, and developing countries will account for most of the emissions within the next several decades.

2.3 An Engineering Problem

It is essentially certain that as CO_2 levels in the atmosphere increase the Earth will become warmer. The results of climate model studies indicate that a CO_2 doubling would raise temperatures globally by 1.5–4.5 °C, and evidence that the Earth is already warming appears to be firm. Although climate models disagree in the details of their predictions, they nearly all predict more summer dryness in continental interiors (Houghton, 1996). It will in all likelihood be at least a decade and possibly much longer before models are able to predict details of regional climate change. In the meantime the skeptics will no doubt remain skeptical.

Yet the evidence that global warming is already occurring is strong, and it seems reasonable that we should act to prevent it from becoming extreme. But this is certainly no time to abandon our trust in technology. Abundant energy is necessary for modern society. It will be necessary in the developing world if the populations there are to achieve some amount of prosperity. There are ways to reduce CO_2 emissions by increasing end use efficiency (Rosenfeld *et al.*, 1997). However, the real key to supplying carbon free energy more likely lies in development of affordable and safe nuclear (fission and fusion) and renewable energy. In a very real sense global warming has become an engineering problem. Given the rather daunting challenge of supplying the world with 10 to 30 TWt of carbon free energy within the next 50 to 100 years, and even more if atmospheric CO_2 is to be stabilized, as implied by the various scenarios, it appears inevitable that we should explore a wide range of possibilities without preconceived notions. The range of options under serious consideration by most researchers is perhaps too limited.

2.3.1 Adaptive decision strategies

Policy choices are often based on "optimal" responses to what is likely to happen in the future: to forecasts. In the next chapter Lempert and Schlesinger remind us that "the one thing that we know for sure about forecasts is that most of them are wrong". This is consistent with our general philosophy that we need to keep open a wide range of policy options and possible energy sources.

This is not to say that forecasts themselves are not useful. The real goal of the Framework Convention on Climate Change is to stabilize the atmospheric concentration of CO_2 at a level that is safe in some, currently unspecified and (really unknown), sense. There are, of course, an infinity of possible paths for accomplishing this. Climate change presents a problem of decision making under conditions of great uncertainty. Lempert and Schlesinger argue that robust strategies can result from adaptive-decision processes that change over time in response to observations of changes in climate and as economic conditions evolve. Robust strategies are those that will work reasonably well over a broad range of plausible future scenarios. As an example they examine the impacts of climate variability on the choices of near-term policies, and find that under many circumstances a combined strategy involving direct technology incentives designed to encourage early adoptions of new emission-reducing technologies in addition to price-based incentives such as carbon taxes and emission trading is more robust than strategies involving either of these policies alone. The various technologies examined in later chapters of this volume can be useful, along with the tool described in this chapter, in shaping future policy actions that strike a balance between these two types of incentives and appeal to many stakeholders.

2.3.2 *The prospects of energy efficiency*

In Chapter 4 Hassol, Strachan and Dowlatabadi argue that although improving energy efficiency will play an essential part in reducing CO_2 emissions, it cannot by itself reduce emissions to a reasonable target during this century while meeting global energy demands at a reasonable cost. Energy efficiency (or the lack of it) is intricately tied to structural changes in the economy and also to sector-specific technological change. Many modelers treat technological learning as exogenous, and fixed by model assumptions, while it is more realistic to consider it endogenous, and to allow it to be affected by other parameters in the model, responding to interacting factors such as policies, prices, and research and development. This parallels the discussion in Chapter 3 in the sense that it evolves and is affected by feedback processes over time.

In the next several decades greenhouse emissions from developing countries will surpass those from developed countries. Therefore, energy intensity trends (E'/GDP) in those countries will be crucial in determining the amount of carbon free energy needed to stabilize the atmospheric loading of CO_2. However, lack of data and modeling experience along with dynamically changing economies make prediction of future E'/GDP difficult. Some question even its relevance. Although E'/GDP seems to be falling in China, purchasing power parity might be more nearly relevant than *GDP*, and when this is used as a measure of economic gains energy intensity gains (based on E'/PPP) disappear.

The authors go on to discuss the path dependence of shaping the future availability of various energy efficient technologies and their relations to policy instruments, including case studies of transportation, buildings, industrial production and electricity generation.

2.3.3 *Earth-based renewables*

Chapter 5 focuses on terrestrially based renewable energy options. Short and Keegan discuss costs and benefits of renewable energy technologies along with the anticipated cost reductions over the next 20 years. Hydroelectricity and biomass have been cost competitive for many years, and already provide substantial energy globally. Their discussions of market opportunities and barriers, including the problem of "lock in", are particularly helpful.

Renewable energy resources, of course, exist in enormous quantities, but are generally dispersed. Places where a given renewable resource (e.g. solar, wind, geothermal) is most concentrated are generally not places where many people live. The problem of intermittency must also be addressed. In this regard the authors discuss briefly the problem of energy carriers in the form of hydrogen, superconductivity and electric vehicles. Short and Keegan conclude that the potential for terrestrially based renewable energy is huge, pointing to IPCC figures that show a potential some 21 times global energy use today.

2.3.4 *Transportation and storage*

Berry and Lamont further address the problems of transport and storage in Chapter 6. The transportation sector will be a particularly difficult one, and the authors provide an extensive and useful discussion of hydrogen powered vehicles. Gasoline and diesel fuel are used almost exclusively as transportation fuel, and as various countries develop the demand for personal transportation vehicles will grow. They conclude that the key technology is the one that links electricity and transportation fuel, efficient hydrogen production through electrolysis.

2.3.5 *Nuclear fission*

In Chapter 7 Krakowski and Wilson discuss the possible contribution of nuclear fission reactors to the future fossil free energy mix. Following World

War II there was a rapid growth in the commercial application of nuclear energy. Since that time, however, there has been little growth, although a few industrialized countries (notably France) have continued to sustain a meager growth. The costs of building nuclear power plants have increased, while those for fossil fueled plants have decreased. In addition, the idea of environmentally clean, safe and inexpensive nuclear energy has given way to the prospects of accidental releases of dangerous radioactive materials, the generation as byproducts of breeder reactors materials of military interest, and dangers and difficulties associated with radioactive waste. Krakowski and Wilson thus identify the four cardinal issues that need to be addressed before nuclear energy can become acceptable as an energy source: radioactive waste, proliferation of materials of military interest (plutonium), cost and safety.

It is clear from this (as well as other) chapters that there is sufficient economically recoverable ^{235}U for nuclear fission reactors to contribute as much energy as that postulated in the various scenarios until at least 2100, that is, ramping up to 5 TWe within 100 years. Depending on assumptions about ^{235}U resource availability, breeder reactors will have to be employed after 2050 if they are to continue to contribute to CO_2 stabilization unless some inexpensive method is found to extract uranium from seawater. Nevertheless, any viable future for nuclear energy will depend largely on the rate at which barriers to public acceptance are lowered. Even so, substantial increases in renewable energy resources will be required if atmospheric CO_2 is to be stabilized, absent very large improvements in E'/GDP.

2.3.6 Nuclear fusion

The other arm of nuclear energy is nuclear fusion. Fusion occurs when the positively charged nuclei of two low atomic number elements approach closely enough for the attractive short-range nuclear forces to overcome the repulsive Coulomb force. The mass of the fused nuclei is smaller than the combined masses of the separate nuclei, and the mass deficit is released as energy. Getting the two nuclei close enough requires that they have very high energy. Temperatures must be at least ten million Kelvins and the density must be very high. The density–time–temperature combination must be at least 10^{21} keVm^{-3} s^{-1}. The basic problem with the fusion option is that we have not yet been able to reach this level. Nevertheless, Molvik and Perkins believe that we are near enough that fusion has a chance to become the energy source of choice early in this millennium. Unlike fission, fusion reactors are in some sense inherently safe because the fuel in the core at any time is sufficient for at most only a few seconds of operation. Molvik and Perkins discuss several different fusion fuels and reactor configurations in Chapter 8. They conclude that "An innovative fusion energy research program, focused on the critical issues, will maximize the probability of achieving the full potential of fusion energy. The development of a virtually limitless energy source will provide a profound benefit to future humanity."

2.3.7 Solar power from space

In Chapter 9 Criswell summarizes the problems and prospects of energy efficiency and of the energy sources discussed in earlier chapters, often coming to somewhat different conclusions. The potential for a variety of possible future energy sources is given in his Tables 9.2 to 9.6. These include many of those considered in other chapters: wind, bio-resources, peat, gas hydrates, hydroelectric, tides, waves, ocean thermal and geothermal energy. He concludes that most renewable sources discussed in Chapter 5 cannot by themselves, and perhaps even in combination, supply all of the energy needs of the world in the 21st century. Chapters 5 and 9 will no doubt provide plenty of opportunity for a spirited discussion between proponents of Earth-based renewables and those who doubt that they can supply our future energy needs. Close examination of the details and comparing numbers from Chapter 5 with those in Tables 9.2, however, reveals that the differences between the two chapters lie mostly in the fact that Criswell is basing his conclusions on a world which consumes energy at a rate of 60 TWt, a level that will probably not be reached for 100 years. Criswell addresses the question of which power sources have the potential for supplying 20 TWe by 2050, this being a level consistent with many of the future scenarios that is necessary to provide the global economic growth implied in the scenarios. His major emphasis, however, is on the possibility of supplying renewable (solar) energy from space by locating solar collectors outside the Earth's atmosphere and beaming power down by microwaves. This will certainly be the most controversial chapter in this book. Criswell argues (convincingly, in my opinion) that lunar solar power (LSP), in which solar energy is collected by large solar panels located on the Moon can become feasible and relatively inexpensive if we have the will to invest in it.

2.3.8 Geoengineering

Geoengineering is defined in various ways by different scientists. Many scientists define it as large-scale intentional engineering of the environment for the primary purpose of controlling or counteracting changes in the chemistry of the atmosphere. This definition is consistent with the idea that it is used to remove the CO_2 from the atmosphere as rapidly as fossil fuels or other sources emit it, so that the atmospheric loading does not increase (at least beyond "safe" levels). However, other forms of geoengineering involve controlling the climate itself by, for example, reflecting away sunlight to cool the climate and counteract the greenhouse effects of increasing CO_2 (Flannery *et al.*, 1997). It is also not explicitly a "what if nothing else works" idea. Sequestering carbon, for example, can certainly be imagined as one more way, along with increasing energy efficiency, of limiting the buildup of atmospheric CO_2.

In Chapter 10 Keith examines many possible geoengineering options. Most of them involve huge engineering enterprises, from radiation shields to sequestering carbon by fertilization of the ocean to aforestation. In many of these schemes one must ask the obvious question "and what else happens?" Keith discusses the risk of side effects and unintended consequences as well as political and ethical considerations.

References

Alcamo, J. *et al.*, 1995. An evaluation of the IPCC IS92 emission scenarios. In Houghton, J. T. (ed.) *Climate Change 1994*, Cambridge University Press.

Flannery, B. P., H. Kheshgi, G. Marland, and M.C. MacCracken, 1997. Geoengineering climate. In R. G. Watts (ed.) *Engineering Response to Global Climate Change*, CRC Lewis Publishers.

Hammond, A. L. (ed.), 1994. *World Resources 1993–94*, Oxford University Press.

Hoffert, M. I., K. Caldeira, A. K. Jain, L. P. Harvey, E. F. Haites, S. D. Potter, S. H. Schneider, R. G. Watts, T. M. L. Wigley, and D. J. Wuebbles, 1998. Energy implications of future stabilization of atmospheric CO_2 content, *Nature*, **395**, 881–4.

Houghton, J. (ed.), 1996. *Climate Change 1995: The Science of Climate Change*, Cambridge University Press.

Houghton, J., B. A. Callander, and S. K. Varney (eds.), 1992. *Climate Change 1992*, Cambridge University Press.

Jain, A. K., H. S. Kheshgi, M. I. Hoffert, and D. J. Wuebbles, 1995. Distribution of radiocarbon as a test of global carbon cycle models, *Glob. Biogeochem. Cycles*, **9**, 153–66.

Kaya, Y., 1989. *Impacts of Carbon Dioxide Emission Control on GNP Growth: Interpretation of Proposed Scenarios* (IPCC Response Strategies Working Group Memorandum, 1989).

Leggett, J. *et al.*, 1992. In *Climate Change 1992*, J. Houghton (ed.), Cambridge University Press.

Manne, A. and R. Richels, 1995. The greenhouse debate – economic efficiency, burden sharing and hedging strategies, *Energy Journal*.

Nakicenovic, N., 1996. Freeing energy from carbon, *Daedalus*, **125**, 95–112.

Nakicenovic, N., A. Grubler, and A. McDonald (eds.), 1998. *Global Energy Perspectives*, Cambridge University Press.

Nordhaus, W. D., 1979. *The Efficient Use of Energy Resources*, Yale University Press.

Pendergraft, C. A., 1999. Managing planet earth: adaptation and cosmology, *The Cato Journal*, **19**, 69–83.

Pepper *et al.*, 1992. Emission scenarios for the IPCC – an update: background documentation on assumptions, methodology and results, US Environmental Protection Agency, Washington, D.C.

Raskin, P. and R. Margolis, 1995. *Global Energy in the 21st Century: Patterns, Projections, and Problems*. Polestar Series Rep. No. 3, Stockholm Environmental Institute, Boston, MA.

Richels, R. and J. A. Edmonds, 1994. In *Integrated Assessment of Mitigation, Impacts and Adaptation to Climate Change*, (ed. N. Nakocenovic), International Institute for Applied Systems Analysis.

Rosenfeld, A., B. Atkinson, L. Price, R. Ciliano, J. I. Mills, and K. Friedman, 1997. Energy demand reduction. In Watts, R. G. (ed.), *Engineering Response to Global Climate Change*, CRC Lewis Press.

Wigley, T. M. L., R. Richels, and J. A. Edmonds, 1996. Economics and environmental choices in the stabilization of atmospheric CO_2 concentrations, *Nature*, **379**, 240–3.

3

Adaptive Strategies for Climate Change

3.1 Introduction

This book presents forecasts for a range of innovative energy technologies that might help stabilize atmospheric concentrations of greenhouse gases during the course of the 21st century. These forecasts provide a wealth of important information for those who wish to inform their view of the climate-change problem and the actions governments or the private sector might take to address it. But these chapters nonetheless present a fundamental dilemma – for the one thing we know for sure about forecasts is that most of them are wrong. How then should we use the information in this book to shape policy?

The difficulty resides not so much in the forecasts themselves as in the methods that we commonly employ to bring the information they contain to bear on adjudicating among alternative policy choices. Generally, we argue about policy by first settling on our view of what will happen in the future and then by using this understanding to decide what actions we should take in response. For instance, if we came to believe, through arguments such as those in this book, that there were cost-effective technological means to stabilize greenhouse-gas emissions, we might be more likely to support policies that sought to achieve such a stabilization. We often make these arguments non-quantitatively, even if systematically. There are also a host of powerful mathematical tools, based on the mathematical techniques of optimization, that help us systematize and elaborate on this style of thinking about the future. These methods encapsulate our knowledge about the future in probability distributions that then allow us to rank the desirability of alternative strategies. These prediction-based analytic approaches work extraordinarily well in a wide variety of cases. So much so that they strongly affect our notions of the criteria we ought to use to compare our policy choices and the way we ought to use forecasts to support these choices. We often speak of choosing the optimum or best strategy based on our predictions of the future. In this approach then, the purpose of forecasts is to shape our views of what is likely to happen, as a means of shaping our decisions about how to act.

In this chapter we argue for a different approach to climate-change policy and thus a different use for forecasts. We argue that climate change presents a problem of decision-making under conditions of deep uncertainty. We begin with the premise that while we know a great deal about the potential threat of climate change and the actions we might take to prevent it, we cannot now, nor are we likely for the foreseeable future, answer the most basic questions, such as is climate change a serious problem and how much would it cost to prevent it? We argue that in the face of this uncertainty, we should seek robust strategies. Robust strategies are ones that will work reasonably well no matter what the future holds. Not only is this desirable in its own right, but a robust strategy may provide a firm basis for consensus among stakeholders with differing views about the climate-change problem, because all can agree on the strategy without agreeing on what is likely to happen in the future. Rather than using forecasts to specify a particular path into the future, in this alternative approach forecasts describe a range of plausible scenarios and sharpen our sense of what any particular future, if it comes to pass, might look like.

In this chapter we argue that robust strategies for climate change are possible by means of adaptive-decision strategies, that is, strategies that evolve over time in response to observations of changes in the climate and economic systems. Viewing climate policy as an adaptive process provides an important reconfiguration of the climate-change policy problem. The long-term goal of the Framework Convention on Climate Change calls for society to stabilize atmospheric levels of greenhouse gases at some, currently unknown, safe level, and the protocol negotiated in Kyoto in December 1997 commits the world's developed countries to specific, binding near-term emissions reductions. There is currently much debate as to how to best implement these reductions and whether or not they are justified. Viewed as the first steps of an adaptive-decision strategy, however, the Kyoto targets and timetables are but one step in a series of actions whose main purpose is to increase society's ability to implement large emissions reductions in the future. Contrary to most of today's assessments, the real measure of the Framework Convention's success a decade hence should not be any reductions in atmospheric concentrations of greenhouse gases, but rather the new potential for large-scale emissions reductions society has created for the years ahead.*

In this context, the potential for new technologies, the processes of innovation and diffusion by which these technologies come into widespread use, and

---

* See Lempert (2001) for an application of robust adaptive-decision strategies to the diplomatic situation that followed the United States withdrawal from the Kyoto Protocol.

the government policies needed to encourage such processes, play a central role in society's response to climate change. We will suggest in this chapter that the processes of technology diffusion might provide key indicators for monitoring the progress of an adaptive climate-change strategy, particularly in the face of significant climate variability that will make it difficult to observe a reliable signal about the extent of human-caused impacts to climate change. There is also much debate over the type of policy instrument – carbon taxes or tradable permits – that should be used to encourage emissions reductions. We find that one does not need to have particularly high expectations that these innovative technologies will achieve the potential described here in order to put in place an adaptive-decision strategy that also makes important use of a third type of policy instrument, technology incentives. Forecasts such as those in this book can provide a great deal of information that can help policy-makers design such robust adaptive strategies, but in order to be most useful, such forecasts need to provide a broader range and different types of information than is often provided.

We begin this chapter with a discussion of the importance of adaptive-decision strategies and how they differ from other views of the climate-change problem. We will then describe some first steps in employing an analytic method, based on a multi-scenario simulation technique called exploratory modeling, for designing and evaluating adaptive-decision strategies for climate change, and provide a simple example of its application. We will then turn to the design of adaptive-decision strategies, examining the conditions under which such strategies are appropriate, the tradeoffs between actions and observations in the design of such strategies, and some initial steps in examining climate-change policy as a process. We will present two example analyses – the impacts of climate variability on the design of such strategies and the choice of policy instruments in the presence of the potential for significant technology innovation. During the course of the discussion, we will show the role of innovation in emissions-reducing technologies in such adaptive-decision strategies, as well as the type of information from forecasts like those in this book that can prove most useful in designing them.

3.2 The Case for Adaptive-Decision Strategies

Much of the current policy and political debate about climate change focuses on the question of how much to reduce emissions of anthropogenic greenhouse gases. The Framework Convention on Climate Change, in particular the protocol negotiated in Kyoto in December 1997, now calls for the world's developed countries (Annex I nations in the protocol) to commit to binding reductions that would bring their emissions back to roughly 1990 levels by 2010. It does little damage to the nuances of the debate to divide most of the responses to this protocol into two opposing camps. Supporters argue that the risk of severe impacts from climate change is sufficiently high, and the prospects of inexpensive abatement due to technology innovation and social change sufficiently good, that immediate and significant reductions in emissions are the only prudent course. Opponents point to the potentially high economic costs of emissions reductions and the large uncertainties surrounding the predicted damages, and argue that little or no action be taken at this time.[1]

Numerous quantitative policy studies have attempted to determine the relative validity of these competing arguments. Many take the Kyoto targets as a given and attempt to predict the economic cost of meeting them. These estimates cover a wide range, depending on assumptions about diverse factors, such as how the protocol is implemented and the market acceptance of efficiency-enhancing and emissions-reducing technologies (Repetto and Austin, 1997). In addition, many analytic efforts have been made to determine the best arrangements for the emissions trading and other policies that may be used to implement the protocol. For instance, economic studies generally argue that allowing countries to trade emissions permits among themselves will lower the costs relative to requiring each country to meet its reduction target individually.

More broadly, other studies address the question of whether or not the Kyoto targets represent the most cost-effective first steps towards the Climate Convention's ultimate goal of long-term stabilization of atmospheric concentrations of greenhouse gases. The influential study by Wigley, Richels, and Edmonds (Wigley *et al.*, 1996 – hereafter WRE) argued that the Convention's goals can be met at least cost with near-term emissions reductions much smaller than those required by Kyoto. In response, a number of studies have questioned the assumptions underlying WRE. In particular, these rebuttals argue that alternative, plausible assumptions about the speed and characteristics of technology innovation suggest that, given the large time scales associated with changing energy-related capital infrastructure and the feedback effects whereby early investments in new technology can reduce subsequent costs, Kyoto's emissions reductions would be beneficial.

It is not at all clear that the question of the best level of near-term emissions reductions can be resolved by this type of argument. Ultimately, the predicted benefit of such reductions depends on a broad range of assumptions about the seriousness of potential impacts due to climate change, the potential of new

---

[1] There are, of course, additional debates over topics such as the extent to which the fast-growing developing countries ought to bear the burden of emissions reductions, the extent to which developed countries can pay others to make emissions reductions for them, and the best means to reward those firms that voluntarily commit to early actions.

emissions-reducing technologies, the value people in the future will place on the preservation of nature, the extent to which market reforms succeed in many developing countries, and a host of other factors that are impossible to predict with any accuracy. Thus, the studies that favor and oppose particular levels of emissions reductions often differ, not so much because of any disparities in the underlying methodology, but in the basic assumptions they make about the future. Because a wide range of such assumptions is plausible, the studies cannot distinguish among a wide range of potential policy proposals. Furthermore, such studies will provide little basis for agreement among competing sides in the climate-change-policy debate, since each stakeholder will, not surprisingly, tend to gravitate towards those assumptions that justify the level of emissions-reductions they would otherwise support on ideological, financial, or other grounds.

The concept of adaptive-decision strategies offers a way out of this impasse. Rather than base prescriptions for the proper level of emissions reduction on a judgment that WRE is or is not correct, or that some new emissions-reducing technology will or will not achieve substantial market penetration, we look to strategies that begin with a certain suite of actions, monitor the external environment, and then modify the suite in response to these observations. Rather than predict what the level of emissions will be in future decades, we determine what type of adaptive-decision strategy will perform best across a wide range of potential futures, each of which, in retrospect, clearly requires some different level of emissions reduction.

Simple and obvious as this point may seem at first, it is actually a substantial reconfiguration of the policy problem because it suggests that policy-makers consider a wider set of potential policy actions and a different set of information about the world than they are encouraged to do in the standard analytic framework. Perhaps this can best be seen by comparing the concept of adaptive-decision strategies to the related, analytic approach of sequential-decision strategies which more commonly appears in the climate-change-policy literature (Manne and Richels, 1992; Nordhaus, 1994; Hammitt, Lempert, and Schlesinger, 1992 – hereafter HLS). The sequential-decision approach finds an optimum strategy assuming that policy-makers begin with uncertainty but will receive a specific dollop of uncertainty-reducing information at some fixed time in the future. For instance, HLS consider two-period decision strategies that start with imperfect information about the future costs of climate change and the costs of abatement, with the uncertainty resolved completely in ten years. Generally such a study suggests some modest adjustment to the strategy one would pursue without uncertainty, such as a slightly increased level of near-term emissions reductions.

A sequential-decision strategy assumes that the rate at which policy-makers learn is independent of the seriousness of the problem or of actions they take. But the dependence of what we learn tomorrow on both what we do today and the possible paths the world offers to us may be important factors in choosing a response strategy. For instance, a key determinant of the proper level of near-term emissions reductions ought to be whether or not society can achieve sufficient warning time to respond to signals of adverse climate change. With sufficient warning time, near-term emissions can be lower; without it they should be higher. If climate change is modest, it may take decades to extract the signal from the noise. But if climate change is severe, the signal may well be revealed more quickly. Similarly, near-term emission reductions will likely generate information about abatement costs that can help us better determine the proper level of emissions reductions. Including such considerations in the calculations will affect analysts' recommendations as to the best level of near-term emissions reductions.

Considering strategy as an evolutionary process can also significantly affect the types of actions we consider and the criteria we use to judge those actions. For instance, a key question in the design of adaptive-decision strategies is the information to which decision-makers ought to pay attention. For example, the US Federal Reserve follows an adaptive-decision strategy as it periodically adjusts interest rates to regulate the growth of the US economy. Much attention is paid to the information and indicators the Fed governors use to make their decisions and, when the Fed chairman makes comments that additional information will be deemed important, the markets can react strongly. In fact, sometimes indicators alone can be a large step towards the solution. One of the most striking examples was the US EPA's policy of requiring companies to publish their toxic-release inventories. This communication of information itself reduced emissions, without any other regulation, because companies felt shamed by high numbers.

Perhaps most importantly, the response of real-world decision-makers to deep uncertainty is often not any particular action, but rather on building processes that can respond to uncertainty. This in the broadest sense is what we mean by an adaptive-decision strategy. In his review of risk assessment, William Clark (1980) emphasizes the importance of rules of evidence and other such processes in determining how we perceive and respond to risk. The business literature often recommends organizational changes, designed to increase the firm's ability to respond to new information, as the best response to uncertainty and rapid change. This theme of information flows and how to use them is central to the idea of an adaptive-decision strategy. It may be that the most important responses to the threat of climate change are the establishment of the institutional mechanisms that allow society to respond more quickly and less expensively to opportunities and threats in the decades ahead.

3.3 Assessing Adaptive-Decision Strategies

Despite the fact that adaptive-decision strategies are likely to be the better approach to climate change, and the approach policy-makers are likely to follow in practice, they have been infrequently considered in the policy literature (see Dowlatabadi, 1999, for one of the few such treatments). One important reason is that the analytic-policy literature has been largely dominated by the approach of finding the optimum response to climate change. This optimization approach is not conducive to assessing adaptive-decision strategies for two reasons. First, the analytic demands of optimization calculations usually require neglecting key feedbacks. Often these feedbacks are precisely those involved in an adaptive-decision strategy.

More profoundly, optimization is the wrong criterion for assessing adaptive-decision strategies because it assumes a level of certainty about the future that, if truly present, would obviate most if not all of the need for such strategies. Optimization finds a unique, best strategy based on specific assumptions about the system model, the prior probability distributions on the parameters of that model, and a loss function which represents society's values. But adaptive-decision strategies are most useful when there is deep uncertainty, that is, when we do not know with confidence the model, probabilities, or societal values, or where different stakeholders disagree about these things. In such cases, disagreements about optimum strategies can quickly reduce to arguments about alternative, unprovable assumptions or differences in goals.

Rather than optimization, the criterion for assessing adaptive-decision strategies ought to be robustness (Lempert and Schlesinger, 2000). Robust strategies are ones that will work reasonably well, at least compared to the alternatives, over a wide range of plausible futures. Robust strategies are advantageous because we can often identify them without specifying, or requiring stakeholders to agree on, specific models, priors, or societal values. In general, we can always identify, post hoc, models, priors, and values that make any given adaptive-decision strategy optimal. But in practice, the robustness criterion is useful precisely because it avoids the need for any prior agreement in those cases where a range of plausible scenarios is the best available representation of the information we have about the future.

The concept of robust strategies has a strong theoretical and practical pedigree. The idea is closely related to Simon's (1959) satisficing strategies and Savage's (1954) idea of minimizing the maximum regret. There is also much evidence that actual decision-makers in practice search for robust strategies rather than optimal ones (March, 1994). While these ideas have been familiar in the decision-analysis literature (Watson and Buede, 1987; Matheson and Howard, 1968), in practice they are infrequently employed on the climate-change problem because they are often difficult to implement for problems of any complexity. In recent years, however, an emerging school of what we might call multi-scenario simulation approaches (Lempert, Schlesinger, and Bankes, 1996, henceforth LSB; Morgan and Dowlatabadi, 1996; Rotmans and de Vries, 1997; Casan, Morgan, and Dowlatabadi, 1999), has begun to exploit the capabilities of new computer technology – fast computer processing; extensive, low-cost memory; and powerful, interactive visualization tools – to consider strategies under deep uncertainty. These methods use simulation models to construct different scenarios and, rather than aggregate the results using a probabilistic weighting, instead make arguments from comparisons of fundamentally different, alternative cases.

Our approach, which we call exploratory modeling (Bankes, 1993), is a multi-scenario simulation technique that explicitly implements these classic ideas about robust strategies. In exploratory modeling we use simulation models to create a large ensemble of plausible future scenarios, where each member of the ensemble represents one guess about how the world works and one choice among many alternative strategies we might adopt to influence the world. We then use search and visualization techniques to extract from this ensemble of scenarios information that is useful in distinguishing among policy choices. These methods are consistent with the traditional, probability-based approaches to uncertainty analysis because when such distributions are available, one can lay them across the scenarios and thus calculate expected values for various strategies, value of information, and the like. However, in situations characterized by deep uncertainty, the exploratory modeling method allows us to systematically find strategies that are robust against a wide range of expectations about the future and a range of valuations of that future.

We now provide a simple example that demonstrates how this concept of robustness and these exploratory modeling methods can be used to assess adaptive-decision strategies. This example, based on our 1996 work in LSB, addresses whether and under what conditions adaptive-decision strategies are a reasonable response to climate change. We begin by comparing the performance of the strategies advocated by the two competing camps in the debate over the Kyoto protocol. We define two alternative strategies, "Do-a-Little" (DAL) and "Emissions-Stabilization" (ES), to represent these camps. Each strategy reflects a prediction-based approach to climate change and is expressed as a given emissions path over the course of the 21st century. The former has little near-term reduction and is similar to the results of many economic analyses that assume relatively small long-term damages due to climate change (less than 2% of GWP) and relatively high abatement costs (more than 2% of GWP) to stabilize atmospheric concentrations of greenhouse gases. The latter, which returns global emissions to their 1990 levels by 2010 and holds them there until mid-century, is similar (though slightly more aggressive) than the reductions paths mandated by the Kyoto protocol. While the proponents of these camps do not explicitly reject adaptive-decision approaches, and the Framework Convention on Climate Change explicitly calls for policies that adjust over time in response to new information, the debate in practice is very much characterized by support or opposition to specific targets for the reductions of greenhouse gases.

We compare these DAL and ES strategies using a linked system of simple climate and economic models. Emissions of greenhouse gases determine their atmospheric concentrations, which in turn determine the change in global-mean surface temperature. These temperature trajectories determine the trajectory of damage costs, while the emissions trajectories generate a trajectory of abatement costs. We work in a cost–benefit framework, and measure the performance of each strategy as the present value of the sum of the hundred-year time series of damage and abatement costs. In particular, we focus on comparing the performance of these strategies as a function of three key dimensions of uncertainties we face about climate change (Lave and Dowlatabadi, 1993; Nordhaus, 1994), which we express as: (i) the sensitivity of the climate system to increasing concentrations of greenhouse gases, (ii) damages resulting from an increase in global-mean surface temperature, and (iii) the ability of innovation to significantly reduce the costs of abating greenhouse-gas emissions. In each case we define our plausible range of estimates, wherever possible, as the extreme range found in the published, refereed literature (a similar screening was used in Rotmans and de Vries, 1997).

We simulate the impact of uncertainty about the climate sensitivity on the change in global-mean surface temperature due to anthropogenic emissions of greenhouse gases (GHGs) with our energy-balance-climate/upwelling-diffusion-ocean (EBC/UDO) model (Schlesinger and Jiang, 1991; Schlesinger and Ramankutty, 1992, 1994a,b, 1995). We allow the climate sensitivity to vary between 0.5 °C and 4.5 °C. The upper value is the high value from the IPCC; the low value is from Lindzen (1990). We express uncertainty about the damages due to climate change using a simple phenomenological impacts function of changes in global-average temperature, as in the practice of many integrated assessments (Nordhaus, 1994a; Manne and Richels, 1992; Peck and Teisberg, 1993, 1992). We consider cases where total aggregate damages at the end of the next century for a 3 °C temperature rise range from 0.5% to 20% of gross world product (GWP), based on a survey of experts conducted by Nordhaus (1994b).

The crude innovation model used in LSB to simulate the consequences of uncertainty about the impacts of innovation, first used in HLS, assumes base-case greenhouse-gas emissions (Houghton *et al.*, 1990) are reduced as non-emitting "fuel-switching" technologies diffuse through the economy with an S-shaped, logistic diffusion curve at some policy-determined rate $1/R$. (We will describe a more detailed innovation model below.) The model builds the least-cost mix of emitting and non-emitting technologies to meet the exogenous energy-demand and policy-imposed emissions constraint. The model parameters are chosen so that our basecase innovation case reproduces the results of more detailed models (Manne and Richels, 1991; Nordhaus, 1991). We then assume that innovation can reduce the incremental costs of the non-emitting technologies at some fixed, but currently unknown, annual rate. Abatement costs also depend on the rate of reductions because, in those cases where emissions are reduced sufficiently quickly, emitting capital must be prematurely retired (assuming a normal lifetime of 30 years). This formulation captures in a crude manner the idea of inertia that may affect the choice between early and late action (Grubb, Chapuis, and Ha-Duong, 1995).

Figure 3.1 compares the performance of the DAL and ES strategies as a function of society's expectations about our three, key, climate-change uncertainties: the likelihood of extreme climate sensitivity, the likelihood of extreme damages, and the likelihood of significant innovation. For instance, the lower left-hand corner represents the expectation that the climate is likely to be insensitive to increasing greenhouse-gas concentrations, damages due to any climate change are likely to be small, and that innovation is unlikely to reduce the costs of abatement. In contrast, the upper right-hand corner represents the expectation that the climate system is very sensitive to greenhouse gases, damages are likely to be large, and innovation is likely to radically change the costs of abatement. The curved surface labeled A represents the boundary where we should be indifferent between the two choices, DAL and ES. Formally, this is the indifference surface where the strategies have equal expected values. To the left of the curve, we should prefer DAL; to the right we should prefer ES.

We created this visualization by running a large number of scenarios, each with one of the two strategies and a particular set of assumptions about the three parameters describing our uncertainties. We then laid probability distributions across these parameters as a function of these three expectations, and conducted a computer search to find the surface on which the expected values of the two strategies were equal. Note that this process, similar to the policy-

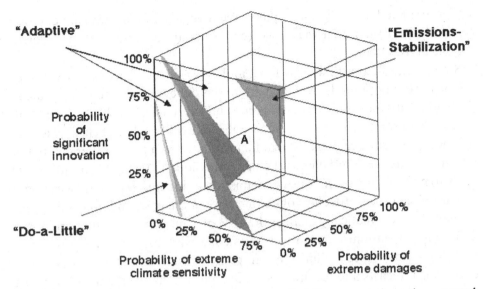

Figure 3.1 Surfaces separating the regions in probability space where the expected value of the "Do-a-Little" policy is preferred over the "Emissions-Stabilization" policy, the adaptive strategy is preferred over the "Do-a-Little" policy, and the adaptive strategy is preferred over the "Emissions-Stabilization" policy, as a function of the probability of extreme damages, significant innovation, and extreme climate sensitivity.

region analysis of Watson and Buede (1987), inverts the standard use of probabilities in decision analysis. Commonly, analysts begin with assumptions about the probability distributions of key uncertainties, which much be gleaned from methods such as expert elicitation (Morgan and Henrion, 1990), and then use these probabilities to find some optimum strategy. Instead, we report key probabilities as outputs of the calculations. That is, we report the probabilities that are implicit in advocating one strategy over another.

There are two important arguments we can make based on the comparison shown in Figure 3.1. First, the figure provides what we call a *landscape of plausible futures* (Park and Lempert, 1998), because it provides an explicit overview of the implications of the full range of uncertainties. Such landscapes are useful because they provide a common framework in which contending stakeholders to the climate-change debate can position themselves. Each stakeholder can find a portion of the landscape that reflects their view of the world and thus agree that the models we are using capture their assumptions and can reproduce the arguments they make from those assumptions. This process helps stakeholders find a language in which to express their disagreements. It also helps them buy-in to the analysis. We have shown this "climate cube" to audiences including both oil-company executives and ardent environmentalists and convinced them that their divergent views are captured within different regions in the cube.

Second, this figure shows why DAL and ES provide an unsatisfactory set of options. We cannot predict which point in this cube best represents the state of the world. Little scientific evidence is currently available, or is likely to be available anytime soon, to convince a reasonable individual whose beliefs place her on one side of surface A that she in fact should switch to the other camp. For instance, someone who believes the new, emissions-reducing technologies are likely to make future greenhouse-gas emission reductions inexpensive can offer dozens of examples of technologies that have in the past, and others that may in the future, dramatically change the cost of emissions reduction and other pollution prevention. Alternatively, someone who believes such emissions-reducing technologies are unlikely can point to dozens of stories about over-optimistic forecasts and technologies that failed to change the world. Because the technological future is fundamentally unpredictable, there is no evidence we can gather nor theorems we can evoke that will definitely sway this argument one way or the other. Thus, we cannot really know which side of the surface A in Figure 3.1 we are on. We can show, however, that guessing wrong can be very costly. That is, if we chose ES in a world where it turns out DAL would have been best, or DAL in an ES world, the costs can be significant. Thus, it should be no surprise that framing the climate-change problem around competing predictions of an unpredictable future should foster a hostile and unresolvable debate.

We next consider how an adaptive-decision strategy performs in comparison to these two static strategies. We posit a very simple, threshold-based adaptive-decision strategy that observes the damage and abatement costs once a decade and can make one correction based on these observations. The strategy begins with slow emissions reductions, as shown in Figure 3.2. If the damage exceeds a particular threshold after ten years, the strategy switches to draconian emission reductions. If not, the strategy waits another ten years. If the damage then exceeds the threshold, or if the abatement costs are below a threshold, then the strategy switches to rapid reductions. If not, it continues checking the damage and abatement costs every decade until the mid-twenty first century. We assume that once the strategy makes a mid-course correction, it makes no further observations nor reduction-rate adjustments. This particular adaptive-decision strategy is one, very simple exemplar, not necessarily the best such strategy, but sufficient for our purposes here.

Note that we express this strategy, not as any particular outcome for emissions-reductions, but as a specific process that will produce different levels of

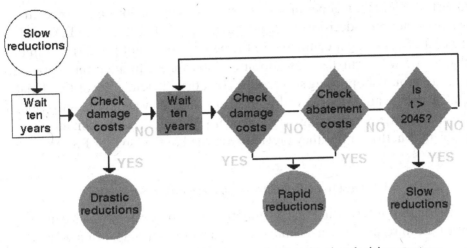

Figure 3.2   Flow chart describing a very simple adaptive-decision strategy.

emissions reductions in alternative scenarios. Thus, in a crude way, we can examine the type of process the decision-makers ought to use rather than the particular outcomes they ought to achieve. This question is different from those generally addressed by optimization approaches which, by definition, assume that the decision-maker conducts a particular, idealized process. Our claim, however, is that modeling the process of decision-making not only helps answer questions more relevant and credible to decision-makers faced with deep uncertainty, it also (and probably not coincidentally) provides solutions that perform better across a wide range of scenarios than those based on optimum outcomes.

We compare the performance of this adaptive-decision strategy to that of the DAL and ES strategies, with the surfaces labeled B and C in Figure 3.1. Surface B, in the lower left-hand corner of the cube, is the indifference surface between the adaptive-decision strategy and DAL. If society's expectations about the future put us in this lower left-hand region, we should prefer DAL; otherwise we should prefer adaptive-decision. Surface C, in the corner of the cube closest to the viewer, is the indifference surface between the adaptive-decision strategy and ES. If society's expectations about the future put us in this upper-right-hand region, then we should prefer ES; otherwise we should prefer adaptive-decision. In addition, the adaptive-decision strategy never makes large errors, so the costs of choosing it incorrectly – where DAL or ES in fact turn out to be the right answer – are not large.

The adaptive-decision strategy performs better than DAL and ES over a wide range of plausible expectations about the future, and thus provides a more

robust response than either of the two static strategies. For those expectations
where the adaptive-decision strategy does not perform better than DAL or ES,
the cost of choosing it compared to the best option is not large. (Later in this
chapter we will formalize this notion of robustness.) In addition, if we convinced a group of opposing stakeholders that the different poles of the climate-
change debate are captured within this cube, then they should now agree that
the adaptive-decision strategy is the proper way to address the climate-change
problem, whether or not they previously resided in the DAL or ES camp.

3.4 Design of Robust Adaptive-Decision Strategies

It is perhaps not surprising that an adaptive-decision strategy which can evolve
over time in response to new information is more robust against a wide range
of uncertainty than strategies which do not adapt. In order to shape policy
choices, however, we need to determine the best adaptive-decision strategies
from among the options available. In the above example, we examined only a
single, threshold-based, adaptive-decision strategy, which started with slow
near-term emissions reductions and used particular trigger levels for its observations of damage and abatement costs. In order to design and use adaptive-
decision strategies, we need to answer question such as: should such strategies
begin with fast or slow emission reductions? What observations should suggest
a change in the initial emissions-reduction rate?

Several academic literatures offer insights into the design of adaptive-
decision strategies. Economics and financial theory suggest that a successful
strategy will employ a portfolio of different policy instruments and the mix and
intensity of these instruments may change over time. Control theory suggests
that there is an intimate relation between our choice of instruments and the
types and accuracy of the observations we can make. Scenario planning offers
a useful language for the components of an adaptive-decision strategy: *shaping
actions* intended to influence the future that comes to pass, *hedging actions*
intended to reduce vulnerability if adverse futures come to pass, and *signposts*
which are observations that warn of the need to change strategies (Dewar *et
al.*, 1993; van der Heijden, 1996).

We can use our exploratory-modeling methods to systematically combine
these seemingly disparate elements and find robust adaptive-decision strategies. Generalizing on the application described above, we compare the performance of a large number of alternative adaptive-decision strategies across the
landscape of plausible futures. In order to make sense of this information, we
calculate the regret of each strategy in each of many plausible states of the
world, defined as the difference between the performance, for that state of the

world, of the strategy and of the optimal strategy. We then search for robust strategies, defined as those with relatively small regret over a wide range of states of the world. In some cases we can find strategies that are robust across the landscape of plausible futures. In other cases, we find particular states of the world that strongly influence the performance of robust strategies. We then might report tradeoff curves which show, for instance, the choice of robust strategy as a function of the probabilistic weight a decision-maker might assign to some critical state of the world.

In this section we will use these methods to address two important questions in the design of adaptive-decision strategies. First, we will examine the impacts of climate variability on the choice of near-term policy choices (Lempert, Schlesinger, Bankes, and Andronova, 2000 – henceforth LSBA). Variability, one of the most salient features of the earth's climate, can strongly affect the success of adaptive-decision strategies by masking adverse trends or fooling society into taking too strong an action. Interestingly for this book, our preliminary analysis suggests that the most robust adaptive strategies in the face of climate variability begin with moderate near-term emission reductions which are more likely to shift to more rapid abatement based on observations of innovation-reduced abatement costs than of increased climate damages. Second, we examine the mix of policy instruments an adaptive-decision strategy ought to use to encourage the diffusion of emissions-reducing technologies (Robalino and Lempert, 2000 – henceforth RL). We find that in many circumstances a mix of price-based mechanisms (e.g., carbon taxes or tradable permits) and direct technology incentives is more robust than a single price-based mechanism alone. This analysis provides some direct lessons for the type of information that technology forecasts can most usefully provide.

3.4.1 Impact of variability on adaptive-decision strategies

In addressing the impacts of climate variability on adaptive-decision strategies, it is important to first note that an adaptive strategy is not always the best strategy to follow if you take into account the costs of adapting. For instance, there might be expensive monitoring equipment needed to gather information, adjustment costs every time a strategy is changed, and/or the observations used to inform the adaptive strategy might be ambiguous so that mistakes are possible. If the costs of adapting are greater than the expected benefits, it is best to just ignore any signposts and make a permanent, for the foreseeable future at least, choice of shaping and hedging actions. This is the message in the lower-left and upper-right hand corners of Figure 3.1. In these regions the costs of waiting to observe new information outweigh the expected benefits of acting on that information.

The risk of adapting incorrectly is a key cost of an adaptive-decision strategy. Observations are often ambiguous, especially if the observed system has noise and other fluctuations. Decision-makers must balance between waiting until the information becomes more clear and the risk of acting on erroneous information (often called Type I and Type II errors). Given the degree of variability in the climate system, these dangers may be acute for climate-change policy. We need to ask if adaptive-decision strategies can still be robust in the face of climate variability, and if so, what these strategies are.

Our work in LSBA is among the first studies of the impacts of variability on climate-policy choices. While a simple, preliminary treatment, it nonetheless provides some important insight into the design of adaptive-decision strategies. We begin by making two changes to the models used in LSB in order to represent climate variability and its impacts. First, we treat climate variability as a white-noise component to the radiative forcing, so that the change in forcing is given by

$$\Delta Q(t) = 6.3334 \ln \left(\frac{ECD(t)}{ECD(1765)} \right) + \Delta F_{SO4} \left(\frac{E_{SO_2}(t)}{E_{SO_2}(1990)} \right) + g(t), \quad (3.1)$$

where $ECD(t)$ is the effective carbon dioxide concentration that would give the same radiative forcing as the actual concentration of carbon dioxide, methane, and other greenhouse gases; $ECD(1765)$ is the pre-industrial value; $E_{SO_2}(t)$ is the emission rate of SO_2 which is converted to sulfate (SO_4) aerosols in the atmosphere; ΔF_{SO4} is the sulfate-aerosol radiative forcing in 1990; and $g(t)$ is Gaussian-distributed noise with mean zero and standard deviation σ_Q (e.g., Hasselmann, 1976) which generates a red-noise-like temperature variability.

We represent our uncertainty about the climate system with a range of plausible values for the climate sensitivity, ΔT_{2x}, and the sulfate forcing, ΔF_{SO_4}. We choose the range of climate sensitivities, $0.5°C \leq \Delta T_{2x} \leq 4.5°C$, as in LSB. In practice we can find a best-guess estimate of the climate-sensitivity-dependent sulfate forcing parameter by using the EBC/UDO model to reproduce the instrumental temperature record (Andronova and Schlesinger, 2001). However, there are many sources of error, such as the influence of volcanoes and solar-irradiance variations, that could affect these estimates. Thus, we characterize our uncertainty about the sulfate forcing by the percentage it deviates from the best-estimate value as a function of climate sensitivity. For any pair of values for the sulfate forcing parameter and the climate sensitivity, we find the best estimate of the associated white-noise climate forcing, σ_Q, by regressing our EBC/UDO climate model using each pair against the instrumental temperature record from 1856 to 1995.

We represent our uncertainty about the impacts of climate change using a simple, phenomenological damage function designed to capture, in aggregate, some of the impacts of climate variability and the ability of society to adapt to changes in variability. We write the annual damage in year t, as percentage of gross world product (GWP), as

$$D(t) = \alpha_1 \left[\frac{\Delta \overline{T}_5(t)}{3°C} \right]^{\eta_1} + \alpha_2 \left[\frac{\Delta T(t) - \Delta \overline{T}_5(t)}{0.15°C} \right]^{\eta_2} + \alpha_3 \left[\frac{\Delta T(t) - \Delta \overline{T}_{30}(t)}{0.35°C} \right]^{\eta_3}, \quad (3.2)$$

where $\Delta T(t)$ is the annual global-mean surface temperature change relative to its pre-industrial value, and $\Delta \overline{T}_5(t)$ and $\Delta \overline{T}_{30}(t)$ are the 5-year and 30-year running averages of $\Delta \overline{T}(t)$.[2] The second term on the right-hand side of Eq. (3.2) represents impacts due to changes in the variability of the climate system that society and ecosystems can adapt to on the timescale of a year or two. The third term represents those impacts that society and ecosystems adapt to more slowly, on the order of a few decades, and thus are sensitive to both the year-to-year variability and the secularly increasing trend in temperature. The first term represents the damages due to a change in the global-mean surface temperature and is similar to the power-law functions used in most simple damage models in the literature, which we can view as representing impacts that society and ecosystems adapt to on century-long timescales. The coefficient α_1 represents the damages due to a 3°C increase in the global-mean temperature and the coefficient α_2 and α_3 represent the maximum damages due to climate variability in 1995 at the 90% confidence level.

We use a variety of empirical data to fix the damage-function parameters in Eq. (3.2). Because there are many gaps in the available information, and much of it that exists is heavily debated, we define a wide range of plausible parameter combinations rather than any best estimate. In LSBA we used only the requirement that the model be consistent with past observations to constrain the input parameters, though in general, we could also use future forecasts to generate constraints as well (as in the work on innovation described below). The wide range of plausible parameters supports, rather than hinders, our goals in this study since in the end we show that a simple adaptive-decision strategy can be robust against both very small and very large damages.

We constrain the parameters for the first term in Eq. (3.2) by noting that whatever damages due to climate change have occurred in the last few years and decades, they cannot have been more than a few tenths of a percent of GWP, otherwise we would have observed unambiguous evidence of damages

---

[2] The N-year running average is given by $\Delta \overline{T}_N(t) = \sum_{\tau=t-N+1}^{t} \Delta T(\tau)$.

to date. With $\Delta \overline{T}_5 (1995) = 0.5°C$, we can write $\alpha_1 \leq 0.1\% \cdot 6^m$, which corresponds to a range for the damage coefficient $\alpha_1 \leq 20\%$ GWP for cubic damages. We constrained the parameters in the second and third terms using time series data on economic losses from large-scale natural disasters (Munich Re Reinsurance, 1997). This is an imperfect data source because these extreme-event damages (which in 1996, excluding earthquakes, caused \$60 billion, 0.2% GWP, in damages) are due at least in part to trends in society's vulnerability to natural disasters rather than to any change in the size or frequency of natural disasters themselves and, conversely, are only one component of the damages due to climate change. Lacking better sources of available information, we used these extreme-event data to define three bounding cases: "Low Variability" with parameters $(\alpha_2, \alpha_3, \eta_2, \eta_3,) = (0.2\%, 0\%, 1, na)$; "High Variability" with parameters $(0.4\%, 0\%, 2, na)$; and "Increasing Variability" with parameters $(0\%, 0.33\%, na, 3)$. (Note that the "Low" and "High" damages due to variability cases drop the third term in Eq. (3.2), while the "Increasing Variability" case drops the second.) The "High" and "Increasing" cases differ in that damages in the latter grow with increasing concentrations of greenhouse gases while the former do not. Thus damages due to climate variability can be affected by emissions-abatement policy in the latter case but not the former.

This damage model has important shortcomings. Among the most important, the white-noise forcing, the driver of the variability in our model, is fitted to the instrumental temperature record and does not change as we run our simulations into the future. Thus the damage distribution due the variability terms change over time only due to increases in the rate of change in the average temperature, though we expect that in actuality changes in variability are more important than changes in the mean (Katz and Brown, 1992; Mendelsohn *et al.*, 1994). Nonetheless, the crude phenomenological damage function in Eq. (3.2) provides a sufficient foundation to support initial explorations of alternative abatement strategies and the impacts of variability on near-term policy choices.

We can now use this model to examine the potential impacts of climate variability on near-term policy choices. Figure 3.3 shows the difficulties an adaptive-decision strategy might have in making observations of the damages due to climate change. The thin lines show the damages due to climate change generated by Eq. (3.2) for two distinct cases. In the first there is a large trend but low variability. In the second there is a large variability, but no trend. The thick line in each case shows the trend a decision-maker (that is, a decision-making process that includes the making and reporting of scientific observations) might reasonably infer from the respective damage time series, calculated using a linear Bayesian estimator that rapidly detects any statistically

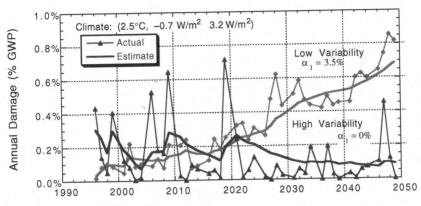

Figure 3.3   Actual (thin lines with markers) and estimated (thick lines) damage time series for two plausible climate impact futures. The lower pair of time series has high damages due to variability and no trend ($\alpha_1 = 0\%$). The upper pair has low damages due to variability and significant trend ($\alpha_1 = 3.5\%$). Both are calculated using the climate parameters $(\Delta T_{2x}, \Delta F_{SO4}, \sigma_Q) = (2.5\,^\circ C - 0.7\ W/m^2,\ 3.2\ W/m^2)$.

significant trend in the damage time series. The estimator does a reasonably good job of tracking the damage time series in both cases, but because of the high variability, the estimates for the trend and no-trend cases do not diverge until about 2020. Thus, an adaptive-decision strategy attempting to distinguish between these two cases based on observations of the damage time series would have to wait at least two decades before being able to act. While there are many cases that the estimator can distinguish more quickly, the question nonetheless remains, how can we design an adaptive-decision strategy that can perform successfully when the observations have the potential for such ambiguity?

In order to address this question, we posit a simple, two-period, threshold-based adaptive-decision strategy, similar to the one used in LSB. The strategy can respond to policy-makers' estimate of any trend in damages based on annual observations of the noisy time series $D(t)$ or to any observed changes, neglecting noise, in abatement costs, represented here by the incremental cost of non-emitting capital. As shown in Figure 3.4, our adaptive strategies begin with a pre-determined abatement rate $1/R_1$ and can switch to a second-period abatement rate $1/R_2$ in the year, t_{trig}, when either the damages exceed, or the abatement costs drop below, some specified target values. The logistic half-life R represents the years needed to reduce emissions to one-half the basecase. The damage target (in % GWP) is given by $D_{est}(t) > D_{thres}$ ($D_{est}(t)$ is the decision-makers' estimate shown in Figure 3.3) and the abatement-cost target (in $/ton carbon abated) is given by $K(t) < K_{thres}$. The second-period rate depends on the year t_{trig}. If $t_{trig} < T_{near}$, then the second-period abatement is given by $R_2 = R_2^{near}$.

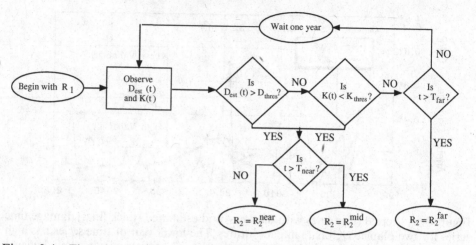

Figure 3.4   Flow chart describing a set of simple adaptive-decision strategies designed
to address climate variability.

If $T_{near} \leq t_{trig} < T_{far}$, the second-period abatement is $R_2 = R_2^{mid}$. If neither condition is met by the year T_{far}, the strategy switches to a second-period abatement $R_2 = R_2^{far}$. We express the decision facing policy-makers as a choice among the eight parameters defining these adaptive-decision strategies.

We now compare the performance of a large number of alternative adaptive-decision strategies. Figure 3.5a compares the expected regret of alternative strategies to the expected regret of the static DAL and ES strategies as a function of our expectations about the future. The horizontal axis gives the likelihood we ascribe to the possibility that DAL, as opposed to ES, is the better response to the climate-change problem. On the left-hand-side of the graph we are sure that DAL is better. Not surprisingly, the expected regret of the DAL strategy at this point is small ($0.2 billion/ year) while the regret of the ES strategy is high ($91 billion/ year). On the right-hand-side we are sure that ES is better, and not surprisingly the expected regret of ES is small ($0.9 billion/year) here and the expected DAL regret is large ($230 billion/year). In the middle of the graph, we ascribe even odds that DAL or ES is the best strategy.[3]

This figure is similar to the cube in Figure 3.1, except that we have collapsed our expectations about the future – the independent variables – into a single

---

[3] Formally, we divide the states of the world into exclusive sets depending on whether DAL, ES, or drastic reductions (DR) is the better policy. (Drastic reductions eliminates anthropogenic greenhouse gas emissions over the first half of the 21st century.) We assign a probabiliy $(1-p_{ES})(1-p_{DR})$ to the states where DAL is better, $p_{ES}(1-p_{DR})$ to the states where ES is better, and p_{DR} to the states where DR is better. The horizontal axis in Figure 5a spans $0\% \leq p_{ES} \leq 100\%$, with $p_{DR}=0\%$. The static DAL and ES policies do not necessarily have zero expected regret at $p_{ES}=0\%$ and 100%, respectively, because one of the adaptive-decision strategies may perform better across the DAL or ES states of the world.

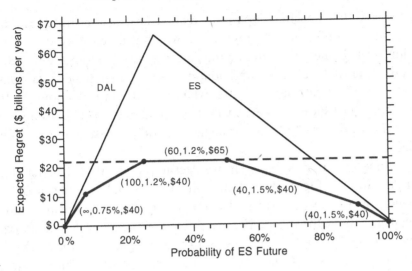

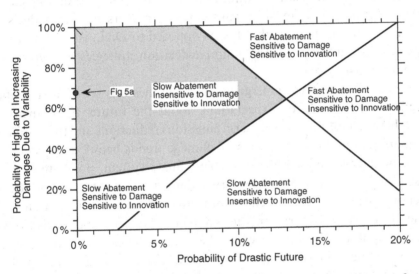

Figure 3.5   (a) Performance of alternative adaptive-decision strategies as a function of the probability ascribed to climate-change futures that require an "Emission-Stabilization" as opposed to "Do-a-Little" response. Strategies are labeled by a triplet representing rate of near-term abatement (in years until emissions are reduced by half), the estimated damages that will trigger a change of strategy (in % of gross world product), and the observed marginal cost of abatement that will trigger a change in strategy (in $/ton-carbon). (b) The most robust adaptive-decision-strategy as a function of expectations about the probability of a drastic future and of high and increasing damages due to variability.

dimension so that we can compare the relative cost of the strategies. There is another major difference. In Figure 3.1 we considered only one adaptive-decision strategy; here we examine the performance of thousands (5120 to be precise). Note that we have found adaptive-decision strategies that dominate the DAL and ES strategies no matter what our expectations about the future. That is, this chart shows that even with the potential of potentially ambiguous observations, adaptive-decision strategies are still the best response to climate change.

Of course, not all possible adaptive-decision strategies are better than ES and DAL. In fact, Figure 3.5a shows only a very small number of dominant strategies, those that perform best compared to the others as a function of expectations about the future. For each of many points across the horizontal axis we find the strategy with the lowest expected regret and label each of these dominant strategies with the three parameters – the rate of near-term emissions reductions, the damage threshold, and the innovation threshold – $(R_1, D_{thres}, K_{thres})$, that are most relevant to policy-makers' near-term decisions. As our expectations increase that ES, as opposed to DAL, is the better strategy, we pass through five different adaptive-decision strategies, with R_1 ranging from infinity to 40 years.

Two interesting patterns emerge as one examines this set of best adaptive strategies as a function of expectations about the future. First, there is a tradeoff between the rate of near-term emissions reductions and the confidence one should require in observations of damage trends before acting on them. That is, in the face of variability in the climate system, policy-makers can choose a response threshold for observed damages that can compensate, to a greater or lesser extent, for any choice of near-term emissions-reduction target. Second, the rate of near-term emissions reductions depends most sensitively on expectations about the future, while the aggressiveness with which society ought to respond to innovations are the least sensitive. That is, the policy choice at the focal point of the current negotiations may be the most controversial component of a robust strategy, in part because stakeholders with different expectations will have the most divergent views as to the proper target, while the least controversial components may be at the periphery of the negotiations.

Figure 3.5a also shows the most robust strategy across the range of expectations, which we find by searching for the strategy with the least squares regret across the points on the horizontal axis. This most robust strategy, given by (60 yrs, 1.2%, $65), begins with moderate emissions reductions, is relatively insensitive to observations of climate damages, and is very sensitive to observations of decreasing abatement costs. Its annualized, present-value expected regret is $22 billion/year, independent of our expectations about the ES future. While this low number does not mean climate change is costless (the regret measures the difference between the performance of a given strategy and that of the best strategy for that state of the world), it does suggest that even faced with deep uncertainty, we can find adaptive-decision strategies that perform nearly as well as the strategies we would choose if we knew the future with clarity.

The adaptive-decision strategies shown in Figure 3.5a still contend with a rather narrow range of uncertainty. For instance, in making this figure, we have assumed equal likelihoods of each of our variability cases – low, high, and increasing – that sulfate emissions are low, as in the scenarios of the Special Report on Emissions Scenarios (SRES; Nakicenovic *et al.*, 2000), and zero probability of climate impacts so severe as to require immediate and rapid phase-out of anthropogenic greenhouse gas emissions. In Figure 3.5b, we examine the sensitivity of the most robust strategy to these assumptions. To make this figure, we find the most robust strategy as a function of: (i) our expectations that immediate and draconian emissions reductions (we call this the Drastic Reductions strategy) are the best response to the threat of climate change and: (ii) the probability that the damages due to climate variability are high or increasing. Our finding that the most robust adaptive-decision strategy has moderate near-term emissions and is more sensitive to observations of abatement costs than to observations of damages, is true in the shaded region on the left of the figure where the likelihood of drastic future ranges from about 0% to 12%, depending on the likelihood of low climate variability.

3.4.2 *Promoting innovation*

If climate change is indeed a serious problem, society will have to make significant reductions in greenhouse-gas emissions, on the order of 80% below extrapolations of current trends, by the end of the 21st century. Technology innovation will likely play a major role in any changes of this scale. A number of modeling studies have made this point simply by treating innovation as an exogenous influence on key parameters, such as the rate of autonomous energy efficiency improvements, the cost of low-carbon emitting technologies, or the rate of technology spillovers from developed to developing countries (see for instance, Dowlatabadi, 1998 and Edmonds and Wise, 1998). By considering a range of assumptions about these parameters, these studies confirm that over the long term, sustained rates of either fast or slow innovation can make virtually any emissions-reduction target either inexpensive or impossibly costly to meet.

This fact, and our argument that the most robust, adaptive-decision strategies may emphasize technology innovation, raises a key question – what should policy-makers do to encourage such innovation? In practice, governments pursue a wide range of technology policies designed to improve the technology options for emissions reductions in the future. Such policies include supporting research and development, training individuals, funding demonstration projects, building infrastructure, disseminating information, and implementing a variety of tax credits and subsidies to encourage the use of new technologies. These programs often appear attractive both on substantive and political grounds, but their record in practice is mixed. In addition, economic theory suggests that in the absence of market failures, such policies are inefficient compared to policies designed to "get the prices right", such as carbon taxes or tradable permits. While there is wide agreement that governments ought to fund research and development (though less agreement on the extent to which this funding ought to focus on specific areas such as emissions-reducing technologies), it is not clear the extent to which climate-change policy ought to focus on getting the price of carbon right, developing technology policies designed to improve future options for emissions reductions, or some combination of both.

Our judgements about this question will depend on our expectations about the ability of policy to change the dynamics of technology diffusion. Many recent studies have begun to address these issues (Grübler, Nakicenovic, and Victor, 1999; Azar and Dowlatabadi, 1999), focusing on both the effects of learning-by-doing on technology diffusion (Gritsevskyi and Nakicenovic, 2000; Mattsson and Wene, 1997; Anderson and Bird, 1992) and on the accumulation of knowledge that can lead to new innovation (Goulder and Schneider, 1998, 1999; Goulder and Mathai, 2000). In addition, much empirical work (e.g., Newell, Jaffe, and Stavins, 1999; Grübler, Nakicenovic, and Victor, 1999) is becoming available to inform these modeling efforts. The modeling efforts to date, however, have largely focused on creating scenarios of the future and showing their sensitivity to a variety of assumptions about technology innovation. Few have used this information to adjudicate among alternative policies and, in particular, examine the desirability of technology policies. In our recent work, we have examined the conditions under which technology incentives should be a key building block of a robust, adaptive-decision approach to climate change. This is an interesting question in its own right, but it also sheds important light on the design of adaptive design strategies and on the types of information about technology futures that, in the absence of an ability to forecast, should prove most useful to policy analysis and policy-makers.

We have addressed these questions using our exploratory modeling methods and a model of technology diffusion which focuses on the social and economic

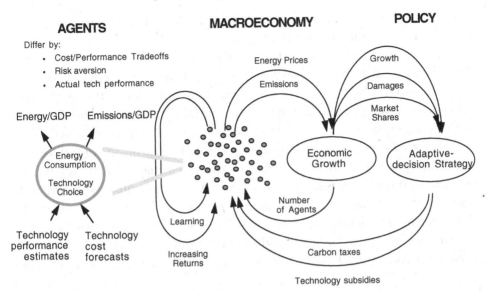

Figure 3.6   Agent-based model of technology diffusion used to compare alternative climate-change abatement strategies. Economic agents choose among alternative technologies on the basis of forecasts of cost and performance, which in turn are influenced by learning among the agents and potential price decreases due to increasing returns to scale. The agents have heterogeneous initial expectations about technology performance and heterogeneous preferences for technology cost/performance tradeoffs. The agents' choices influence the level of energy prices and of greenhouse-gas emissions, which both influence the rate of economic growth. Policy decisions about the level of carbon taxes and technology incentives, which depend on observations of economic growth, damages and technology diffusion, also influence the agents' technology choices.

factors that influence how economic actors choose to adopt, or not to adopt, new emissions-reducing technologies. As shown in Figure 3.6, we use an agent-based model of technology diffusion coupled to a simple macro model of economic growth. Used in this fashion, an agent-based model is merely a stochastic mathematical function representing the dynamics of factors in the macro-model, such as energy intensity. Like any time series model, its parameters can be calibrated to reproduce real-world data. Such agent-based models are particularly useful, however, because they conveniently represent key factors influencing technology diffusion, and thus policy choices, such as the heterogeneity of economic actors and the flows of imperfect information that influence their decisions. Each agent in our model represents a producer of a composite good that is aggregated as total GDP, using energy as one key input. In each time period the agents first choose among several energy-generation technologies and second, given their chosen technology, choose how much energy to consume. (That is, agents choose a production function and where to operate on that production function.)

We assume that agents pick a technology in order to maximize their utility, which depends on each agent's expectations about the cost and performance of each technology. The agents have imperfect information about the current performance of new technologies, but can improve their information based on their own past experience, if any, with the technology, and by querying other agents who have used it (Ellison and Fudenberg, 1995). The agents are also uncertain about the future costs of new technologies, which may or may not decline significantly due to increasing returns to scale. Agents estimate these future costs based on observations of the past rates of adoption and cost declines. Thus, the diffusion rate can depend reflexively on itself, since each user generates new information that can influence the adoption decisions of other potential users.

We write each agent's utility function using a risk-adjusted, Cobb–Douglas form

$$\langle U_{i,g,j}(t) \rangle = \langle Perf_{i,g,j}(t)^{\alpha_i} \rangle \langle Cost_{i,g,j}(t)^{1-\alpha_i} \rangle - \lambda_i [\mathrm{Var}_{Perf}(t) + \mathrm{Var}_{Cost}(t)], \quad (3.3)$$

where the first term, $\langle Perf_{i,g,j}(t)^{\alpha_i} \rangle$, is the expectation of agent i, in region g, at time t of the performance it will get from technology j. This term depends on information flows among the agents, which we model crudely as a random sampling process, where each agent queries some number of the other agents each time period. The second term, $\langle Cost_{i,g,j}(t)^{1-\alpha_i} \rangle$, is the expected cost of using the technology over its lifetime, derived from observations of past trends in usage and cost of the technology. This term depends on the potential for cost reductions due to increasing returns to scale. The third term represents the agent's risk aversion, taken as a function of the variance of the estimates of technology performance and future costs.

Equation (3.3) focuses on the heterogeneity of the agents' preferences, which is important in creating the possibility for early adopters and niche markets that can strongly influence the course of technology diffusion. First, different agents obtain different levels of performance from the same technology because technologies differ according to characteristics such as size of the equipment and ease of maintenance that matter to some users more than others (Davies, 1979). We represent this in our model with a distribution of performance factors for each technology across the population of agents. Economic actors also have different cost/performance preferences for new technologies. Potential early adopters are generally much more sensitive to performance than price, while late-adopters are often more price sensitive. We represent this in our model by allowing agents to have different values for the exponents α_i, representing the cost/performance tradeoffs, and in the risk aversion coefficient λ_i. Finally, in a world of imperfect information, different economic actors will have different expectations about the performance and cost of each technology. At the beginning of each of our simulations, each agent is assigned an expectation about the performance of each technology, which it can update as time goes by. Each agent's expectations of performance are private; that is, they apply only to that agent, since in fact each agent will gain a unique performance from each technology. The cost forecasts are public, that is, shared in common by all the agents, but each agent in general will have different planning horizons, determined by the remaining lifetime of the technology they are already using. Each type of heterogeneity considered in our model can significantly affect market shares and the dynamics of the diffusion process (Bassanini and Dosi, 1998).

This model requires data on the social and economic context of technology diffusion, quite different from that generally demanded by other climate-change-policy models and often supplied by technology forecasts. Each of the technologies in our study is represented by three factors: the cost (which can drop over time due to increasing returns), the carbon intensity (the quantity of CO_2 emitted when generating one unit of energy), and the performance. The cost and the carbon intensity are intrinsic to the technology, and are treated in our model similarly to other models. However, the performance, as noted above, depends on the agents using it. In addition, we need information on the different preferences and current expectations of the users and potential users of the new technologies. Thus, we are interested in data which show how the performance of a technology varies across many different types of users and, in particular, how different key segments of the market along the technology adoption life cycle (Moore and McKenna, 1995) – the early adopters, early and late majority, and laggards – may perceive, judge, and use the technology. In our model we treat these factors crudely, assuming in each case that preferences, expectations, and performance are distributed normally across our population of agents. Using these simple assumptions we can draw what seem to be important, but general, policy conclusions. More detailed policy recommendations would probably require better information about such factors.

In addition to considering the agents' individual technology and energy consumption choices, we also need to consider the aggregate impact of the agents' actions. We capture these effects with a very simple representation of economic growth. Assuming a world economy in a *steady state*, we write the Gross Domestic Product (*GDP*) in each of two world regions, the OECD and the rest of the world (ROW), with the difference equation, where output per capita grows at some exogenous rate that can be modified by changes in the price of energy and any damages due to climate change. Thus,

$$GDP_g(t) = GDP_g(t-1)[1 + \gamma_g - \phi_{xg} \cdot \Delta C_{eng}(t) - \phi_{sg} \cdot C_{sub}(t)][1 - impact\,(t)], \quad (3.4)$$

where γ_g represents the exogenous growth rate in the regions $g = $ OECD and ROW; C_{eng} is the growth rate of the average cost of energy per unit of output in region g, including the costs of energy-producing technologies, any carbon tax imposed in order to reduce CO_2 emissions, and any subsidies on new low-emitting technologies; C_{sub} is the per unit of output cost of the subsidy; ϕ_{xg} and ϕ_{sg} are the corresponding elasticities in energy prices, including the costs of energy-producing technologies, any carbon tax imposed in order to reduce CO_2 emissions, and the cost of the subsidy to changes in economic growth; and impacts due to climate change are given by a simple polynomial function of the atmospheric concentrations of greenhouse gases, $impact\,(t) = \kappa_o\,[Conc\,(t)/Conc\,(1765)]^{\kappa_1}$, using a simple difference equation to relate concentrations to emissions (Cline, 1992; Nordhaus and Yang, 1996). The aggregate decisions of the population of agents affect the cost of energy, the emissions, and thus, the climate impacts. In turn, the rate of GDP growth affects the number of new agents with new expectations and no sunk capital cost and, thus, the decisions of each individual agent.

Having defined our model, we can use it to create the landscape of plausible futures by defining ranges for the model inputs and constraints on the model outputs. On the micro (bottom-up) level, we confine ourselves to parameters describing only three generic types of technologies: *high-emission-intensity* systems, such as coal-fired power plants, which at present provide the bulk of the world's energy; *medium-emissions-intensity* systems, such as natural-gas-powered combustion, which provide a significant minority of the world's energy; and *low-emissions-intensity* systems, such as renewable, biomass, and/or new nuclear power facilities, which at present are not in widespread use, but may be significant energy sources in the future. This is clearly a very coarse grouping, similar to that used by Grübler, Nakicenovic, and Victor (1999), but it is sufficient to draw policy conclusions about the appropriate mix of carbon taxes and technology subsidies. We choose a range of cost and performance parameters for these three generic energy technologies, as well as parameters describing the behavior of the population of consumers of these energy technologies, based on averages of data describing technology systems currently with significant market penetration and standard forecasts of technologies with potential significant future market penetration (Tester *et al.*, 1991; Manne and Richels, 1992).

On the macro (top-down) level, we use data for a variety of parameters describing the growth of the economy and its response to changes in the cost of energy and damages due to climate change (World Bank, 1996; Dean and

Hoeller, 1992). These parameters include the exogenous rate of economic growth, the elasticity of economic growth with respect to the cost of energy, the elasticity of economic growth with respect to the cost of the subsidy, the parameters characterizing damages, and the concentration of carbon in the atmosphere, the parameters defining the energy-demand functions, and those used to simulate exogenous improvements in energy efficiency.

We can now search across the thirty-dimensional space of model inputs that plausibly represent the micro-level data, looking for those combinations that give plausible, macroscopic model outputs. Given the simplicity of the model and the types of data readily available, we choose three constraints on model outputs: current (1995) market shares for energy technologies; current levels of carbon emissions, energy intensities, and carbon intensities; and diffusion rates no faster than 20 years (from 1% to 50% penetration). The first two constraints guarantee that our model is consistent with current data. The third, limiting diffusion to a rate no faster than the rates historically observed for energy technologies (Grübler, 1990), forces the model to be consistent with one of the historically observed patterns of technology diffusion.

We next generate the most expansive ensemble we can produce of model input parameters consistent with these constraints. Using a genetic search algorithm, we generated 1611 such sets of input parameters, covering a wide variety of assumptions about key parameters such as the level of increasing returns to scale, agents' heterogeneity – represented by the distribution of values for the coefficients in Eq. (3.3) across the population of agents, uncertainty regarding new technologies, and future damages due to climate change. This set does not include points with very small levels of uncertainty and heterogeneity regarding expectations about the performance of new technologies (such points do not satisfy the constraint on initial market shares), sets of points where the agents' utility functions are largely insensitive to costs, and sets of points with both very high learning rates and levels of increasing returns to scale.

We can now ask the question as to whether an adaptive-decision strategy ought to use technology incentives and carbon taxes, or carbon taxes alone, in order to address the threat of climate change. We posit two strategies, "Taxes Only" and "Combined Strategy", whose performance we compare across the landscape of plausible futures. (We also considered no response and a strategy using only technology subsidies. Neither of these were very attractive compared to the other two.)

As the name implies, the "Taxes Only" strategy employs only a carbon tax, whose level can change over time in response to observations of the rate of economic growth and the damages due to climate change, as shown in Figure 3.7.

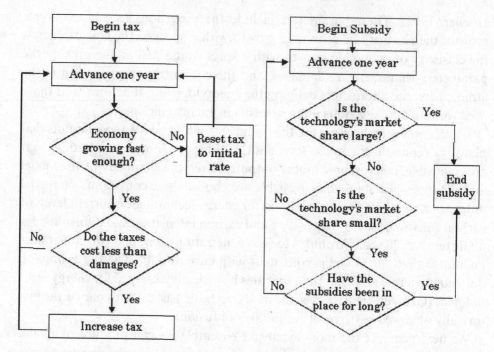

Figure 3.7 Adaptive decision strategy for adjusting carbon taxes and technology
incentives over time.

The tax begins at some initial level per ton of emitted carbon and grows at a
fixed annual rate. However, if in any year the cost of the tax is greater than the
marginal cost of emissions of carbon dioxide (expressed as a percentage of
GDP), the tax remains constant. Similarly, if in any year the global economy
growth rate is below some minimum rate, the tax returns to its initial level, from
which it can begin to grow again. This description of a steadily growing tax is
consistent with the optimum taxes described in the literature (Goulder and
Schneider, 1999), and the stopping condition represents a way in which political conditions may force a tax to terminate.

The "Combined Strategy" uses both the carbon tax and a technology
subsidy, which can change over time in response to observations of the market
share of low-emitting technologies, as shown in Figure 3.7. The subsidy begins
at some initial level, expressed as percent of the cost of the subsidized technology. This subsidy stays at a constant level over time until either the market share
for low-emitting technologies goes above a threshold value or the market share
fails to reach a minimum level after a certain number of years. If either of these
conditions is met, the subsidy is permanently terminated. Thus, we assume the
subsidy is terminated once policy-makers observe that the technology succeeds
or that it fails to diffuse by some deadline.

These tax and incentive policies are together characterized by a total of seven parameters – the beginning levels of the tax and subsidy, the annual increase in the tax, the minimum level of economic growth needed to maintain the tax, the maximum market share which terminates the subsidy, and the minimum market share over what time period that terminates the subsidy. We choose the particular value of these parameters used to define our "Taxes Only" and our "Combined Strategy" by searching for the best tax and the best subsidy strategy at the point in uncertainty space characterized by the average value for each of the model input parameters. The tax starts at an initial value of $100/ton carbon and grows at 5% per year. The initial subsidy is 40% of the cost of the low-emitting technology. The subsidy terminates if the low-emitting technology reaches 50% of market share, or fails to reach 20% of market share after 15 years. This process is a crude approximation to the procedure used in LSBA, in which we found the best strategy for many different states of the world. We are currently applying the LSBA procedure to a comparison of tradable permits and technology incentives. The general conclusions presented here seem to hold.

In order to compare the performance of the alternative strategies, we calculate their performance for each of a large number of different states of the world, looking for those that distinguish one policy choice from another. There are too many dimensions of uncertainty (30) in this model for an exhaustive search, so we used knowledge about a system and the goals of the analysis to summarize a very large space with a small number of key scenarios. We performed a Monte Carlo sample over the thirty-dimensional space, used econometrics (Aoki, 1995) to look for those input parameters most strongly correlated with a key model output of interest (GHG emissions after fifty years), and used variations across these key parameters to define scenarios. (This is a version of critical factors analysis, as in Hope *et al.* (1993).) After making hypotheses about policy recommendations based on this reduced set of scenarios, we then tested our results by launching a genetic search algorithm (Miller, 1998) across the previously unexamined dimensions looking for counter examples to our conclusions.

Figure 3.8, a typical result of such comparisons, shows the regret of the Taxes-Only strategy (dashed line) with the regret of the Combined strategy (solid line), as a function of the heterogeneity of the agent population. For this figure we have assumed moderately increasing returns to scale, a moderate level of social interactions, and moderate damages due to climate change. The figure shows that the Taxes-Only strategy is preferable in a world where the agents are homogeneous. As the heterogeneity of the agents' preferences increases, the Combined tax and subsidy strategy quickly become more attractive.

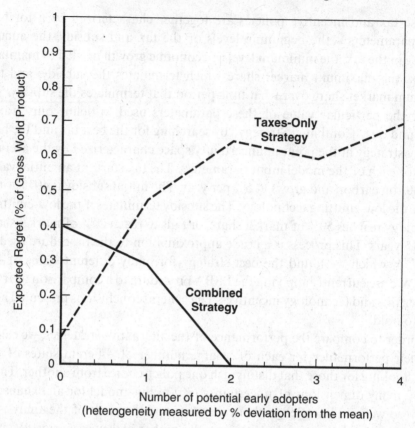

Figure 3.8 Expected regret of the taxes only (dashed line) and combined tax and technology incentives (solid line) adaptive-decision strategies. The lines are displayed as a function of heterogeneity of the agents' population. All other input parameters are held constant at their mean values.

The Combined strategy becomes less costly than Taxes-Only due to the competition between two effects. Carbon taxes reduce emissions by slowing economic growth and inducing agents to switch to low-emitting technologies. Subsidies slow growth slightly, and may or may not induce agents to switch to low-emitting technologies. When the agent population has low heterogeneity, there are few potential early adopters and the tax is more efficient than the subsidy. Increasing heterogeneity favors the Combined strategy, because it creates a number of potential early adopters that the subsidy will encourage to use the new, low-emitting technology which, in turn, generates learning and cost reductions that benefit society as a whole. But the Subsidy-Only strategy never dominates in our simulations, because when the heterogeneity is large, there are always a substantial number of late-adopters who will reduce emissions only when faced with a tax.

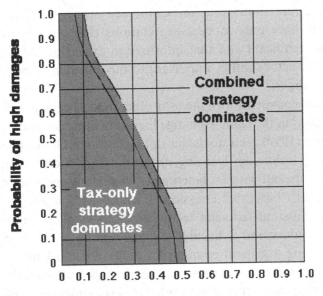

Figure 3.9 Regions in probability space where the expected gross world product resulting from the Taxes-Only strategy is greater than that for the Combined tax and technology incentives strategy as a function of the probability of high climate damages and of the potential for significant cost reductions for new technologies due to increasing returns to scale, reduction of uncertainty about the performance of new technologies due to learning among agents, and heterogeneity in the cost/performance preferences for new technologies among agents ("non-classical world").

We have considered a large number of results such as those in Figure 3.8, which we summarize in Figure 3.9. The figure shows the expectations about the future that should cause a decision-maker to prefer the Taxes-Only strategy to the Combined strategy. The horizontal axis represents the range of expectations a decision-maker might have for how likely it is – from very unlikely on the left to very likely on the right – that factors such as the potential number of early adopters and the amount of increasing returns to scale will significantly influence the diffusion of new technologies.[4] The vertical axis represents the range of expectations a decision-maker might have that there will be significant impacts due to climate change (greater than 0.3% of the global economic

---

[4] The horizontal axis shows a linear scaling factor applied to four key model inputs: the level of increasing returns for low-emitting technologies, the rate at which agents learn from one another about the performance of new technologies, the agents' risk aversion in Eq. (3.3), and the heterogeneity of the agents' preferences defined by the α_i parameter in Eq. (3.3). On the far left (a value of zero on the horizontal axis), the four input parameters take on their minimum plausible values. On the far right (a value of 1 on the horizontal axis), the four input parameters take on their maximum plausible values.

product). The figure shows that the Combined strategy dominates even if decision-makers have only modest expectations that impacts from climate change will be significant and that information exchange and heterogeneity among economic actors will be important to the diffusion of new, emissions-reducing technologies.

Our results are consistent with those of other models of induced technology change introduced in the climate-change literature in recent years. Gritsevskyi and Nakicenovic (1999) use a stochastic optimization technique to examine the effects of induced technology learning, uncertainty, and increasing returns, and technology clusters, but without heterogeneous preferences. They find a wide, bimodal range of "basecase" emissions scenarios and argue for near-term investment in emissions-reducing technologies focused on clusters of such technologies. Goulder and Schneider (1998) examine greenhouse-gas-abatement policies using a general equilibrium model for the United States that takes into account incentives to invest in research and development, knowledge spillovers, and the functioning of research and development markets. They find that the tax should be accompanied by a subsidy *only* when there are spillover benefits from research and development. The work presented here shows that this result can be more general. Many types of spillovers, such as those resulting from increasing returns to scale, network externalities, or *non-R&D knowledge spillovers* from users to non-users of new technologies may justify a subsidy. However, we find that the level of spillovers that justify the subsidy depends on the degree of heterogeneity of agents' preferences and their attitude towards risk, and that, all other things being equal, the critical level to justify the subsidy decreases as heterogeneity and risk aversion increase.

3.5 Conclusions

When viewed as an adaptive-decision strategy, one that can evolve over time in response to observations of the climate and economic systems, climate-change policy has important implications for new energy technologies and the role of forecasts of these technologies. In the work described here, we examined several facets of this broad question. First, we argued that an adaptive-decision approach to climate change will be significantly more robust in the face of deep uncertainty than an approach that does not adapt. We found that the near-term steps in such a robust, adaptive-decision response to climate change should likely include actions explicitly designed to encourage the early adoption of new, emissions-reducing technologies. Such actions are justified if there are significant social benefits of early adoption beyond those gained by the adopters themselves, as is the case when there is heterogeneity in economic actors' cost/performance preferences for new technologies, when new technologies have increasing returns to scale, or when potential adopters can learn from others about the uncertain performance of new technologies. These results are consistent with most other climate-policy analyses which include both significant uncertainty and technological change that responds to policy choices and economic signals.

In addition, we examined the impacts of climate variability on near-term policy choices and found that the future course of emissions-reducing technologies may be one of the key indicators that should shape the evolution of an adaptive-decision strategy. The ability of future innovation to radically reduce emissions of greenhouse gases during the 21st century is one of the key uncertainties facing climate-change policy, perhaps more so than our uncertainty in the response of the climate system to those emissions. Particularly in the presence of climate variability, over time, policy-makers may gain more definitive information about the potential of emissions-reducing technology than they will about the damages due to anthropogenic influences on the climate. Over the next decades then, the most reliable information policy-makers can use to adjust emissions constraints on greenhouse gases and other policy actions may come from indicators of society's ability to reduce such emissions at low cost.

Combining these findings supports the broad view that in the near-term, climate-policy ought to be more focused on improving society's long-term ability to reduce greenhouse-gas emissions than on any particular level of emissions reductions. As a long-term goal, the Framework Convention on Climate Change aims to stabilize atmospheric concentrations of greenhouse gases at some safe, currently unknown and unspecified, level. As a first step, the Kyoto Protocol requires the developed countries to reduce their greenhouse-gas emissions to specific targets by the end of the next decade. Both advocates and opponents usually see these emissions caps as hedging actions, designed to decrease the severity of any future climate change by reducing the concentration of greenhouse gases in the atmosphere. Thus, much of the climate-change debate, particularly in the United States, takes the particular level of these goals very seriously, revolves around whether or not such targets are justified, and seeks to determine the least expensive means to meet them.

An adaptive-decision approach, combined with the uncertain, but large potential of the various technologies described in this volume, suggests that the Kyoto targets are in fact shaping actions, whose most significant impact may be making future emissions reductions, if necessary, less costly and easier to implement, rather than reducing atmospheric greenhouse-gas concentrations. Goals and forecasts often serve this type of shaping function. For instance, the leadership of organizations, such as private firms, often set goals for sales or profits intended to motivate their employees. The leaders would be disappointed if the goals were consistently met, for that would be an indication they had not asked their employees to strive hard enough. Conversely, the firm's financial planners will likely produce conservative sales and profit forecasts and be disappointed if reality does not exceed their expectations. Similarly, the Kyoto targets may be most important as a signal to numerous actors throughout society that they should take the climate problem seriously and plan their actions and investments accordingly.

To some extent, Kyoto has already been effective in this regard. For instance, many firms are beginning to reduce their own emissions, factor potential climate change into their internal planning, and invest in research that would help them reduce emissions more in the future. Traders have begun to implement markets for trading carbon-emissions rights. Governments have begun to construct the institutions necessary to implement international regulation of greenhouse-gas emissions. It is unclear and, for an adaptive-decision strategy, perhaps irrelevant, whether or not these actions will result in Kyoto emissions-reductions targets being met. Rather, the question becomes what are the means to induce technological change that best balance between the possibility that such innovations will soon be needed to address severe climate change and the possibilities that they will not be needed or that the technologies will not meet their promise.

In this context, technology forecasts, to be most useful for informing the design of adaptive-decision strategies, should provide a broader range and different types of information than is often the case. Typically, an energy-technology forecast describes the future costs of the power produced by the system, estimates the maximum amount of society's energy needs the technology might satisfy, and reviews the technology's pros and cons. Often these predictions stem from an analysis of fundamental physical or engineering constraints and/or from analogies from similar systems. For instance, forecasts of the potential generation capability of future photovoltaic energy systems might be based on comparisons of the current and theoretical efficiencies of various types of cells and a survey of the solar insolation and land available for photovoltaic installations. Forecasts of the future costs of these systems might compare past rates of cost reductions to limits imposed by the basic costs of materials (that is, setting the design, production, and installation costs to zero). Sometimes these technology forecasts are given as best estimates; sometimes as a range of scenarios. Generally they are conceived as an answer to the question: how important to society will this technology be in the future?

Such information is useful to policy analyses of climate change. But it is also contrary to much of what we know about processes of technology change, especially over the timescales relevant to the climate problem. In the long term, we know costs and market shares of particular technologies are impossible to predict with any accuracy. We also know that social and economic factors, generally under-emphasized in technology forecasts, are among the primary determinants of the fate of new technologies. The unpredictability of such social and economic factors is a reason that technological change itself is impossible to predict. As one example, a recent study found that cost projections for renewable-energy systems had been reasonably accurate over the last twenty years, while predictions of market share have proved significantly too optimistic. The reasons are not entirely clear, but cost reductions may have been helped by rapid advances in materials and information technologies occurring outside of renewable-energy development, and the market shares have been hindered by the fossil-energy prices being much lower than past forecasters imagined possible.

Nonetheless, these social and economic factors impose common patterns on the processes of technology diffusion and can play an important role in constraining future scenarios, thus helping us choose the best near-term steps in an adaptive-decision strategy. For instance, we know that new technologies generally experience lifecycles that begin with periods of rapid improvement in cost and performance, continue with a period of mature incremental improvements, and end with a period of replacement by newer technologies. A technology's early stages are characterized by many firms experimenting with alternative designs, followed by a period of consolidation over a single design and competition of better means to produce and distribute that design (Utterbeck, 1994). We know there are common patterns of entry and exit as technologies compete against one another in the market place, although radical new technologies sometimes replace established competitors, the old technologies can sometimes crush the new technologies with a burst of incremental improvements. We know there are network effects in which improvements, or the lack thereof, in one technology area can impact improvements in other areas.

Our analysis gives one example of how such patterns, and information about the social and economic factors that influence them, can be used to help design adaptive-decision strategies. We find that expectations about factors such as the heterogeneity of the potential adopters of a new technology – the availability of niche markets – and the speed with which information about experience with a new technology travels through a society – the networks and industry structure among these niche markets – are important in determining the best policy choices. In our analysis we used only crude estimates for such processes. Technology forecasts that provide information on factors such as the size of niche markets, the types of networks in which these potential early adopters reside, and the types of cost/performance preferences they display, could greatly enhance analysis of this type. Forecasts that examine how these factors might play out in a variety of diffusion scenarios would be even more useful. Such information might be seen as less rigorous than the hard, physical constraints on predicted cost and total deployment generally offered in technology forecasts. However, good descriptions of early markets and potential diffusion paths are a more realistic goal than predicting the long-term course of new technologies and, as described in this chapter, provide a more solid basis on which to design responses to climate change that adapt over time and, thus, are robust across a wide range of plausible futures.

References

Anderson, D. and C. D. Bird, 1992: Carbon accumulation and technical progress – a simulation case study of costs, *Oxford Bulletin of Economics and Statistics*, **54**, 1–29.

Andronova, N. G. and M. E. Schlesinger, 2001: Objective estimation of the probability density function for climate sensitivity, *J. Geophys. Res.* **106 (D19)**, 22605.

Aoki, M., 1995: The Japanese Economic Review, **46**, 148–65.

Azar, C. and H. Dowlatabadi, 1999: A review of technical change in assessments of climate policy, *Annual Review of Energy and Environment*.

Bankes, S. C., 1993: Exploratory modeling for policy analysis, *Operations Research*, **41**, no. 3, 435–49, (also published as RAND RP-211).

Bassanini, A. P. and G. Dosi, 1998: *Competing Technologies, International Diffusion and the Rate of Convergence to a Stable Market Structure*, Interim Report from the International Institute for Applied Systems Analysis, (1998).

Casan, E. A., M. G. Morgan, and H. Dowlatabadi, 1999: Mixed levels of uncertainty in complex policy models, *Risk Analysis*, **19**, 1, 33–42.

Clark, W. C., 1980: Witches, floods, and wonder drugs: historical perspectives on risk management, in *Societal Risk Assessment: How Safe is Safe Enough?* eds. R. Schwing and W. Albers, Jr., Plenum Press, New York.

Cline, W., 1992: *The Economics of Global Warming*, Institute for International Economics, Washington, DC.

Davies, 1979: *The Diffusion of Process Innovations*, Cambridge University Press, Cambridge.

Dean A. and P. Hoeller, 1992: *Costs of Reducing CO_2 Emissions. Evidence From Six Global Models*, Working Papers, Department of Economics, no 122. OECD, Paris.

Dewar, J. A., C. H. Builder, W. M. Hix, and M. H. Levin, 1993: *Assumption Based Planning: A Planning Tool for Very Uncertain Times*, RAND, Santa Monica, CA, MR-114-A.

Dowlatabadi, H., 1998: Sensitivity of climate change mitigation estimates to assumptions about technical change, *Energy Economics*, **20**, 473–93.

Dowlatabadi, H., 1999: Adaptive management of climate change mitigation: a strategy for coping with uncertainty, paper presented at the Pew Center

Workshop on the Economics and Integrated Assessment of climate change, July, Washington, D.C.

Edmonds, J. and M. Wise, 1998: The value of advanced energy technologies in stabilizing atmospheric CO_2, in *Cost-Benefit Analyses of Climate Change*, F. L. Toth (ed.), Birkhauser.

Ellison, G. and D. Fudenberg, 1995: *Quarterly Journal of Economics*, **440**, 93–125.

Goulder, L. H. and K. Mathai, 2000: Optimal CO_2 abatement in the presence of induced technological change, *Journal of Environmental Economics and Management*, **39**, 1–38.

Goulder, L. H. and S. H. Schneider, 1999: Induced technological change and the attractiveness of CO_2 abatement policies, *Resources and Energy Economics*, **21** (3–4) August, 211–53.

Gritsevskyi, A. and N. Nakicenovic, 2000: Modeling uncertainty of induced technological change, *Energy Policy*, **28**, 907–21.

Grubb, M., T. Chapuis, and M. Ha-Duong, 1995: *Energy Policy*, **23**, 417–31.

Grübler, A., 1990: *The Rise and Fall of Infrastructures*, Physica Verlag, Heidelberg.

Grübler, A., N. Nakicenovic, and D. G. Victor, 1999: Dynamics of energy technologies and global change, *Energy Policy*, **27**, 247–80.

Hammitt, J. K., R. J. Lempert, and M. E. Schlesinger, 1992: A sequential-decision strategy for abating climate change. *Nature*, **357**, 315–18.

Hasselmann, K., 1976: Stochastic climate models. Part 1: Theory, *Tellus*, **28**, 473–85.

Hope, C., J. Anderson, and P. Wenman, 1993: Policy analysis of the greenhouse effect: an application of the PAGE model, *Energy Policy*, **21**, 3.

Houghton, J. T., G. J. Jenkins, and J. J. Ephraums (eds.), 1990: *Climate Change, The IPCC Scientific Assessment*. Cambridge University Press, Cambridge.

Katz, R. and B. G. Brown, 1992: Extreme events in a changing climate: variability is more important than averages, *Climatic Change*, **21**, 289–302.

Lave, L. and H. Dowlatabadi, 1993: Climate change policy: the effects of personal beliefs and scientific uncertainty, *Environ. Sci. Technol.*, **27**, 1962–72.

Lempert, R. S., 2001: Finding transatlantic common ground on climate change, *The International Spectator*, vol. **XXXVI**, No. 2, April–June.

Lempert, R. J. and M. E. Schlesinger, 2000: Robust strategies for abating climate change, *Climatic Change*, **45**, 387–401.

Lempert, R. J., M. E. Schlesinger, and S. C. Bankes, 1996: When we don't know the costs or the benefits: adaptive strategies for abating climate change, *Climatic Change*, **33**, 235–74.

Lempert, R. J., M. E. Schlesinger, S. C. Bankes, and N. G. Andronova, 2000: The impact of variability on near-term climate policy choices and the value of information, *Climatic Change*, **45**, 129–61.

Lindzen, R. S., 1990, Some coolness about global warming, *Bull. Amer. Meteorol. Soc.*, **71**, 288–99.

Manne, A. S. and R. G. Richels, 1991, Global CO_2 emission reductions: the impact of rising energy costs, *The Energy Journal*, **12**, 88–107.

Manne, A. S. and R. G. Richels, 1992: *Buying Greenhouse Insurance: The Economic Costs of Carbon Dioxide Emissions Limits*. MIT Press, Cambridge, MA.

March, J. G., 1994: *A Primer on Decision-Making*, Free Press, New York.

Matheson, J. E. and R. A. Howard, 1968: *An Introduction to Decision Analysis*, SRI.

Mattsson, N. and C. O. Wene, 1997: Assessing new energy technologies using an energy system model with endogenised experience curves, *Journal of Energy Research*, **21**, 385–93.

Mendelsohn, R., W. Nordhaus, and D. Shaw: 1994, The impact of global warming on agriculture: a Ricardian analysis, *American Economic Review*, **84**, 753–71.

Miller, J. H., 1998: Active nonlinear tests (ANTs) of complex simulations models, *Management Science*, **44**, 6, 820–30.

Moore, G. and R. McKenna, 1995: *Crossing the Chasm*, Harperbusiness.

Morgan, M. G. and H. Dowlatabadi, 1996: Learning from integrated assessments of climate change, *Climatic Change*, **34**, 337–68.

Morgan M. G. and M. Henrion, 1990: *Uncertainty: A Guide to Dealing with Uncertainty in Quantitative Risk and Policy Analysis*, Cambridge University Press, 332 pp.

Munich Re, 1997: (Muncherner Ruckversicherungs-Gesellschaft, D-80791 Munchen, Germany (www.munichre.com).

Nakicenovic, N. and R. Swart (eds.), 2000: *IPCC Special Report on Emissions Scenarios, a Special Report of the Intergovernmental Panel on Climate Change*. Cambridge University Press, Cambridge, 570 pp.

Newell, R. G., A. B. Jaffe, and R. N. Stavins, 1999: The induced innovation hypothesis and energy-saving technological change, *Quarterly Journal of Economics*, **114**(3), 941–75.

Nordhaus, W. D., 1991, The cost of slowing climate change: a survey, *The Energy Journal*, **12**, 37–65.

Nordhaus, W. D., 1994a: *Managing the Global Commons: The Economics of Global Change*, MIT Press, 213 pp.

Nordhaus, W. D., 1994b: Expert opinion on climate change, *Amer. Sci.*, **82** (Jan.–Feb.), 45–51

Nordhaus, W. D. and Z. Yang, 1996: *The American Economic Review*, **86** N4.

Park, G. and R. J. Lempert, 1998: The class of 2014: preserving access to California higher education, RAND, MR-971–EDU.

Peck, S. C. and T. J. Teisberg, 1992: CETA: a model for carbon emissions trajectory assessment, *Energy J.*, **13**(1), 15, 55–77.

Peck, S. C. and T. J. Teisberg, 1993: Global warming uncertainties and the value of information: an analysis using CETA, *Resource and Energy Economics*, **15**, 71–97.

Repetto, R. and D. Austin, 1997: *The Costs of Climate Projection: A Guide for the Perplexed*, World Resources Institute.

Robalino, D. A. and R. J. Lempert, 2000: Carrots and sticks for new technology: crafting greenhouse-gas reductions policies for a heterogeneous and uncertain world, *Integrated Assessment*, **1**, no. 1, 2000, 1–19.

Rotmans, J. and H. J. M. de Vries, 1997: *Perspectives on Global Change: The TARGETS Approach*, Cambridge University Press, Cambridge.

Savage, L.J., 1954: *The Foundations of Statistics*, New York, Wiley.

Schlesinger, M. E. and X. Jiang, 1991: Revised projection of future greenhouse warming, *Nature*, **350**, 219–21.

Schlesinger, M. E. and N. Ramankutty, 1992: Implications for global warming of intercycle solar-irradiance variations, *Nature*, **360**, 330–3.

Schlesinger, M. E. and N. Ramankutty, 1994a: An oscillation in the global climate system of period 65–70 years, *Nature*, **367**, 723–6.

Schlesinger, M. E. and N. Ramankutty, 1994b: Low frequency oscillation. Reply, *Nature*, **372**, 508–9.

Schlesinger, M. E. and N. Ramankutty, 1995: Is the recently reported 65–70 year surface-temperature oscillation the result of climatic noise?, *J. Geophys. Res.*, **100**, 13, 767–13, 774.

Simon, H. A., 1959: Theories of decision-making in economics and the behavioral sciences, *American Economic Review*, June.

Tester, J. W., D. O. Wood, and N. A. Ferrari, (eds), 1991: *Energy and the Environment in the 21st Century*, MIT Press.

Utterbeck, J. M., 1994: *Mastering the Dynamics of Innovation*, Harvard Business School Press.

van der Heijden, K., 1996: *Scenarios: The Art of Strategic Conversation*, Wiley and Sons, 305 pp.

Watson, S. and D. Buede, 1987: *Decision Synthesis*, Cambridge University Press, Cambridge.

Wigley, T. M. L., R. Richels, and J. A. Edmonds, 1996: Alternative emissions pathways for stabilizing CO_2 concentrations. *Nature*, **379**, 240–3.

World Bank, 1996: World Development Report. Washington DC.

4

Energy Efficiency: A Little Goes A Long Way

4.1 Introduction

4.1.1 Why is energy efficiency important?

Four issues will be crucial in determining the amount of energy needed in the future, and hence the amount of carbon-free energy to meet climate targets:

1. Population growth or how many people will be using energy,
2. GDP growth or how rich these people will be,
3. The fuel mix and the potential emissions, and
4. Energy efficiency or how well the fuels are used.

This chapter examines the role energy efficiency can play. We will argue that energy efficiency alone can not provide the solution to meet targets for abatement of greenhouse gas (GHG) emissions. This is because the simultaneous and interacting effects of changes in global population, growth or shrinkage in the world economy, and the proportions of various fuels employed to meet energy needs, have a much greater potential to raise or lower GHG emissions. However, we will also argue that energy efficiency is an essential part of the solution to abating GHG emissions to meet global energy demands at a realistic cost.

Changes in energy efficiency are affected by a host of technological, social, economic, and political factors. This necessitates that the processes of technological, social, economic, and political change be made an endogenous feature of any model of energy use, whether for specific activities or aggregated for the whole economy. This chapter will endeavor to elaborate on this complex problem and pay particular attention to structural changes in the economy, endogenous technical change, the social aspects of energy technology adoption, path dependencies for attaining lower levels of energy intensity, and the consequences of energy policy instruments.

Table 4.1 *An illustration of the importance of energy intensity*

Yearly change in energy intensity of the global economy	Energy requirements in the year 2100 (year 2000 at a base level of 100)		
	No change in population or income	Double population at same income	Double population and double income
−3%	4.8	9.6	19.2
−2%	13.3	26.6	53.2
−1%	36.6	73.2	146.4
No change	100	200	400
+1%	270	540	1080

4.1.2 Scenarios for future energy efficiencies

Measuring changes in energy efficiency over long time scales is very difficult. Firstly, technological paradigms change radically over long time horizons. Secondly, there is great uncertainty as to whether energy use is improving or economic output is increasing. Thirdly, country comparisons are fraught with potential for misinterpretation due to reliance on measures of market GDP instead of GDP at purchasing power parity (PPP), and lack of knowledge of a country's energy use and economic performance. This last point is crucial as the majority of expected energy growth will be in the developing world, where the most uncertainty exists concerning both future overall development and expected energy intensities, and the vast majority of analysis and research has been carried out on developed countries.

Table 4.1 illustrates a simple numerical representation of the importance of energy intensity.[1] These five scenarios produce radically different requirements for energy when extrapolated over 100 years. Also shown is the importance of growth in population and income. A more realistic calculation would show how income rising from a low base results in much larger increases in energy use than income rising from a higher base, as energy demand flattens out as we move up the income curve. This is further evidence of the importance of the future energy intensity of developing countries.

There is a wide variety of opinion regarding how much efficiency gains can offset future power needs. Estimates of changes in energy efficiency are based on both empirical and historical data, as well as modeling attempts based on theoretical constructs of the demand for energy and the resources and technologies available to provide it. The authors of the IPCC central "Business-As-Usual" scenario (IS92a) believed that an improvement in energy intensity of

[1] Energy intensity describes the aggregated level of energy use for economic output. Its use and limitations will be discussed in more detail in the next section.

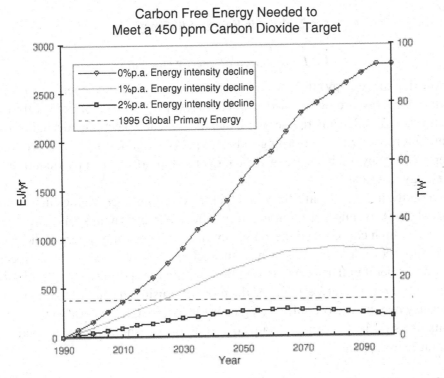

Figure 4.1   Model predictions of required carbon free energy for a 450 ppm CO_2 target (Source: Azar and Dowlatabadi, 1998).

about 1% per year would be sustainable over the next century employing only those emission control policies internationally agreed to at the 1992 Rio Climate Treaty negotiations. Various studies model or otherwise forecast long-term rates of energy intensity decline ranging from no change to greater than 2% per year. Others point to historical precedents for rapid efficiency improvement over shorter time scales of about 3% per year in the US from 1979–1986 (Lovins, 1998).

Therefore, based on a range of answers for how energy efficient the future will be, we may need as little as 5 TW of carbon-free power in 100 years (half as much total energy as the world uses today) or more than 90 TW. In Figure 4.1, Azar and Dowlatabadi (1998) provide estimates at the 50th percentile for the amounts of carbon-free power needed in the future assuming energy intensity declines of 0%, 1% and 2% per year. This study, in agreement with Hoffert *et al.* (1998), indicates that sustained efficiency improvements in the range of 2% per year lead to modest requirements for carbon-free power in the coming century, while smaller efficiency gains mean an increasingly large need for such power.

4.2 What is Energy Efficiency?

4.2.1 Definition of energy efficiency

Before defining energy efficiency it is instructive to define what energy provides. Units of energy are not valued in themselves; rather the economic value of energy is derived from the services that it provides: keeping the lights on, heating a room or transporting goods and people to a destination. Therefore, energy efficiency is a measure of the energy used in providing a particular level of energy services.

Secondly, a general paradigm in which to consider energy efficiency is not one whereby technological innovation offers energy savings that are then limited by high costs, social inertia or political maneuvering. Energy efficiency is instead a complete analysis of how the technical, political and social aspects of some societal undertaking interact both within the specific activity and with other activities. Therefore, some of the most promising mechanisms for energy efficiency may be video-conferencing over the Internet, a firm's drive to ensure lights are switched off in unoccupied rooms or a fashion for individuals to commute by bicycle.

4.2.2 Energy efficiency vs energy intensity

It is easy to confuse energy efficiency and energy intensity. Energy efficiency as we describe it above is a bottom-up view applied to individual activities. We describe energy intensity as a top-down or aggregated look at energy use in an economy. For consistency and clarity, the rest of this chapter will use the term efficiency for specific activities and intensity for aggregated energy use. However, the relationship between the two is far more complex and controversial than a simple aggregation.

One measure in common usage to define energy intensity is primary energy supply divided by GDP (E/GDP). For most OECD countries, this is between 7 and 14 mega-joules per US dollar (MJ/US$) with the US and Sweden in the upper part of this range and Japan in the lower range (Azar and Dowlatabadi, 1998). However, energy intensity is not simply the inverse of energy efficiency. Energy intensity is also affected by a nation's climate, heating and cooling requirements, amount of indoor space, lifestyles, population density, economic structure (industrial versus service-based, strong dependence on primary materials or not, etc.), and various other factors. For example, Norway has a very high ratio of energy use to output in manufacturing because a high concentration of its industrial sector is in energy-intensive processes (paper/pulp, refining, non-ferrous metals, and ferro-alloys) (Schipper *et al.*, 1998). Pure energy efficiency is estimated to account for between one-third and two-thirds of energy intensity, with the other factors making up the rest.

These difficulties in comparing energy intensities between countries (when defined as E/GDP) has led some leading commentators to call for this measure not to be used at all. This objection in using E/GDP is both for spatial and temporal comparisons but also in aggregation from technology-specific energy efficiencies to a measure of national or international energy efficiency. It is complex and difficult to equate particular technology-specific measures of energy efficiency with the aggregated measure of energy intensity because intensities reflect behavior, choice, capacity or system utilization and other factors besides just engineering considerations.

Researchers at the International Energy Agency (IEA) suggest that the ratio of energy use to GDP is not the measure upon which we should be focusing. They explain, "since the denominator represents many diverse activities, the ratio cannot really measure efficiency. Since the numerator aggregates many fuels and stirs electricity into the mix as well, even the notion of 'energy' is confused. Moreover, the mix of activities generally varies from country to country and over time. Since the energy intensities of these activities vary widely, variations in the mix of activities can cause significant variations in the ratio of energy to GDP over time, or explain enormous differences among countries. Since efficiencies refer either to specific physical processes or to specific economic activities, it is hard to take seriously an aggregate ratio of energy use to GDP as an indicator of either energy efficiency or economic efficiency. Thus we are left with a quantity that does not tell us much more than how much energy is used relative to GDP. In response we have decided to disaggregate energy uses and activities and to calculate intensities where numerators and denominators match as closely as possible." (Schipper *et al.*, 1998). This has been the main focus of the IEA's "Energy Indicators" effort.

The IEA has thus defined its own version of "energy intensity" which differs from the general use of the term (E/GDP). They explain: "measuring 'efficiency' is far more difficult than it seems to be. This is because we rarely observe the physical quantities that define an 'efficiency' in the engineering sense. And we rarely measure or estimate the economic inputs and outputs that define economic efficiency. To avoid this confusion we introduced energy intensities, defined as energy use per unit of activity or output for a large number of economic and human activities. In economics, intensities are often used to measure how much of one (of many) resources is used to produce a given output. These intensities can be aggregated under certain conditions, but

should not be confused with the ratio of energy use to GDP, which unfortunately is still used widely to measure 'efficiency'."(Schipper et al., 1998).

Another confounding issue is that how GDP is measured differs across countries, compromising the accuracy of international comparisons. For example,
developing countries are often judged energy inefficient on the basis of their
energy intensity because the GDP figures used are based on their relatively
weak market exchange rates. But if purchasing power parity (PPP) exchange
rates are used instead, the energy intensity of developing countries is similar to
that of industrialized nations (Azar and Dowlatabadi, 1998; WEC/IIASA,
1995).

Although the measure of E/GDP is an imperfect measure of energy intensity, and aggregation of individual energy efficiencies is fraught with
difficulties, the rest of this chapter will discuss energy intensity for aggregated
energy use and energy efficiency for individual technologies and activities. This
is for two reasons. Firstly, this chapter examines the role of energy efficiency in
determining how much carbon-free power is required for an abatement policy
to respond to anthropogenic global climate change. For this purpose, a molecule of carbon is the same whether emitted from a steel mill or a residential
Stirling engine in either Arizona or Angola, and thus some measure of aggregation is needed for comparison, prediction and discussion. Secondly, the use
of E/GDP, with its limitations clearly understood, allows flexibility to expand
on the importance of lifestyles, economic factors, political decisions, and social
processes in determining how energy is used and how vital the evolution of
these interacting processes will be in shaping future energy consumption and
hence emissions of greenhouse gases.

4.3 Historical Trends and Future Predictions

4.3.1 Energy intensity trends

Looking at general scenarios is useful to give some understanding of the role
of energy efficiency for future energy requirements. But to understand the
process of efficiency improvements and how they come about, we must recognize and understand why energy efficiency change does not proceed in a linear
fashion. Analysis by Nakicenovic (1997) of historical technological change
illustrates how economic activity undergoes paradigm shifts based on successive long waves of key technologies. The paper illustrates how the process of
economic growth and technological change is not smooth and continuous and
that the timing of the transition from one dominant cluster to another is consistent with the pattern of Kondratieff long waves. Energy efficiency has been

shown to respond in a Hicksian relationship to the cost of energy (Newell *et al.*, 1999): when energy prices have risen, then energy efficiency has also risen. Lastly, there is an inherent difficulty in assigning energy use to technologies and activities that have not yet been developed or even conceived. For example, what was the energy efficiency of personal computing technology 20 years ago, never mind 100 years ago?

A key distinction between the high and low estimates of future energy intensity is the projected cost of energy. Simply put, when energy is cheap, there is little incentive to save it. But price is clearly not the only factor. Tax structure, R&D investments, efficiency standards, and innovative efforts to commercialize new technologies are also important elements. Public policies that correctly price energy by internalizing externalities and help to overcome market failures can go a long way towards leveling the playing field for efficiency improvements to compete with supply options. However, the history of energy policy has often seen subsidies of less-than-optimal energy supply technologies and fuels supported for a variety of largely political motivations. Also important in energy efficiency is the market and regulatory structure of energy supply and information flows concerning energy options. Lastly and crucially, social processes provide inertia against changing energy-intensive practices, and social dynamics can result in unanticipated increased energy usage. The classic US (and other parts of the developed world) example of this is the dependence on the automobile, and the related exodus of people from inner cities to suburbia and onto exurbia.

4.3.2 Empirical analysis

Empirical analysis of historical energy use has tended to focus either on aggregated energy intensity in a top-down approach, or technology-specific energy efficiency developments in bottom-up studies. As described in the previous section, there is considerable difficulty in comparing and translating results on intensity to efficiency and vice versa. Generally, bottom-up analyses have shown larger improvements in the effective use of energy. This is at least in part due to the fact that technical efficiency gains have been devoted to enhancing energy service attributes rather than reducing energy inputs needed to secure a given level of service. For example, technology gains in automobile efficiency have gone to providing more vehicle power and other amenities rather than to getting more miles to the gallon.

We have stressed the caution needed when considering long-term trends. Energy efficiencies for specific technologies or activities can be defined. But aggregating these measures to an economy's energy intensity is further

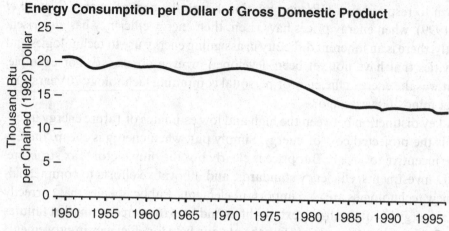

Energy Consumption per Dollar of Gross Domestic Product

Figure 4.2   Estimated US energy intensity 1950–1996 (Source: EIA, 1997).

complicated as the shifting mix of economic activities in any country changes over time.

With these difficulties in mind, we begin by looking at estimates of past energy intensities in the US. The World Energy Council reports the long-term rate of energy intensity improvement in the US, averaged over the past 200 years, was about 1% per year (WEC/IIASA, 1995). However, analysis on decadal time scales (Dowlatabadi and Oravetz, 1997) has shown that energy intensity responds to changes in energy prices beyond the price elasticity of demand. Thus for the 20 years from 1954 to 1974, energy intensity rose by close to 2% per annum. This can be attributed to technological drift as energy efficiency was not incorporated into the design of goods and services. Following the 1970s oil shocks, the energy intensity of the US economy declined in response to price signals; from 1979–1986, energy intensity fell an average of 3.1% per year. However, as energy prices tumbled post 1986, the decline in energy intensity steadily slowed as the US failed to carry on the energy efficiency experiment long enough to see continued improvement (Rosenfeld, 1999). Figure 4.2 shows another estimate of US energy intensity trends.

In other industrialized nations, a similar story is seen. Following the two oil crises in the 1970s, energy intensity over the years 1973–1993 declined by 1.4%/yr in the IEA countries (IEA, 1995). Following the collapse of energy prices post 1986, the rate of decline slowed, reaching 0.6%/yr on average for the IEA countries over the period 1989–1993. Thirteen IEA countries actually report increasing E/GDP trends (WEC/IIASA, 1995).

So what drives trends in long term energy intensity? The authors of

WEC/IIASA (1995) conclude that "energy intensity improvement rates are related to per capita GDP growth rates. The faster an economy grows in per capita terms, the faster is productivity growth, the rate of capital turnover and the introduction of new technologies, and the faster energy intensities improve." However, an alternative result of economic growth can be greater investment in energy intensive industries, and subsequent increases in energy intensity as long as saturation for energy intensive goods has not been reached. Conversely, in cases of negative per capita GDP change, such as the recent experiences in the former Soviet Union, energy intensities decline, even though overall energy use may also be declining. This latter effect dominates for Kazakhstan, as is shown in Figure 4.4.

Dowlatabadi and Oravetz (1997) also find that intensity trends have varied widely through time and responded strongly to prevailing economic conditions. They postulate that scrapping of old capital plays a key role. This can often come about as a precursor to the growth phase of the economic cycle and is associated with deep recessions. The deep recession leads to the death of activities and behavior that is economically of marginal value (this is despite public policy attempts to save jobs, industries, or communities for political motivations), and is often characterized by energy inefficient capital in sunset industries. This structural change in economies opens the path for new ways of meeting demands and new behaviors demanding new energy services.

Given that declines in energy intensity are non-linear and respond to price changes and structural shifts in the economy, what evidence is there for individual energy efficiency improvements to take advantage of new opportunities? Ausubel and Marchetti (1996) discuss ongoing historical trends in energy efficiency improvements from an engineering perspective. They point to the 300-year quest to develop more efficient engines, from 1%-efficient steam engines in 1700 to today's best gas turbines which approach 50% of their theoretical efficiency limit (Figure 4.3). Fuel cells, which Ausubel (1998) says may power our cars in 20 to 30 years, will increase that efficiency to about 70%, as fuel cells do not incur the inevitable efficiency limits of combustion systems imposed by the laws of thermodynamics. Similarly, Ausubel and Marchetti (1996) point to the dramatically increasing efficiency of lighting technology over the past 150 years (Figure 4.3, analyzed as a sigmoid (logistic) growth process, shown in a linear transform that normalizes the ceiling of each process to 100%).

Of course, technological change can also lead to worsening energy efficiencies. For instance, in the post-WWII era, rising incomes and falling electricity prices led to the marketing of refrigerators with less insulation and larger interior volumes (von Weizäcker *et al.*, 1995).

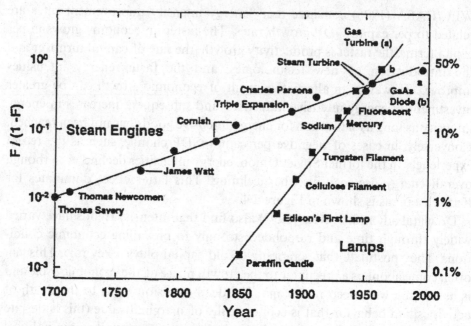

Figure 4.3   Improvement in the energy efficiency of motors and lamps (Source:
Ausubel and Marchetti, 1996).

On the whole, however, empirical evidence exists for considerable energy
efficiency improvements in specific technologies. In some technical areas the
improvement has been remarkable, for example in energy efficient lighting,
fuels cells and photovoltaics. As impressive, and potentially even more important, is the corresponding decrease in costs of these energy efficient technologies. However, it should be noted that there is an inevitable time lag for
adoption of efficiency improvements (in addition to reluctance of potential
adopters as discussed later) due to the delay of capital stock turnover. As
Rosenfeld (1999) explains, capital stock (for a proxy life of 15 years) begins to
turn over the first year, but the improvement in energy efficiency is then just the
fraction of new units multiplied by the fraction of these that were higher
efficiency units, e.g., 5% of 5%. These fractions grow each year until the stock
is completely replaced.

4.3.3 Modeling of possible scenarios

To continue our discussion of how improvements (or lack thereof) in energy
efficiency from individual technologies are intricately tied to structural changes
in the economy, we must make the distinction between models in which tech-

nological learning is exogenous, or assumed in the model, or the far more realistic cases in which it is endogenous, or affected by other parameters in the model.

In traditional energy-economic models, energy intensity is exogenously specified, that is, it is dealt with as an input which is not responsive to anything within the system. In macroeconomic models (Manne and Richels, 1992; Walley and Wigle, 1991), this process is captured by the autonomous energy efficiency improvement (AEEI). The AEEI is assumed to capture all *non*-price induced changes in energy intensity, including structural change in the economy and sector-specific technological change. Thus, it is neither a measure of technical efficiency improvements per se, nor is it autonomous since policy-driven adoption of more energy efficient technologies is included. AEEI is usually specified in a range between 0.5% and 1.0% per annum, although some modelers use even lower values. However, the empirical evidence and theoretical understanding of technological change does not support such a relentless linear efficiency improvement (or decline in energy intensity).

A more realistic categorization for energy intensity makes it endogenous, or responsive to the interacting system factors such as policies, prices, research and development, and path dependence (Goulder and Schneider, 1998; Dowlatabadi, 1998). Generally, results from these models are consistent with high energy prices leading to a *demand pull* model of technological change improving the efficiency of energy use. In times of low energy costs, other characteristics including performance and ease of use may be valued more than efficiency.

4.4 Developing Nations and Uncertainty in Future Energy Use

4.4.1 Developed vs developing countries

Although GHG emissions have historically been overwhelmingly due to developed nations, changing population and economic growth dynamics in developing countries will result in them surpassing developed nations in terms of GHG emissions by around 2030 (Dowlatabadi and Oravetz, 1997). Thus the energy intensities of these developing economies will be crucial in determining the need for carbon-free power. Unfortunately, analysis of energy intensity trends has been concentrated on the developed world, and thus E/GDP as a measure of energy intensity is even less well defined for the developing world, due to difficulties in defining GDP, and poor data on energy use. This is one of the inputs into the large uncertainties in projections of future global energy intensities and hence carbon emissions.

Looking at energy intensity trends in developing countries, these have generally been rising over the past decade (Goldemberg, 1996), reflecting an earlier stage of industrialization where heavy industry and infrastructure expand. The fear of rising energy intensities in populous nations as they industrialize has been a part of the rationale behind the Clean Development Mechanism agreed in the Kyoto Protocol of the UN Framework Convention on Climate Change. However, analysis of technology transfer in developing countries suggests that recipients of such a mechanism are reluctant to use new energy efficient technologies unless they are also widely used in the donor country (Sagar and Kandlikar, 1997).

Figure 4.4 details changes in energy use and growth in GDP from 1950 through 1966 for six developing countries. This figure illustrates the lack of similar trends in energy efficiency for different developing nations.

As energy intensity is measured in terms of energy input per economic output, the recent efficiency gains in China may partially reflect how the world has historically underpriced Chinese economic production. If GDP based on market currency rates has been undervalued (as in the Chinese case), then falling energy intensity could be a reflection of this underestimation. If the historic GDP is given as power purchasing parity (relative to US $), the efficiency gains may be significantly smaller.

Periods of industrialization can be characterized not only by an increase in energy use but an increase in energy intensity. This is shown for the cases for India and Argentina. Therefore, even if developed countries steadily improve their energy efficiencies, an increasingly large proportion of energy consumption will be in developing countries, and thus their changing energy intensities are of great importance to global carbon emissions as their percentage of global energy use increases. Other commentators propose that developing countries will not necessarily have rising energy intensity trends as they industrialize. Several developing countries have experienced long run negative energy intensity trends, including Korea and Brazil (Nilsson, 1993). A later section will discuss how structural changes in the economy are linked to technological change with far reaching consequences for energy efficiency. This was certainly the case for Mexico, where energy intensity declined following its economic troubles of the 1980s.

Will the impact of energy efficiency on the need for carbon-free energy technologies be much reduced due to the importance of developing countries as they industrialize? Could global energy intensities actually rise in the next 100 years? For insights into these questions we need to examine the complex process of energy efficiency changes. However, in any predictions of energy efficiency there are great uncertainties, and this is especially true in developing

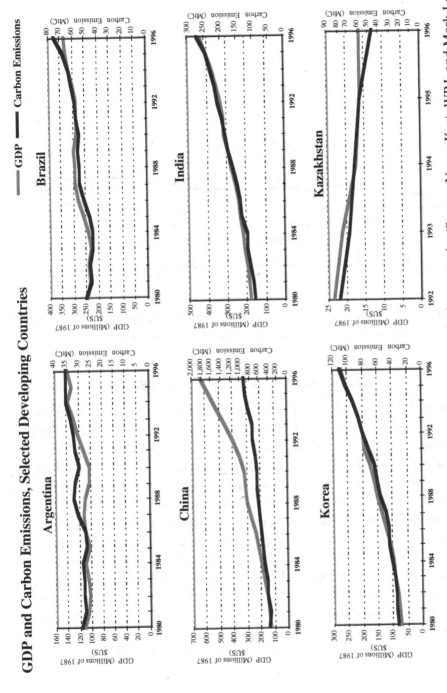

Figure 4.4   Trends in GDP and energy intensities for selected developing countries (Source: Nancy Keat, WRI, and Mark Levine, in Levine *et al.*, 1992).

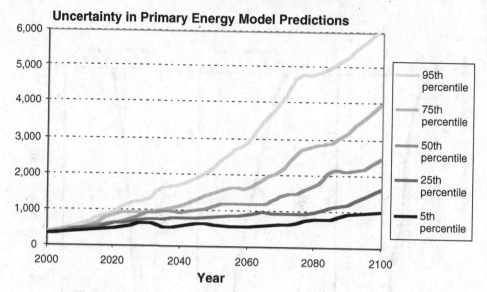

Figure 4.5   Uncertainties in global energy model predictions.

countries where a lack of data, a lack of modeling experience and dynamically changing economies and societies make prediction difficult.

4.4.2 Uncertainty in predicting energy efficiency

So far we have been discussing the impact of changes in future energy use in terms of median values. However, very large uncertainties exist for the impact on the amount of carbon-free power needed for a given scenario of energy intensity change. This is illustrated in Figure 4.5 taken from the ICAM climate model developed by Dowlatabadi and fellow researchers at Carnegie Mellon University. This considerable uncertainty is due to the interacting factors affecting energy use as discussed in previous sections. This large uncertainty in future requirements for carbon-free power to meet specific climate targets has been used by some researchers (e.g., Wigley, Richels and Edmonds, 1996) to suggest a delay in abatement policies, including efforts to improve the energy efficiencies in technologies, and thus, by aggregation, reduce the energy intensity of an economy. However, any delay in instigating energy efficiency policies increases the risk of "lock in" to more energy-intensive technological paradigms. This danger is explored further in the section on path dependence.

4.4.3 Uncertainties in energy intensity trends of developing countries

Energy intensity trends in developing countries will be crucial in determining future global energy use in the long term. However, the largest uncertainty exists in the energy intensity paths of the developing countries. This is due in part to their development being modeled based on the experience of the development of industrialized countries under very different conditions. Predictions, by their nature, are only as strong as their weakest or most uncertain component. Identification of these uncertainties is critical to making better decisions.

There is significant heterogeneity in the extent of knowledge available about different parts of the world. For example, whereas commercial energy markets dominate energy supply in industrialized nations, non-commercial fuels, often collected communally, meet the needs of a sizable fraction of the world's population. Economic models of production are often specified for industrialized nations where capital is plentiful and labor is scarce (or needs to be made less important in order to limit union power). These models are often adopted as being representative of production patterns in less industrialized settings – even where there may be a desire to increase employment for social reasons and there is a dearth of capital. Therefore, insights about the distributional impacts of policies involving first world and other nations are highly uncertain if not fundamentally suspect.

Additionally, when modeling changes in energy use we adopt continuous mathematical forms and consider only small perturbations to the system. When considering long-term global change issues and sustainability, we may need to redefine our representations as discontinuous processes suffering significant changes. This change in how the issues are represented will also entail a different set of questions. This is very different from the more familiar question of "will the benefits of a small intervention outweigh the costs of its implementation?" This is particularly true when predicting future energy efficiencies of developing countries.

4.5 Viewpoints on Energy Efficiency

4.5.1 Energy analyst vs corporate decision maker vs individual

Projections of potential and achievable energy efficiency depend on the viewpoint of the analyst and in part allude to the "bottom-up" versus "top-down" debate. Technology-based "bottom-up" models indicate that enormous opportunity exists to improve energy efficiency with current technology at no net cost (the energy savings equal or outweigh the outlays) and that various "market failures" are to blame for the slow adoption of these technologies. Traditional "top-down" macroeconomic models, on the other hand, assume that all economical options are exploited, so by definition, if a technology has not been adopted, it must not be economically efficient. Lovins (1996) analogizes this to not picking up a $100 bill lying on the sidewalk because if it were real, someone would have picked it up already.

Another way of looking at the debate is by asking the question "Where can the energy savings be made?" Research examining personal energy use (Schlumpf *et al.*, 1998) illustrates that there are relatively modest savings that individuals can make on their own without major changes in their lifestyles. Social theory tells us that there is great inertia to such lifestyle changes, and it is believed unlikely that such changes will arise for environmental reasons in the foreseeable future. A similar process occurs for corporate decision-makers deciding on purchases for their organizations. There is a limited set of commercially available options for rational decision-makers to choose from. Therefore, although an energy analyst may envisage large savings waiting to be exploited, actual decision-makers are constrained by what is available to them at the time of decision. This leads into the next section and a discussion of path dependence or shaping the future "shopping list" of technologies of varying efficiencies that can be selected. Of course, even if these technologies are available at reasonable cost there are other factors important in the adoption decision, and this is discussed in the section on social aspects of energy efficiency.

4.6 The Development of Energy Efficient Technologies

4.6.1 *Technological change as endogenous and interactive with patterns of energy use*

Energy efficiency is *not* technological opportunity limited by social, regulatory or political factors. This bold statement is designed to emphasize the point that technological change and energy efficiency are an integrated and endogenous aspect of the design and operation of economic activities. Future efficient technologies (and their successful commercialization) will be as much a consequence of energy prices, social attitudes about environmental issues and political energy priorities as of envisaged and hypothesized potential technical efficiencies.

In a system, a resource constraint can be solved both by innovation in how a technology is used to derive services from the resource and by innovations making additional resources or equipment available. Forces capable of motivating technological change bringing greater energy efficiency are also capable of motivating technological change making more energy available. There is no inherent reason a system's dynamics automatically work to promote efficiency.

4.6.2 Path dependence

Due to shorter-term considerations, a society can be "locked-in" to a technology that is inferior (e.g., in energy use) in the longer term. Thus, path dependencies are important. To illustrate path dependence or technology lock-in, take the example of a decision-maker considering whether to invest in an old or a new technology which provide the same service but at different costs and efficiencies. The decision-maker may find it more profitable in the short run to invest in "old" technologies. But from a societal perspective, investments in alternative technologies may turn out to be only slightly more costly. Further, switching to the new efficient technologies will result in the accumulation of much experience in these technologies, and consequently much lower costs associated with their use. Future attempts to switch to more efficient technologies will undoubtedly cost significantly more to society, due to the missed opportunity to join the learning curve for these technologies with its associated improvement in product performance and reduction in costs (Arthur, 1987). The risk of technology lock-in in general is analyzed in a paper by Ayres & Martinàs (1992). Additionally, in his analysis of the QWERTY keyboard, David (1985) shows that expectations of future market share and performance become key parameters in the lock-in of technologies. Lastly, the importance of a sponsoring firm or organization of a technology is important, as this can determine whether current or future performance will predict the dominance of the technology (Katz and Shapiro, 1986).

From the viewpoint of a decision-maker, an efficient option can only be chosen if it is commercially available. On this basis, several commentators (see e.g., Harvey, 1998) have argued that climate policy should focus more research into energy efficient technologies. Through a learning process, products can improve their efficiency and reduce their costs as they travel down the experience curve. This would allow future decision-makers to have the option of an efficient technology to cost-effectively meet their needs, and is especially valuable when considering the slow turnover of capital (50 years for residential housing, for example). Such a proactive strategy would have wide-ranging implications for energy efficiency over the next 100 years.

Another illustration of path dependence and efficient technologies is given in a series of studies by Jaffe and colleagues where they estimate the role of performance standards and subsidies in the range of technologies offered to consumers (Jaffe and Stavins, 1995; Newell *et al.*, 1999). They find that performance standards can motivate producers to remove their least efficient models, thus speeding the adoption and further development of new, efficient technologies. In combination with price-induced efficiency improvements, performance standards illustrate the importance of improving individual technologies for overall reductions in energy use.

4.7 Adoption of Energy Efficiency

4.7.1 Economics of energy efficiency adoption

The discussion so far has illustrated the bumpy route of economic and operational factors that leads to new efficient technologies and stressed the importance of institutional support to create a cycle of continued use with resultant technical improvement and cost reductions. But all of these insights into the process of increasing energy efficiency lead to naught if consumers and firms do not adopt these energy efficient technologies. Of course costs play a pivotal role in the uptake of energy efficiency. Individual efficiency investments have been shown to be induced by rising energy prices (Newell *et al.*, 1999).

When analyzing the uptake of energy efficiency innovations, these technologies have historically poor levels of adoption despite the high projected rates of returns. This has been called the "Energy Paradox" (Jaffe and Stavins, 1993). Explanations have included the additional costs that it poses to organizations to change their method of operation (Cebon, 1992), or that for consumers at least, the idealized engineering-economic projected savings are never approached (Metcalf and Hassett, 1998). Another possibility is that consumers are simply not aware of the technology and potential savings (Brill *et al.*, 1998). Morgenstern (1996) has found that the provision of free technical information is important, but not nearly as important as the effects of energy prices, although it may be that firms need a good information network to keep up with developments in energy efficiency (DeCanio and Watkins, 1998).

A promising avenue of research examines investment under the realistic conditions of uncertainty and irreversibility (Dixit and Pindyck, 1994). This avenue of research is centered on the reality that investments are "lumpy" through time, and are generally irreversible. Thus the uncertainty inherent in decision making assumes a much greater importance and investors are more likely to wait until uncertainty is reduced. Studies looking only at substitutions under conditions of rising prices and increasing uncertainty found that higher net present values were required and the optimal strategy was to delay investment (Hassett and Metcalf, 1993; Farzin *et al.*, 1998).

Another analysis tool is diffusion theory (e.g., Griliches, 1957; Mansfield, 1961). Insights from diffusion theory include the categorization of adopters as innovators and laggards, illustrations of the uptake of individual innovations as an exponential curve with a rapid take-off following early adoption, and the necessity of both mass and personal communications media for successful adoption.

Some proponents of energy efficiency feel that it has such high cost effectiveness that they cannot believe adopters would have any reluctance to investing in it, if only certain market barriers were corrected. For example, as mentioned above, Lovins (1996) likens the non-adoption of energy efficiency measures to pedestrians not picking up a $100 bill on the sidewalk. This analogy certainly holds resonance with engineering economic studies of energy efficiency investments that offer simple payback periods of less than two years, and sometimes even months (e.g., simple energy management measures including automatic sensors for turning off lights in unoccupied rooms). A separate study (Strachan and Dowlatabadi, 1999) shows that adopters do invest in energy efficiency, even though these investments turn out to be poor in comparison to the existing alternative of centrally generated grid electricity and on-site boilers.

Others believe that potential adopters may be inherently reluctant to invest in energy efficiency for a variety of reasons, including expectations of rapid technological change and uncertainty about theoretical returns. Under any of these circumstances, we turn to the non-economic consequences (perceived or real) of investing in energy efficiency to explain adoption or non-adoption behavior, and these are discussed in the next section. It should be noted that potential investors in energy efficiency opportunities compare it not only to existing energy use but also to the range of other spending opportunities for the firm or individual. In addition, energy investment is often discretionary and typically only accounts for a small fraction ($\sim$5%) of operating costs. Thus, even if energy efficiency is considered to be a good investment, other (non-energy) investments may be preferred at the expense of efficiency improvements.

4.7.2 Social factors in energy efficiency adoption

Some of the common social barriers to energy efficiency adoption include the lack of information, the cost of making and implementing a decision to change practices, and any difference in the energy services that the new technology offers (for example, reduced lighting output from early compact fluorescent bulbs). However, social processes can equally work to promote energy efficiency. Examples of this (which also reinforce the observation that social feedbacks are as important as economics and technology in determining efficiency levels) include the move towards fitter lifestyles, the construction of bicycle lanes in European cities, and the role of increasing Internet communication including video-conferencing in reducing transportation. Of course, it is possible that more electronic communication could mean more communication in general and perhaps lead to higher total energy use.

The acknowledged limitations of diffusion as a social theory explaining adoption has been taken up by Gruebler (1997), who presents an empirical examination of diffusion as a process of imitation and homogenization, but with clusters and lumps. These discontinuities may be considered as inherent features of the evolutionary process that governs social behavior. Technology clusters have historically been instrumental in alleviating many adverse environmental effects, and the emergence of a new cluster could hold the promise of a more environmentally compatible technological trajectory (Freeman et al., 1982).

Other studies of organizational behavior have shown that organizations follow standard operating procedures (Cyert and March, 1992) and these processes determine which decisions are considered and hence made. Theory on bounded rationality (Simon, 1982) led to the term "satisficing" being coined for when decision-makers settle not for the optimal solution to a problem but on the first solution that suffices a given criterion. Perhaps sadly, in business decision making, especially when energy efficiency investments are discretionary, energy efficiency is unlikely to be that criterion.

Finally, it should be stressed that efficiency gains do not always mean reductions in energy use. Some amount of the gains in energy efficiency may be offset by increasing uses of energy. Increasing wealth often leads to a more energy-intensive lifestyle. As reported in the *New York Times* (1998), even as the average number of people per household is shrinking, the average size of new homes is growing. And these homes are loaded with energy-consuming features, from central air-conditioning to home theaters. So even if the amount of energy used to heat a square foot of living space is falling, the number of square feet, and thus the total energy use, is growing. Energy use on the roads is rising even faster due to suburban sprawl that increases miles driven and the popularity of sport utility vehicles, minivans and pick-up trucks. Instead of being used to increase fuel efficiency, technological advances have gone largely to increase vehicle power (average horsepower rose from 99 in 1982 to 156 in 1996, with the 0-to-60 mph-time falling from 14.4 to 10.7 seconds).

4.8 Policy Aspects of Energy Efficiency

4.8.1 Policy instruments

One instructive way of categorizing policy options is to make the distinction between a set of market instruments (such as taxes and permits) designed to boost the motivation for higher efficiency, and a second set of instruments designed to set production on a path of reducing costs and improving energy efficiency of products. The debate between the two centers around which policy options would be most effective and which are politically feasible.

An underlying reason for energy efficiency improvements not having greater penetration is the historic low price of energy. Proponents of boosting energy efficiency via market instruments see the price internalization of all the costs of using energy as the most effective route to spurring reductions in energy use. This strategy recognizes that the bottom line that motivates most users to adopt commercially available energy efficient technologies is economic savings. The motivation to adopt would increase following an internalization of all costs and the consequential rise in the price of energy. Other observers point out the overwhelming political difficulties in achieving this desirable goal of raising energy prices in the near term.

Therefore, rather than pursue the hope of political action to increase the costs of fuels and electricity, other policy options to increase the numbers and performance of energy efficiency technologies are proposed. These include performance standards to eliminate the least efficient models and promote the trend towards higher efficiencies, and increased research and development on energy efficient technologies. Most analysts recognize the desirability of both policy options and suggest a combination of the two. A collection of views from the debate on the best policy instruments is given in the next section on policy prescriptions.

Tietenberg (1992) provides a good summary of market based instruments. If the costs to the economy of these instruments are deemed too great, a far smaller levy could be used to stimulate R&D into energy efficient technologies. This would be the starting off point on a path of increasing energy efficiency and reducing costs. Such R&D would cover the spectrum from basic research to commercialization, and is already being pursued in major programs under the auspices of the US Department of Energy and the European Union's DG 12 and 17. But as an instrument, R&D only provides half the impetus, as new technologies must be adopted and continue to be refined and improved. Instruments to achieve these goals include information campaigns such as the numerous US Federal programs, the UK EEBPP,[2] or the international CADDET[3] organization. Insights from diffusion theory tell us that these information sources must offer both mass media and individual sources of information for successful implementation of energy efficiency.

Another mechanism to promote an energy efficiency path is minimum standards to which suppliers must adhere. Examples of these include building codes, CAFE[4] standards for vehicles, and energy standards for appliances. An advantage of standards is that they remove the least efficient options from adopters' possible list of purchases, and successive applications of standards can steer an industry down a more efficient path. However, these standards are usually based on the best available technology (BAT) and thus can be technology limiting as suppliers might see no incentive in carrying out research into more efficient models if they believe that this level of efficiency will then become mandatory.[5] An additional option is for government to stimulate economies of scale and resultant lower costs by itself adopting many units of a new technology.

Other mechanisms for promoting less energy intensive technologies include tax breaks or credits. The Californian experience of tax breaks to support wind energy production was deemed by some to be technology limiting and a subsidy of expensive power under guaranteed favorable economic circumstances. One solution to these objections has been used in the UK where quotas for renewable energy have been allocated on an auction basis with the lowest bids gaining a subsidized rate. This program (NFFO[6]) has seen the prices of renewable technologies fall by a factor of three to the point that they are now competitive with peak time electricity prices (Mitchell, 1997). Lastly, energy efficiency savings have been claimed through utility demand side management programs, although the potential for abundant future savings from such programs has been called into question (Sonnenblick, 1995).

Governments' investment in energy efficiency has been extremely successful, at least by their own estimates. In the US, $8 billion (1997$) in federal outlays for research, development and deployment of energy efficiency technologies between 1978 and 1996 helped spur private sector investment, achieving $150 billion in annual savings. This is a tremendous return on investment, not counting environmental and health improvements (PCAST, 1997). Although these figures are extremely impressive, they may be overestimated due to the free rider problem of government paying for research that private firms might have

---

[2] Energy Efficiency Best Practice Program.
[3] Center for the Analysis and Dissemination of Demonstrated Energy Technologies.
[4] Corporate Average Fuel Economy.
[5] The CAFE performance standards were designed to ensure this did not happen.
[6] Non Fossil Fuel Obligation.

done anyway. In addition, there are cases of energy savings and cost reductions being serendipitous results of government programs designed to support at-risk industries or technologies for political motivations. (An example of this was the successful NFFO program in the UK that included renewable energy to make public support of the UK nuclear program allowable under EU rules).

4.8.2 *Policy prescriptions*

The "Energy Innovations" (EIA, 1997) study argues that "sector-specific market mechanisms would guide consumer and business decisions toward greater efficiency. Examples include an electricity generation emissions allowance and tradable permit system; a revenue-neutral industrial investment tax credit, which can offset increase in energy costs by reductions in the cost of capital; and transportation pricing reforms such as pay-as-you-drive insurance, which would shift a portion of auto insurance premiums to a cost that varies with miles driven. Each such policy reallocates costs in order to motivate higher energy efficiency and lower emissions while avoiding a net increase of taxes or fees within the sector." The study also includes policy initiatives for advanced vehicle development and deployment, market introduction incentives to help move prototype energy efficient technologies into mass production, and minimum efficiency standards.

Harvey (1998) suggests a strategy that would include realistic pricing of energy, but he takes issue with the widely held notion that price is the only key variable. He points out that gasoline purchases account for only 11% of the cost of running a car – considerably less than insurance. Doubling the price of gasoline would increase fuel mileage by a third at most, he says, which is far short of what is technically feasible, not to mention the fact that it is politically intractable. More broadly, most economists estimate that it would require a carbon tax of some $100 per ton to cap US carbon emissions. Changing energy consumption through price signals, when energy represents only a small fraction of the cost of doing business, is very difficult, and it may be politically impossible. Therefore, he recommends that modest taxes – $4 or $5 per ton of carbon – be assessed against environmentally damaging fuels, and the proceeds of those taxes be used to transform energy technologies. "Promoting the commercialization of energy efficiency and renewable energy technologies is a far cheaper way to achieve the same effect as raising taxes, and requires a much more modest political effort."

Advances in science and engineering (through public and private R&D efforts) have already done much to reduce the costs of many energy efficiency technologies. But in many cases, making technologies commercially competitive requires bringing their prices down further by producing and deploying them on a significant scale. Bulk purchases provide both the economies that result from large-scale production and the savings from improvements discovered when a technology is applied in the field. There are few programs to bring key technologies down the last part of the cost curve, leaving important advances in science and engineering adrift. Auctions are one option for directly supporting commercialization. An auction selects, through a bidding process, the most competitive technologies; a subsidy (paid for through a modest carbon tax as described above) makes up the difference between the winning bid and the market energy price. As mentioned earlier, such a system used in the UK cut the price of renewable energy offerings in half in just six years. The subsidy was funded through an increase in electricity rates that amounted to less than 0.5% (PCAST, 1997).

Harvey (1998) concludes that a national program of this type in the US of $3 billion to $6 billion per year, guaranteed to last at least five to ten years, should be adequate to make half a dozen key technologies (for energy efficiency and renewable energy) commercially viable. While not a trivial sum, compared to our national energy bill or the consequences of continued CO_2 production and local air pollution, it is a modest investment indeed. A carbon tax of only $4 per ton would support a fund of $5.6 billion per year. The total amount needed should be contrasted, perhaps, to a carbon tax required to reduce carbon production through elasticities of demand, which most economists project to be in the range of $50 to $100 per ton. A tax only 5 or 10% as large, efficiently applied to reducing the cost of key technologies, represents an intelligent commercialization strategy which makes a broader carbon cap much easier to achieve and much less costly.

Messner (1997) endogenized technology dynamics into a bottom-up energy systems model to examine the relationship between declining specific investment in energy technologies and overall experience or capacities installed. Her results show the importance of early investment in new technology developments. "New technologies will not become cheaper irrespective of research, development, and demonstration (RD&D) decisions; they will do so only if determined RD&D policies and investment strategies enhance their development." WEC/IIASA (1995) adheres to this position, and in their modeled scenarios, technology is not assumed to "fall like 'manna from heaven' but instead is the result of deliberate search, experimentation, and implementation, directed by both social and political policies as well as economic opportunities."

Azar and Dowlatabadi (1998) recommend a policy of investment credits to help gain experience with new technologies or deploy them more widely, thus inducing cost savings. They further recommend technology performance standards to "limit the choice consumers face to those technologies that meet consumer needs and embody progress towards severing the link between energy services and carbon emissions."

A study by the Alliance to Save Energy and Dale Jorgenson, Chair of Harvard's Economics Department (ASE, 1998), recommends an energy tax that is distinct from the carbon, *ad valorem* (market value), or Btu (energy content) taxes commonly proposed. Instead, "the taxes are based on damage estimates associated with costs from air pollution and global climate change mitigation not currently internalized through existing regulation." Another unique feature of the ASE study is that it provides results for the combined effects of fiscal and energy tax reform rather than the more common focus solely on the latter. ASE (1998) finds that "correctly pricing energy and taxing consumption – by taxing fossil energy to reflect air pollution and climate change damages and by replacing the present income tax system (at federal, state and local levels) with a consumption tax, all on a revenue-neutral basis . . . increases investment, consumption, exports, and the GDP; decreases fossil-fuel energy use, fossil fuel energy intensity, and carbon emissions; and increases aggregate social welfare."

4.9 Efficiency Case Studies

The four sectors of transportation, buildings, industry, and electricity generation dominate our use of energy. For each we describe a key energy efficiency technology and identify opportunities and barriers to their use in the overall framework of economic, technical, social, and political change.

4.9.1 Transportation

Transportation poses perhaps the world's greatest energy challenge, and presents the best example of the social and economic dynamics affecting a change to a more efficient system. This is certainly the case in the US. The US transportation sector is 97% dependent on oil and is the dominant user of oil in the US, accounting for 60% of the national demand – more than is extracted domestically (PCAST, 1997). With low gasoline prices (Figure 4.6), there is little incentive for consumers to demand more efficient vehicles. Schipper (1999) stresses the importance of fuel prices on efficiency as gasoline costs make up a large fraction of the *variable* costs of driving, and influence vehicle power and size, miles driven and vehicle fuel efficiency. But aided by low fuel prices, the consistent efficiency gains that manufacturers have achieved have

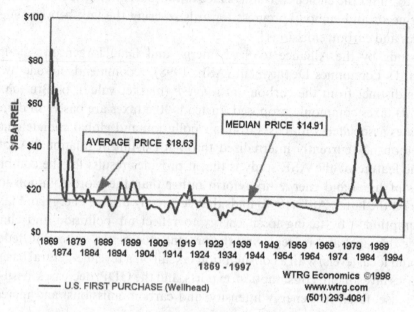

Figure 4.6   US gasoline prices.

not translated into more miles to the gallon but rather more safety and luxury features. Even limited efficiency standards have failed to quench consumers' desire for larger, faster, safer cars, as is evident in the popularity of sport utility vehicles (SUVs). These are not classified as cars and hence are subject to the lower light truck CAFE standard. Figure 4.7 illustrates the higher efficiency of the US vehicle fleet following the high gasoline prices of the 1970s and the recent slowdown in efficiency gains. In addition, population growth and growth in car ownership in the US continues to offset the meager efficiency gains of about 1% per year. In the US, the population of personal vehicles will likely grow from about 200 million to about 300 million during the 21st century (Ausubel *et al.*, 1998).

One frequently mentioned technological solution to increasing fuel efficiency is the fuel cell-powered vehicle in which compressed hydrogen mixes with oxygen from the air to produce electrical current in a low-temperature chemical reaction that emits only water. This electrochemical process is potentially 20–30% more efficient than the thermodynamic process of internal combustion engines (Ausubel *et al.*, 1998). Several manufacturers are already building prototype cars powered by fuel cells. Daimler–Chrysler plans to begin penetrating the market within 10 years, starting at about 100 000 cars per year.

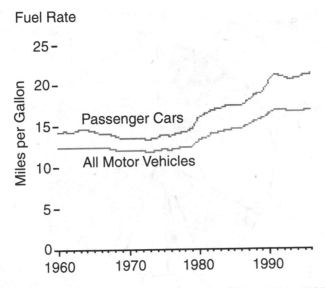

Figure 4.7   Efficiency of US vehicle fleet (Source: EIA, 1997).

Already on the road are hybrid fuel cell models by Honda and Toyota. Toyota's "Prius" (~66mpg) is priced at about $18k (although this is a subsidized cost) and has sold over 15 000 cars in Japan (NPR, 1999). But because of the large, lumpy investments in plant required, the traditional ten-year lifetime of cars, and gradual public acceptance, it will likely take two to three more decades before fuel cell cars dominate the fleet (Ausubel *et al.*, 1998). In addition, there are structural costs associated with such a change in the technological base, such as the restructuring of the automobile servicing industry. Even more difficult would be the provision of enough natural gas or hydrogen in the right places to allow 200 million fuel cell vehicles to operate.

Aside from technological breakthroughs, there are other possible solutions. Different societal norms in Europe in terms of living closer to work (which generally means closer to city centers), the greater availability of public transport and higher taxation on gasoline, have led to the continent having smaller, lighter automobiles that are generally one-third less fuel-intensive than present US vehicles. This is perhaps the clearest example of social and economic dynamics locking an activity into a certain level of energy use. This effect has also been driven by a history of low fuel prices in the US and social acceptance of higher prices in Europe. It is also one of the most intractable problems when considering the inertia of the US population to abandon their love affair with the internal combustion engine.

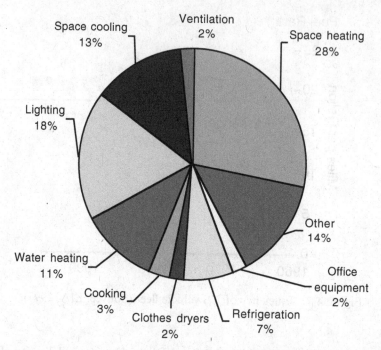

Figure 4.8   Percentage of consumption by end-use in buildings in 1995 (Source: EIA, 1997).

4.9.2 Buildings

Although typically accounting for up to a third of energy use in developed countries, energy use occurs over many separate activities, as is shown in Figure 4.8. It is notoriously difficult to successfully implement energy efficient technologies into the buildings sector. This is because energy use in commercial buildings is usually only a few percent of operating costs and is thus a discretionary investment. From a user perspective, tenants normally only live or work in a building for less than five years, thus greatly limiting their incentive to improve their buildings. Owners and builders of homes market their properties far more on upfront costs rather than long-term living expenses. In addition, the most inefficient (and coldest) homes are often lived in by the poor who have the fewest resources to address this issue. Finally, legislative efforts to improve the building stock are hampered by overlapping building regulations at different levels of government, a lack of information and the long (50 years) capital turnover of the building stock.

Despite these difficulties, there have been some efficiency success stories. Twenty years ago, Lawrence Berkeley Laboratory launched a program to develop advanced, spectrally-selective coatings for windows. The program,

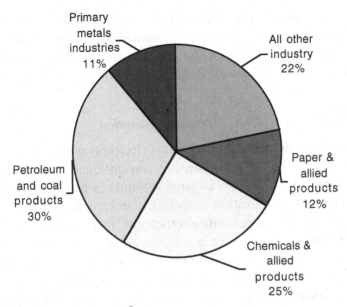

Figure 4.9   Percentage of energy use in the manufacturing sector (Source: EIA, 1996).

which cost some $3 million over the last two decades, transformed the world's window industry. New windows are three times as efficient as double-glazed windows. "Low-E" glass now accounts for 35% of all windows sold in the US. Cumulative energy savings from the diffusion of this technology to date have exceeded $2.1 billion in the US alone, and are projected to grow to $17 billion by 2015. However, this program also illustrates the long time period a new technology may have to wait for wide-spread adoption to occur, as 65% of all windows sold in the US today are still not as efficient as this proven technology.

4.9.3 Industry

The industrial sector is heterogeneous, as is shown in Figure 4.9, although the seven process industries targeted under the DOE Office of Industrial Technology's Industries of the Future program (aluminum, chemicals, forest products, glass, metal-casting, petroleum refining, and steel) are responsible for 80% of energy consumption in the US manufacturing sector.

One crosscut energy efficiency technology that has enjoyed both government support and successful industrial adoption is cogeneration utilizing gas or steam turbines as prime movers. In the UK, installed capacity of cogeneration has doubled in the last 10 years from an installed base of 1850 MWe to 3700

MWe, saving an estimated 20 million tonnes of CO_2 each year (UK-DTI, 1998). This cost effective technology has benefited both from a consistent information program by the UK Government and the deregulation of the UK energy industry, with resultant opportunities for on-site energy generation.

4.9.4 Electricity generation

Many people are unaware that conventional electricity generation occurs only at an efficiency of 30–35%. This is based on conventional steam turbine plants and is largely due to the wastage of large amounts of heat that are inevitably produced during the combustion process. Commercially available combined cycle gas turbines (CCGT) generate electricity at 50% efficiency due to the use of waste heat from the gas turbine in a second-stage steam turbine. Even more impressive are prototype advanced turbines with a target electrical efficiency of up to 70% for a combined cycle (Energy Innovations, 1997). This public–private research effort is being coordinated by the US Department of Energy's Advanced Turbine Systems program. Advanced turbine systems enable higher operation temperatures (up to 2600°F) and thus improve thermal efficiency (from Carnot's law applied to thermodynamic machines). In practice, temperatures are limited by damage to the turbine blades, and this is being addressed by new materials research and new cost effective ways to cool the turbine blades. The private members of the consortium include six turbine manufacturers and 83 universities.

Advanced turbine systems (like conventional CCGTs before them) will have far reaching consequences for efficiency of electrical generation. As they are far smaller than centralized coal or nuclear plants, ATS plants can be built with short lead times, and at low enough costs to compete with the marginal operating costs of existing older plants, thus forcing the retirement of inefficient plant. As the electricity markets in developed countries become deregulated, and social pressure to prevent the construction of transmission lines develops further, the opportunities for highly efficient generation at the point of use grow. This offers the possibility of using the excess heat in a cogeneration scheme and a structural change to a decentralized paradigm of energy supply with great efficiency savings (Patterson, 1998).

4.10 Conclusions

Energy efficiency is a powerful tool as part of the solution to reduce anthropogenic emissions of greenhouse gases. The strength of energy efficiency is that adopters are interested in the services that energy provides, not physical units of energy. However, we have endeavored to show that energy efficiency declines should not be thought of as linear trends stretching out over the next 100 years. Instead the mechanism by which energy efficiency develops is a messy, interactive and complex one, existing endogenously within the framework of human activity.

Energy efficiencies can be defined for specific activities over decadal time scales. Aggregation over economies to compare overall energy intensities is complex and controversial, and is limited spatially by climate, economic and social differences between nations, and temporally by the evolution of social mechanisms and the use of as-yet-undiscovered technologies.

Efficiency improvements occur mainly when structural changes in the economy occur. This may be in the retirement of old capital at the same time a new technology becomes available, or when changes in a sphere of human social and economic activity result in the employment of new energy technology. On this bumpy road to efficiency improvement, technological change is not the driver of efficiency improvement. Rather it should be thought of as endogenous to the changes in the economic, social, regulatory, and technological paradigms. It is these interactions that make energy efficiency trends so difficult to project into the distant future.

When trying to predict long term changes in energy efficiency, the contribution of developing countries will be crucial. This is because the bulk of the world is moving toward greater use of fossil fuels and the movement to industrialized societies will be based on rising energy consumption. Despite the importance of developing countries, most efficiency analysis has been carried out on the industrialized world. Additionally, the future development of industrialized countries and the path of energy intensities that they will follow is very uncertain.

Interestingly, some of energy efficiency's fiercest and longest-standing proponents make the mistake of implying that efficiency is so powerful and the market so effective that policy "trimtabs" to correct "subtle" imperfections are all that's needed to address the needs of climate change policy. Certainly numerous studies support the contention that energy efficiency investments can not only reduce energy use and greenhouse gas emissions but can do so while improving economic performance, increasing jobs, enhancing overall environmental quality, reducing the price of non-energy goods and services, and increasing household wealth. But this will not happen by itself, and various policies to promote energy efficiency have been discussed. Policy options include a realistic price of energy that internalizes all the costs associated with its use and hence increases the motivation to save energy. In light of the political intractability of achieving meaningful energy price increases, another set of policy options that allow the economy to set itself down a path where energy efficiency products are offered by suppliers, bought by adopters and continue to decline in price and improve in performance are particularly important for a long-term climate abatement strategy at reasonable cost. These policies include increased RD&D for energy efficiency technologies, support mechanisms for adoption, and energy standards.

Energy efficiency can be thought of as making a little go a long way. Azar and Dowlatabadi (1998) estimate that to meet a 450ppm CO_2 concentration target, given no energy intensity reduction, 2700 EJ/year of carbon-free energy would be required. As present global energy supply is of the order of 400 EJ/year, this requirement is clearly enormous. Given the nature of energy efficiency development through interactions with social and economic evolvement that we have stressed throughout this chapter, a conservative estimate of long run annual decrease in global energy intensities is 0.5% a year, resulting in the need for only 60% of this estimated carbon-free energy or 1630 EJ/yr. If, through policies to encourage the development of more energy efficient technologies and support for those adopting and using these technologies, an average improvement of 1% per year is achieved, then only 37% or 990 EJ/year is required. If, with determined efforts, the average energy intensity is reduced by 1.5% a year, then only 22% of this carbon free energy or 600 EJ/year is required for atmospheric CO_2 stabilization at 450ppm. And if, through even more aggressive policies, energy intensity is reduced by 2% per year, then only 9% or 250 EJ/year of carbon-free energy would be needed. However, with present global energy supply at around 400 EJ/year, this transition to carbon-free energy sources by 2010 would still be a major endeavor.

These numbers indicate that energy efficiency cannot be the whole solution but is an essential part of a solution to abate greenhouse gas emissions without the need for huge quantities of carbon-free power. A little efficiency does go a long way.

References

Arthur, W. B., 1987. Path-dependent processes and the emergence of macro-structure. *European Journal of Operations Research*, **30**:294–303.

ASE, 1998, *Price It Right: Energy Pricing and Fundamental Tax Reform*, Alliance to Save Energy, Dr. D. L. Norland and K. Y. Ninassi, in cooperation with Professor D. W. Jorgenson, Chairman, Department of Economics, Harvard University. Alliance to Save Energy, Washington DC.

Ausubel, J. H., 1998. The environment for future business: efficiency will win, *Pollution Prevention Review*, Winter 1998, John Wiley & Sons, Inc.

Ausubel, J. H. and C. Marchetti, 1996. Elektron: Electrical Systems in Retrospect and Prospect, *Daedalus*, **125**:139–69.

Ausubel, J. H., C. Marchetti, and P. S. Meyer, 1998. Toward green mobility: the evolution of transport, *European Review*, **6**:137–56.

Ayres, R. U. and K. Martinàs, 1992. Experience and the life cycle: some analytical implications. *Technovation*, **12**(7):465–86.

Azar, C. and H. Dowlatabadi, 1999. A review of technical change in assessments of climate policy, *Annual Review of Energy & Environment*, **24**:513–44.

Brill, A., K. Hassett, and G. Metcalf, 1998. Knowing what's good for you: energy conservation investment and the uninformed consumer hypothesis. In *Climate Change Policy: The Road to Buenos Aires*. Washington DC: American Enterprise Institute.

Cebon, P., 1992. 'Twixt cup and lip: organizational behavior, technical prediction and conservation practice. *Energy Policy*, September: 802–14.

Cyert, R. and J. March, 1992. *A Behavioral Theory of the Firm*. Cambridge, MA, Blackwell.

David, P., 1985. Clio and the economics of QWERTY. *American Economic Review*, **75**(2):332–7.

DeCanio, S. and W. Watkins, 1998. Investment in energy efficiency: do the characteristics of firms matter? *The Review of Economics and Statistics*, **80**(1):95–107.

Dixit, A. and R. Pindyck, 1994, *Investment Under Uncertainty*. Princeton University Press.

Dowlatabadi, H., 1998. Sensitivity of climate change mitigation estimates to assumption about technical change, to appear in a special issue of *Energy Economics*, eds. C. Carrera and J.-C. Hourcade.

Dowlatabadi, H. and M. A. Oravetz, 1997. US long-term energy intensity: Backcast and projection, to appear in *Energy Policy*.

EIA, 1997. *Annual Energy Review 1996*, Energy Information Agency, Washington DC.

Energy Innovations, 1997. *Energy Innovations: A Prosperous Path to a Clean Environment*. Washington DC: Alliance to Save Energy, American Council for an Energy-Efficient Economy, Natural Resources Defense Council, Tellus Institute, and Union of Concerned Scientists.

Farzin, Y., K. Huisman, and P. Kort, 1998. Optimal timing of technology adoption. *Journal of economics dynamics and control*, **22**:779–99.

Freeman, C., J. Clark, and L. Soete, 1982. *Unemployment and Technical Innovation: A Study of Long Waves and Economic Development*. London: Francis Pinter.

Goldemberg, J., 1996. A note on the energy intensity of developing countries. *Energy Policy*, **24**(8): 759–61.

Goulder, L. H. and S. H. Schneider, 1998. Induced technological change, crowding out, and the attractiveness of CO_2 emissions abatement, *Resource and Environmental Economics* (submitted).

Griliches, Z., 1957. Hybrid corn: an exploration in the economics of technical change. *Econometrica*, **25**(4):501–22.

Gruebler, A., 1997. Time for a change: on the patterns of diffusion of innovation. January 1997, *Daedelus*, Journal of the American Academy of Arts and Sciences, from the issue entitled, *The Liberation of the Environment*, Summer 1996, **125**(3):19–42.

Harvey, H., 1998. Essay, *The Energy Foundation Annual Report*, and personal communication, 1/99.

Hassett, K. and G. Metcalf, 1993. Energy conservation investment: do consumers discount the future correctly? *Energy Policy*, **21**(6):710–16.

Hoffert, M. I., K. Caldiera, A. K. Jain, E. F. Haites, L. D. Harvey, S. D. Potter, M. E. Schlesinger, S. H. Schneider, R. G. Watts, T. M. L. Wigley, and D. J. Wuebbles, 1998. Energy implications of future stabilization of atmospheric CO_2 content. *Nature*, **395**:881–4.

Jaffe, A. and R. Stavins, 1993. *The Energy Paradox and the Diffusion of Energy Conservation Technology,* Resources for the Future: Washington DC.

Jaffe, A. and R. Stavins, 1995. Dynamic incentives of environmental regulations: the effects of alternative policy instruments on technology diffusion. *Journal of Environmental Economics and Management*, **29**:S43–S63.

Katz, M. and C. Shapiro, 1986. Technology adoption in the presence of network externalities. *Journal of Political Economy*, **94**(4):822–41.

Levine, M. D., L. Feng, and J. E. Sinton, 1992. China's energy system: historical evolution, current issues, and prospects, *Annual Review of Energy and the Environment*, **17**:405–36.

Lovins, A., 1996. Negawatts. Twelve transitions, eight improvements and one distraction, *Energy Policy*, **24**(4):331.

Manne, A. S. and R. G. Richels, 1992. *Buying Greenhouse Insurance – the Economic Costs of CO_2 Emission Limits.* Boston, MA, MIT Press.

Mansfield, E., 1961. Technical change and the rate of innovation. *Econometrica*, **29**(4):741–66.

Messner, S., 1997. Endogenized technological learning in an energy systems model, *Journal of Evolutionary Economics*, **7**(3), September 1997. Springer-Verlag.

Metcalf, G. and K. Hassett, 1998. Measuring the energy savings from home improvement investments: evidence from monthly billing data. In *Climate Change Policy: The Road to Buenos Aires*. Washington DC: American Enterprise Institute for Public Policy Research.

Mitchell, C., 1997. The renewable NFFO: a review, *Energy Policy*, **23**, No.12, 1077–91.

Morgenstern, R., 1996. *Does the Provision of Free Technical Information Really Influence Firm Behavior?*, Resources for the Future: Washington DC.

Nakicenovic, N., 1997. *Technological Change as a Learning Process*, ILASA: Laxenberg, Austria.

New York Times, 1998. US splurging on energy after falling off its diet, A. Myerson, October 22, 1998, pp. A1, C6.

Newell, R., A. Jaffe, and R. Stavins, 1999. *The Induced Innovation Hypothesis and Energy Saving Technological Change*, Resources for the Future: Washington DC.

Nilsson, L., 1993. Energy intensity trends in 31 industrial and developing countries. *Energy*, **18**:309–22.

Patterson, W., 1998. *Electric Futures: Pointers and Possibilities.* London: Earthscan.

PCAST, 1997. *Federal Energy Research and Development for the Challenges of the Twenty-First Century*, Report of the Energy Research and Development Panel, The President's Committee of Advisors on Science and Technology, November 1997, J. P. Holdren, Chairman, Washington DC.

Rosenfeld, A., 1999, personal communication.

Sagar, A. and M. Kandlikar, 1997. Knowledge rhetoric and power, international politics of climate change, *Economic and Political Weekly*, Dec. 1997, 3139–48

Schipper, L., F. Unander, and C. Marie, 1998. *The IEA Energy Indicators Effort: Extension to Carbon Emissions as a Tool of the Conference of Parties*, International Energy Agency, Energy Efficiency and Technology Policy Office, November.

Schipper, L., 1999. Personal communication.

Schlumpf, C., J. Behringer, G. Durrenberger, and C. Pahl-Worstl, 1998. The personal CO_2 calculator: a modeling tool for participatory integrated assessment model, submitted to *Environmental Modeling and Assessment*.

Simon, H., 1982. *Models of Bounded Rationality*. Cambridge MA: MIT Press.

Sonnenblick, 1995. *A Framework for Improving the Cost Effectiveness of Demand Side Management Program Evaluations.* Ph.D. Thesis, Carnegie Mellon University, Pittsburgh.

Strachan, N. and H. Dowlatabadi, 1999. *The Adoption of a Decentralized Energy Technology: The Case of UK Engine Cogeneration.* Paper read at ACEEE Summer Study on Energy Efficiency in Industry, June 1999, at Saratoga Springs, New York.

Tietenberg, T., 1992. *Environmental and Natural Resource Economics*, 3rd edition, Harper Collins Publishers.

UK-DTI, 1998. *Digest of UK Energy Statistics*, HMSO, London.

von Weizäcker, E. U., A. B. Lovins, and L. H. Lovins, 1995. *Factor Four – Doubling Wealth, Halving Resource Use.* London, Earthscan.

Walley, J. and R. Wigle, 1991. The international incidence of carbon taxes. In *Global Warming: Economic Policy Responses*, ed. R. Dronbush and J. M. Poterba. 233–62.

Wigley T., R. Richels, and J. Edmonds, 1996. Economic and environmental choices in stabilization of atmospheric CO_2 concentrations, *Nature*, **369**:240–3.

World Energy Council (WEC) and International Institute for Applied Systems Analysis (IIASA), 1995. *Global Energy Perspectives to 2050 and Beyond*, N. Nakicenovic, Study Director; WEC, London, UK.

5

The Potential of Renewable Energy to Reduce Carbon Emissions

5.1 Introduction and Overview

Reducing carbon emissions to levels consistent with the stabilization of greenhouse gas concentrations in the atmosphere will require enormous amounts of clean energy supplies to displace current fossil fuel combustion. Renewable energy supplies are one source of such clean energy.

This chapter focuses on terrestrially based renewable energy options. As used here, renewable energy is defined simply as any form of energy that can be used without diminishing the resource. Some forms of renewable energy, like solar and wind, are essentially limitless. Others, like geothermal, may be depleted in the short term, but are eventually regenerated. Some, like hydroelectricity, municipal solid waste, landfill gas, and biomass must be regenerated continually. But none are totally depleted forever.

Renewable energy is generally clean. Most forms of renewable energy do not involve combustion and, therefore, do not emit gases to the atmosphere. In particular, renewables are attractive because they generally emit no net carbon dioxide. What is just as important for those countries experiencing severe local air pollution problems is that non-combusting renewables, like solar and wind, produce energy without emitting sulfur dioxide, nitrogen oxides, particulate matter, or toxic materials. In addition, renewable electric technologies, like photovoltaics and wind that do not use heat to produce electricity, require no water resources for cooling towers, and have no water effluents.

For those renewables that do employ combustion, like biomass power and municipal solid waste, most of the emissions, other than NO_x, would have occurred anyway as a result of natural decomposition processes. In some cases, the natural decomposition may have produced stronger greenhouse gases, like the uncontrolled methane emissions from landfills. Certainly the mix of effluents is different for combustible renewables than for fossil fuels. For example, biomass combustion produces none of the sulfur emissions produced by coal and oil combustion. There is even evidence that cofiring a coal plant with biomass can reduce NO_x emissions per unit of coal combusted. On the other hand, the combustion of municipal solid waste can produce as many particulates, volatile organic chemicals, and NO_x emissions as coal burning.

Renewables provide other benefits as well. For one, they increase the diversity of energy supplies, relieving pressure on fossil fuel supplies and oil prices, and increasing energy security. In the long term, renewables are one of the few sustainable energy supply options available. Eventually they will be essential for replacing diminishing fossil fuel resources, especially given the inequitable distribution of those natural resources.

Many renewable energy technologies can be deployed in small increments. Initial increments of these modular technologies can provide economic returns while other increments are being installed, thereby reducing capital constraints. The modular nature of technologies like wind and photovoltaics also makes them candidates for deployment within an electric distribution system, bypassing the need for additional distribution lines and transformers, reducing transmission losses, and increasing the reliability of the electric supply.

For both developed and developing countries, renewables also represent an indigenous supply with significant employment opportunities in manufacturing, installation, and infrastructure development. It is not the intent of this chapter to focus on all these benefits. They are mentioned simply as co-benefits to the focus of this chapter on the ability of renewables to reduce carbon emissions worldwide.

There are vast renewable resources that can be used to reduce world fossil fuel use and carbon emissions. These resources take many forms – solar, wind, hydro, and geothermal, to name a few. They are spread widely, but non-uniformly, around the globe. While the total resource is several orders of magnitude larger than current world energy consumption, it is doubtful that all fossil fuels could be displaced by renewables. There are issues with respect to cost, access to the resources, technical integration into current energy systems, environmental factors, social barriers, and other institutional concerns.

In this chapter, we explore the potential for renewable energy technologies to address long-term global climate change. On a regional basis, we examine the full set of resources and the applications of these technologies, the issues involved in their deployment, and the prospects for the future. We end with recommendations for more rapid renewable energy deployment, prompted not by the current markets and prices for energy, but by the underlying trends in climate change, local air pollution, depletion of fossil fuel supplies, and each nation's desire to prosper in a secure world environment.

5.2 Characteristics of Renewable Energy Technologies

There are many renewable energy technologies. This chapter does not provide a basic summary description of each of the technologies themselves. Such information can be found in Cassedy (2000), Johansson *et al.* (1993), and on the Internet at www.eren.doe.gov:80/. We will focus here on the characteristics of the technologies that relate to their potential to address climate change. This includes summarizing their cost and performance, looking at the breadth of technologies and applications, and reviewing the size and characteristics of the renewable resources.

5.2.1 Cost of renewable energy

Some forms of renewable energy, like hydroelectricity and biomass, have been cost-competitive for many years in certain applications and provide a substantial energy supply worldwide. Others, like passive solar building design, are cost competitive, but haven't yet overcome all the market factors that currently preclude their widespread use. Technologies like wind and geothermal are currently cost competitive at their best resource sites, but need further improvements and support to reach their full market potential. A few, like photovoltaics, have identified niche off-grid electric markets that the industry is building to the point where it can competitively address retail power markets. Still others, like ethanol from biomass, are evolving both in the laboratory and the marketplace to the point where they will be competitive without price supports. Table 5.1 provides a summary of the general renewable energy technologies and their economics.

In reality, there are a large variety of earth-based renewable energy technologies applicable to a wide range of markets, each with different values, costs, and considerations. Breaking the renewable electric technologies of Table 5.1 down to the next level, Table 5.2 provides some insight as to why it is difficult to speak generically of the future potential for renewables. Table 5.2 shows the different variations that can occur for each of the renewable energy technologies presented in Table 5.1. Even the detail of Table 5.2 does not provide the full technology picture. For example, while there are three major solar concentrating power systems, variants of these exist in terms of the types of engines used to generate power, different forms of storage, and types of concentrators. Obviously the cost of each variant will be different. Nonetheless, the variety exists because there are different applications that take advantage of specific features of a particular variant, and because it is not clear at this point which variant will be the ultimate least-cost option. The cost data presented in Table

Table 5.1 *Renewable energy technology cost*

Technology	Current cost of energy (US cents/kWh)	Reduction in capital cost over last 10 years (%)	Anticipated reduction in cost of energy over next 20 years (%)
Photovoltaics	25–50	40	50
Wind	4–6	30–50	50
Biomass power	7–9	10	15–20
Landfill gas	4–6	10–15	0
Small scale hydro	2–10		
Geothermal	3–10	20	25
Concentrating solar power	12–27	50	25
Active solar heating ($/Mbtu)	3–20	30–60	30–50
Biofuels ($/liter)	0.24–0.37	10	25

Sources: adjusted from EPRI (1997) and IEA (1997).

5.1 represent the least-cost options today. Even for a particular variant, there is substantial spread in today's costs and prices with the size of the system, the particular application, and the region in which it is produced and installed.

5.2.2 R&D potential

The largest single factor precluding more rapid penetration of renewable energy technologies into these markets is the cost of renewables. Table 5.1 shows that the cost of most renewable energy technologies has been declining and is expected to continue to do so. The cost decreases have occurred from industrial learning through the production process as well as through research and development (R&D). While all renewable technologies are expected to have future cost decreases, the potential for cost reductions is not the same for all of them. For example, hydroelectricity is a relatively mature technology where the majority of the research is directed at mitigating environmental impacts, not reducing costs. On the other hand, photovoltaics' current high processing costs of relatively inexpensive materials provide ample opportunity for future cost reductions. Below, we examine the R&D-related cost reduction potential for three renewable energy technologies with significant R&D opportunities.

Table 5.2 *Renewable electric technologies*

Technology	Variations
Photovoltaics	Configuration – flat plate, concentrating Material type – crystalline silicon, thin films Cell type – single junction, multijunction
Wind	Location – land, water Orientation – horizontal axis, vertical axis
Biopower	Resource – agricultural and mill wastes, forest wastes, municipal wastes, dedicated energy crops Output – cogenération, power only Configuration – cofiring of coal plants, direct combustion, gasification Cycle – steam, combustion turbine, combined cycle gasifier
Hydroelectricity	Resource – reservoir, run-of-river
Landfill gas	Output – cogeneration, power only Configuration – combustion turbine
Geothermal	Application – heat pumps for space conditioning, industrial process heat, power only Resource – hot water, steam, hot dry rock Power conversion – flashed steam, binary
Ocean energy	Resource – tidal, wave power, ocean thermal energy conversion
Solar concentrating power	Concentrator – troughs, dishes, central receiver Configuration – hybrid, solar-only Storage

5.2.2.1 Photovoltaics

The two distinguishing features of R&D on photovoltaic systems are the multiple paths available for improvements and the fact that processing costs make up the bulk of today's module costs. By pursuing multiple R&D paths, the probability of finding one or more routes to cost competitive photovoltaic modules is increased. While silicon crystal constitutes more than 85% of production today (Cassedy, 2000), a multitude of thin films are being investigated. These include polycrystalline sheets, amorphous silicon, and a host of semiconducting compounds, such as cadmium telluride, gallium arsenide, and copper indium selenide, all of which absorb wavelengths that predominate in the solar spectrum, increasing their potential efficiency compared to silicon materials. Because these materials can be deposited from the gaseous state, they offer the possibility of multijunction cells with efficiencies potentially greater than 30% resulting from absorption of a wide spectrum of wavelengths. Such high-efficiency cells also lend themselves to the use of reflectors and/or lenses, along with mechanisms, to track the sun to increase the sunlight incident on the photovoltaic cell. Many innovative tracking mechanisms and concentrator assemblies are being investigated for such concentrating systems. Cost reductions are also possible in the manufacturing process itself, through both direct research on manufacturing processes and through learning during the production process. Learning is also reducing the balance-of-system costs for components other than the photovoltaic modules, like the inverter, controls, grid interconnections, tracking mechanisms, and frames.

At issue is whether or not this research and learning will bring the ultimate cost of photovoltaic-supplied electricity from its current levelized cost of $0.25–$0.50/kWh to the $0.03–$0.06/kWh required to be competitive today in grid-connected distributed electric markets. Although this may be extremely difficult, research continues because potential improvements are promising and the market value of a clean modular energy source like photovoltaics is expected to increase.

5.2.2.2 Wind

The current trend is towards taller, more powerful, wind machines. Energy output is greater with taller machines because the energy available rises very quickly with wind speeds (as the cube of the wind speed), and wind speeds generally increase by about 60% for every doubling in tower height. Furthermore, the longer rotors possible on a taller tower greatly increase the area swept by the rotor (square of the rotor length) and, consequently, the energy captured. The advantages of higher towers were sought in the early 1980s through the US-government-sponsored "MOD" series, with MOD –5B machines as large as 3.2 MW with rotors 97 meters in diameter. However, mechanical difficulties with the turbines, support structures, and especially the turbine blades led to a new generation of smaller machines in the 200–400 kW range. Subsequent advances in blade design and variable-speed turbines are now opening new opportunities for machines as large as 1 MW or greater. More flexible blades are also allowing renewed consideration of machines with rotors downwind of the generator and tower. If these can withstand the turbulence introduced by the tower structure, they may allow even larger machines. Other advances being investigated include improved controls and better wind forecasting. These improvements are forecast to bring costs down below $0.03/kWh for the better wind sites.

5.2.2.3 Ethanol from biomass

Currently, fermentation processes are being used to produce ethanol from sugar and starch feedstocks, such as sugar cane in Brazil and corn in the United States. However, in the last few years, hydrolysis processes have been developed with acids and newly discovered enzymes that break down cellulosic material of plants like poplars and switchgrass into fermentable sugars. A simultaneous saccharification and fermentation process has been developed allowing hydrolysis of both cellulosic and hemicellulosic material. Refinements of this process are currently being pursued that should allow high-productivity conversion of the cellulosic material found in fast-growing short-rotation woody crops and herbaceous switchgrasses. R&D is also under way to improve the production rates for these dedicated energy crops. These include breeding and genetic manipulation for resistance to pests and diseases and for higher yields; efficient propagation, harvesting and handling methods; and improved nutrient management systems. Production has already increased to 10 dry tons per acre (Cassedy, 2000) with cost goals of $35/ton. Although commercial plants are not yet in operation for producing ethanol from cellulosic materials, estimates of production costs today are in the range of $1.05–$1.55 per gallon. Early goals were to achieve costs of $0.67/gallon (Johansson *et al.*, 1993), which would be competitive with gasoline at about $1.00 per gallon.[1]

5.2.3 Market opportunities

Not only are there many renewable energy sub-technologies, as shown in Table 5.2, but there are a host of different applications for renewables, each with different costs and values. Table 5.3 shows some of the major markets that renewables can address. Some of these opportunities are still in the laboratory stage, e.g. biofuels, while others have already largely saturated their markets, e.g. large hydro and biomass cooking. However, even the saturated markets present new opportunities. For example, research on more fish-friendly hydro may expand the market for this mature technology. Subsidized corn-to-ethanol biofuels are already pervasive in the United States while research continues on ethanol production from cellulosic material.

The point of Table 5.3 is that many market opportunities exist, and for renewable energy to make its maximum possible contribution to the reduction of greenhouse gases, a number of market and technology issues will have to be

---

[1] In a properly designed engine, a gallon of ethanol will yield about 80% of the driving range of a gallon of gasoline, due primarily to the lower energy content of ethanol.

Table 5.3 *Market opportunities*

Technology	Market/technology	Market size	Value/competitiveness
Photovoltaics	Consumer electronics	Small	High value
	Cathodic protection	Small	High value
	Remote water pumping	Small	High value
	Village power	Large	Battery and diesel competitors are expensive
	Distributed on-grid generation	Large	Provides additional value beyond energy
	Central-station power	Large	Lower value
Wind	Village power	Large	Battery and diesel competitors are expensive
	Distributed on-grid generation	Small, limited sites	Provides additional value beyond energy
	Central-station power	Large	Competitive at best resource sites
Biomass power	Forest/agricultural industry cogeneration	Moderate	Highly competitive; almost 6 GW of wood and pulp liquor power plants exist in the US today
	Village power	Large	Battery and diesel competitors are expensive
	Industrial co-products	Moderate	High-value, non-energy co-products can make it competitive
	Cofiring of coal plants	Small	Economics depend on proximity of quality resource to the coal plants
	Central power	Large	Requires improvements in biogasifiers
Landfill gas	Industrial cogeneration	Small	Limited by co-location requirements for landfill and industry
	Power only	Small	High value in reducing emissions of methane, a greenhouse gas
Hydro	Large hydro	Large	Environmental issues preclude most additional development
	Small-scale hydro	Moderate	Highly competitive, but sites limited

Technology	Application	Size	Comments
Geothermal	Geothermal heat pumps for space heating/cooling	Moderate	Competitive where both cooling and heating space loads exist
	Industrial process heat	Small	Limited by resource/industry co-location
	Industrial cogeneration	Small	Limited by resource/industry co-location and temperature requirements
	Central hydrothermal	Small	Resource limited; economics very site specific
Concentrating solar power	Water and process heat with troughs	Moderate	Requires direct sunlight for concentration; high materials cost
	Distributed generation with dishes	Small	Reliability remains an issue; costs of collector and engines still high
	Central power with troughs and power towers	Moderate	Further cost reductions are required
Active solar heating	Pool heating	Small	Competitive in non-freezing climates; extends swimming season
	Cooking	Moderate	Sunlight availability is an issue
	Hot water heating	Large	Competitive with electric heating, especially in freezing climates
	Space heating	Large	Shell improvements and high-efficiency gas furnaces are currently more economic
Passive solar	Space heating	Large	Solar tempering combined with shell improvements are effective
Biomass heat	Cooking	Moderate	Resource depletion is an issue; decreasing with urbanization
	Space and water heat	Moderate	Resource depletion is an issue; decreasing with urbanization
Biofuels	All transportation fuel needs	Large	Further R&D on crop engineering and conversion process required.

addressed. In many cases, these multiple market opportunities are essential to building up renewable energy markets with initial niche markets, thereby enabling productivity gains, which lead to lower costs and entry to larger bulk markets. In Table 5.3, the more mature niche markets for each technology are listed first, with the less easily penetrated markets near the bottom. Future penetration of these markets will depend heavily on the realization of the anticipated cost/performance improvements summarized in Table 5.1. However, even the mature and stalled markets could expand significantly were heavy reductions in carbon emissions mandated worldwide.

5.2.4 Resources

Unlike fossil fuels that can be mined, pumped, or otherwise collected and transported to their point of use, renewables generally must be consumed at or near the point of the resource or converted there to a transportable form of energy like electricity, gas, or liquid fuel. Fortunately, renewable resources are fairly ubiquitous. Resources like solar radiation, wind, and biomass exist worldwide at varying concentrations. Even geothermal and ocean energy are fairly widely distributed, though more localized than wind and solar.

Not only are renewable resources spread around the globe, they exist in huge, albeit dispersed, quantities. For example, it has been estimated that enough solar radiation reaches the earth every 45 seconds to meet the world's current electricity demand for a year. Similarly, there is enough wind in the state of North Dakota to meet 35% of all current US electricity demand. While such estimates are interesting as upper bounds, they say little about the cost, accessibility, or spatial and temporal variability of these vast resources. We discuss such issues for each of the major renewable resources in the paragraphs that follow. More specific regional resource data is available in the subsequent section on "Regional perspectives".

5.2.4.1 Solar

Sunlight, the most ubiquitous of all the renewable resources, is also the driving force behind wind, biomass, and hydroelectric. Solar energy directly powers many of the technologies of Table 5.1, including photovoltaics, solar concentrators, and active and passive solar thermal. However, for energy purposes, not all sunlight is equal. Solar concentrators can only concentrate direct sunlight; they cannot focus indirect sunlight that has been scattered by aerosols in the atmosphere. Thus, concentrating solar systems are most viable in unpolluted desert regions with high insolation. Non-concentrating solar energy systems work better in these regions also, but do not suffer the same performance penalties in regions where there is relatively more indirect sunlight.

The total amount of solar energy reaching the earth's atmosphere is 170000 terawatts, or about 13000 times current world energy requirements. While the amount of solar power available worldwide is extremely large, the power incident per unit area is small because it is spread over the entire surface of the earth. Figure 5.1 shows the total (direct and indirect) insolation worldwide over the course of an average year as measured in terms of kWh/m^2 per day. Peak values are just over $1 kW/m^2$ at the earth's surface, with daily maximums in the best locations of about $8 kWh/m^2$ per day.

While Figure 5.1 shows the spatial variation of the insolation resource, it does not show the temporal variation. Obviously, the resource is not available at night or during cloudy periods. This means that any application for solar must either require energy only when the sun shines, or be able to convert the sunshine to some form of stored energy, or have another form of energy available as backup. This issue of dealing with the intermittency of solar resources is discussed further below in the "Market issues" section.

5.2.4.2 Wind

Wind sites that are economic today typically have winds that average near the surface (10 m) better than 6.5 m/s (14.5 mph). Below 5 m/s, sites are not considered economic today, but that could change with improvements in technology and changing market conditions. Wind resource quality is not simply a function of wind speed at the ground, but also of turbine height above the ground, air density (cooler, more dense air contains more kinetic energy), and the frequency with which the wind blows. As shown in Figure 5.2, most regions of the world have ample wind resources. It has been estimated that global technical wind potential might be as large as 14000 (IPCC, 1996) – 26000 TWh/yr (IIASA, 1981), or up to twice the 1995 world annual electricity consumption of 13000 TWh (IEA, 1998).

In the last few years, wind systems have made significant advances in penetrating world markets, with installations increasing by more than 35% in 1999 to reach a total of 10 GW of total installed capacity worldwide. Much of this has occurred because of the confluence of technology improvements and cost reductions with increased public concern about local air pollution and global climate change. In particular, the European goal of 4000 MW by 2000 was reached by 1997, doubled in that year to 8000, and exceeded again by the end of 1999 with 8915 MW installed. Consequently, the 2010 goal has now been increased from 25000 MW to 40000 MW (EWEA, 2000).

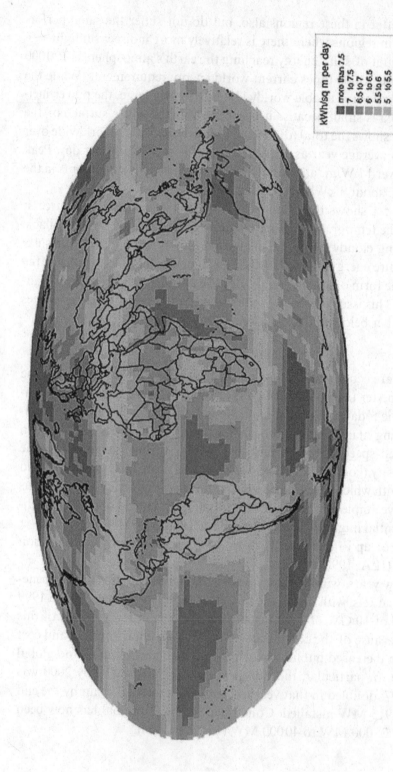

Figure 5.1 Global horizontal insolation.

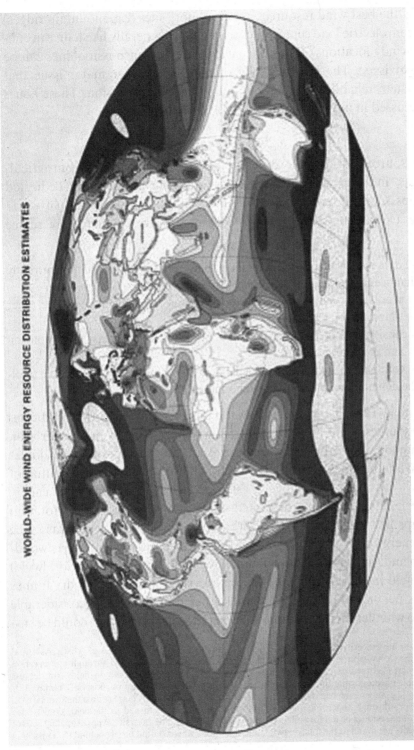

WORLD-WIDE WIND ENERGY RESOURCE DISTRIBUTION ESTIMATES

Darker areas denote higher-speed wind resources

Figure 5.2   Worldwide wind energy resource distribution estimates.

Many of the best wind resources are on remote sites (e.g. mountain ridges) far away from electric load centers (i.e. people don't generally like to live in continuously windy locations). Thus, the availability of transmission lines can be a significant issue. The intermittency of wind is another major issue that impacts system reliability, capacity planning, and transmission. These issues will be discussed in more detail in the "Market issues" section.

5.2.4.3 Biomass

Biomass resources exist in many different forms, including wastes from agriculture, forests, industry, and municipalities; standing growth; and dedicated energy crops. Current biomass use accounts for approximately 15% of world energy use (Johansson *et al.*, 1993). Global resource estimates are scarce because:

1. Resource data is unavailable or uncertain in many areas of the world that currently derive significant fractions of their energy use from biomass.
2. Biomass production can use the same resources as required for food and other crops, e.g., land, water, fertilizer.
3. There is considerable debate over the impact of removing biomass wastes on soil nutrients and erosion.[2]
4. Like any commodity, there is no single resource amount; more will become available if the demand and price are high enough.

The estimate in Table 5.4 of the potential biomass supply worldwide indicates biomass resources could be recovered at rates close to 300 exajoules per year, approaching today's worldwide energy use of all energy forms of 400 exajoules per year. While residues are significant contributors to the total potential of Table 4, 90% of the resource shown is assumed to come from dedicated energy crops. The leading-candidate dedicated energy crops include short-rotation woody crops like poplar trees, and herbaceous grasses like switchgrass. The dedicated energy crop estimates assume 10% of the land now in forests/woodlands, cropland, and pasture (approximately 900 million hectares worldwide) will be planted in dedicated energy crops and yield, on average, 15 dry tonnes of biomass per hectare with 20 gigajoules per tonne. There is considerable debate as to whether this level of production (and energy content) could be sus-

---

[2] These concerns are exacerbated in dry-land agriculture practiced in the poorest areas of Africa, Asia, and Latin America (Johansson *et al.*, 1993). Residue removal can also be increased through conservation tillage and increased fertilizer use (especially nitrogen). Too many residues left on fields can depress growth, inhibit nitrate formation, and spread crop diseases. Many crops are purposefully removed or burned today (e.g. sugar cane bagasse, rice). Dung residues are a potentially large source that can be converted to biogas through anaerobic digestion with residue nutrients returned to the field. Forests also require that some residues be left for soil fertility, erosion control and biodiversity. Approximately 70% of a tree's nutrients are found in the foliage, twigs, and fine roots, which should be left behind (Johansson *et al.*, 1993).

Table 5.4 *Potential recoverable biomass supplies (10^{18} joules per year)*

Region	Crop residues	Forest residues	Dung	Dedicated crops
US/Canada	1.7	3.8	0.4	34.8
Europe	1.3	2.0	0.5	11.4
Japan	0.1	0.2	0	0.9
Australia & New Zealand	0.3	0.2	0.2	17.9
Former USSR	0.9	2.0	0.4	46.5
Latin America	2.4	1.2	0.9	51.4
Africa	0.7	1.2	0.7	52.9
China	1.9	0.9	0.6	16.3
Other Asia	3.2	2.2	1.4	33.4
Total	12.5	13.6	5.2	266.9

Source: Johansson *et al.* (1993).

tained in the long term, even with best practices and extensive fertilizer use (Trainer, 1996). Yet others have estimated the long-term annual potential from biomass to be as high as 1300 exajoules per year (IPCC, 1996), significantly larger than the estimates of Table 5.4.

Figure 5.3 presents a supply curve for biomass resources in the United States that includes all forest, mill, and agricultural wastes, as well as dedicated energy crops (but excludes municipal solid wastes). The annual production cost of biomass increases with increasing production as less attractive (i.e. less fertile or less accessible to markets) resources are employed. The initial points on the curve of Figure 5.3 are associated with different forms of biomass wastes. The latter, more costly, points (above $2.00/GJ) are mostly associated with dedicated energy crops, grown explicitly for biomass energy purposes. Even these costs continue to rise with higher production levels as less attractive (e.g. less fertile, less accessible to markets) land is brought under production. The highest US production points in Figure 5.3 are considerably smaller than the estimates of potential shown in Table 5.4 for the United States. The discrepancy is partially due to the fact that the Table 5.4 entry includes Canada, but is also indicative of the differences between different reference sources. There is a high degree of uncertainty in all such estimates, both in terms of ultimate resource and in terms of the cost of that resource. In the United States, transportation costs can also limit the use of biomass resources when transport distances are more than 200 km, which is usually considered non-economic. Costs in other countries will depend on local economies, climate, and a host of other factors.

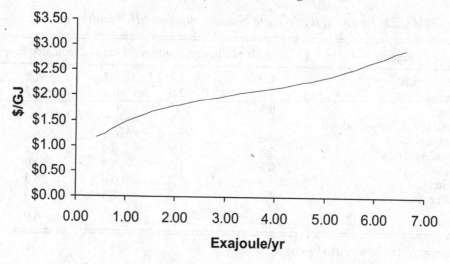

Figure 5.3   US biomass resource costs (Source: Walsh, 1999).

5.2.4.4 Hydroelectricity

Hydropower currently contributes about 6% of the total world primary energy production, second only to biomass among the various renewable energy forms. Worldwide technical potential estimates range as high as 15 000 TWh /yr (Johansson *et al.*, 1993) with only about 14% of that developed to date. As shown in Table 5.5, the economic potential (the amount that is cost competitive) worldwide is about half the technical potential (accessible, but not necessarily cost competitive). But at 8300 TWh/yr, the economic potential is still two-thirds of the current world total electricity use of almost 13 000 TWh/yr.

While existing systems range in size from tens of kilowatts to more than 10 000 Mwe, more than 95% of existing hydro capacity is larger than 10 MWe. Two thirds of the development has been in industrialized countries, leaving significant potential in the developing world for both large and small systems. Even in the United States, where a high fraction of the total potential has been developed, a recent DOE study estimated that there are 5677 sites with more than 30 GWe of capacity that can be developed today (INEEL, 2000).

Although many existing hydroelectric facilities have been located near load centers, the larger remaining sites are typically far from loads and must be connected via long-distance, high-voltage transmission lines. For distances greater than 500 km, the more economic option is high-voltage direct current (HVDC) lines with silicon-controlled rectifiers capable of converting high voltages and large currents from AC to DC power. Such HVDC lines are the basis for proposals to ship power from large Central African hydro projects on the Zaire

Table 5.5 *World hydro* potential*

Region	Economic potential (TWh/yr)	Already exploited (%)	Small hydro potential by 2020** (GWe)	Small hydro potential by 2020** (TWh/yr)
North America	801	72	12.9	59.2
Latin America	3281	12	6.6	27.1
Western Europe	641	63	21.7	90.7
Eastern Europe and CIS	1265	21	6.9	28.3
Mid East and North Africa	257	16	0.3	0.7
Sub-Saharan Africa	711	6	1.1	2.8
Pacific	172	22	0.3	1.3
Asia	1166	45	24.9	98.3
Total	8295	27	74.6	308.4

Notes:
  * Excludes pumped storage.
 ** Based on "favorable" case economic potential; technical potential is estimated to
     be twice as large.
Source: World Energy Council WEC, (1993).

River to loads 6500 km away in Europe. HVDC lines could also be employed for other remote renewable resources like wind.

The worldwide capacity factor for hydro is relatively low (40%) primarily because plants are often used to meet peak loads. Smaller systems are generally run-of-river systems with little or no reservoir storage. Thus, the availability of these systems depends more on the hydrologic cycle. When water is available, they operate more like base-load systems. A third type of hydro-electric system is pumped storage, which, as its name implies, provides for electricity storage. Pumped storage has no potential on its own to displace fossil fuels nor to reduce carbon emissions, but can be used to facilitate the use of intermittent renewable energy technologies.

5.2.4.5 Geothermal

Geothermal energy systems take advantage of the fact that the earth's crust is generally at higher temperatures than the surface temperatures. The crust temperature increases at an average rate of 30° to 35°C per kilometer as one descends from the surface. However, at the edge of the earth's crustal plates, where fluids flow from geysers, natural spas, hot gases, or volcanic lava, the

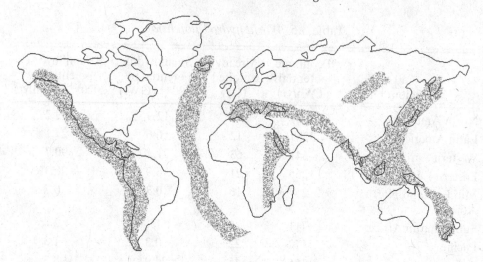

Figure 5.4   World geothermal resource areas.

temperature gradients are generally higher. In these "hot spots", the economics of extracting this natural heat are favorable enough to allow exploitation to a maximum depth of about 5 kilometers. Figure 5.4 shows the regions of the world where plate tectonics create the more favorable geothermal energy opportunities.

It is estimated that the theoretical amount of heat that could be captured within 5 km of the earth's surface is 140×10^6 exajoules. However, much of this is too low in temperature to be useful. Total useful resources have been estimated to be 5000 exajoules (Johansson *et al.*, 1993), an amount roughly equal to 12 years of total world consumption of all energy forms. Annual long-term technical potential has been estimated to be greater than 20 exajoules/year (IPCC, 1996). Today's economic reserves may be only one-tenth as large as the total useful resources, but technology improvements and higher costs for future energy supplies will increase reserve estimates.

As shown in Table 5.6, geothermal resources have been employed in a number of applications ranging from low-temperature bathing (up to 50°C) to high-temperature (above 100°C) electric power generation. Worldwide, the total non-electric use of geothermal energy is almost 14 MWt.

The largest use of geothermal energy is for producing electricity. There is currently more than 8000 MWe of geothermal capacity throughout the world, including more than 2800 MWe in the United States, as shown in Table 5.7. All geothermal electricity generation comes from hydrothermal (hot water/steam) resources. With their continuous operational requirements and high availabilities, geothermal power plants provide largely baseload power.

Table 5.6 *Direct geothermal heat use (MWt)*

Region	Space and water heating	Greenhouses, aquaculture	Industry	Balneology	Other	Total
N. America	936	129	427	284		1776
W. Europe	1770	191	105	682	65	2813
E. Europe	119	761	70	719	446	2115
CIS	429	395	220	360		1404
Pacific	91	52	143	4501	153	4940
China	133	146	91	25		395
Others	122	115	44	179	86	546
Total	3600	1789	1100	6750	750	13989

Source: World Energy Council (1993).

Current power plants for generating electricity from hydrothermal resources can be divided into two general types: steam and binary. Steam plants are typically used with higher-quality, liquid-dominated resources to produce either a liquid or two-phase fluid, which is depressurized to generate steam. With lower-quality, liquid-dominated resources, it is usually more cost-effective and efficient to transfer heat from the geothermal fluid to a volatile working fluid. As a closed-cycle system, dual working fluid or binary plants release no emissions to the atmosphere.

Approximately 1000 MWe of capacity has been curtailed, largely in the United States, due mostly to resource depletion concerns

5.2.4.6 Ocean energy

Ocean energy exists in the form of tides, waves, and thermal and salinity gradients. Each of these resources is extremely large. For example, the solar energy absorbed by the oceans each day is equivalent to 250 billion barrels of oil per day, or enough to meet more than a thousand times the world's current annual energy demands.

Tidal power: Total worldwide potential has been estimated to be 500–1000 TWh/year, but only a fraction of this is likely to be economic in the foreseeable future (Johansson *et al.*, 1993). The distributed nature of this resource and its isolation from large load centers significantly increases the cost of its capture and reduces the size of the useful resource. Today, only the more mature tidal power option is being exploited using commercially available technology, and that at only a relatively modest number of sites, as shown in Table 5.8.

Table 5.7 *Geothermal power capacity by nation (MW)*

Country	1990	1995	1998
Argentina	0.67	0.67	0.00
Australia	0.00	0.17	0.40
China	19.20	28.78	32.00
Costa Rica	0.00	55.00	120.00
El Salvador	95.00	105.00	105.00
France (Guadeloupe)	4.20	4.20	4.20
Greece	0.00	0.00	0.00
Guatemala	0.00	0.00	5.00
Iceland	44.60	49.40	140.00
Indonesia	144.75	309.75	589.50
Italy	545.00	631.70	768.50
Japan	214.60	413.70	530.00
Kenya	45.00	45.00	45.00
Mexico	700.00	753.00	743.00
New Zealand	283.20	286.00	345.00
Nicaragua	70.00	70.00	70.00
Philippines	891.00	1191.00	1848.00
Portugal (Azores)	3.00	5.00	11.00
Russia	11.00	11.00	11.00
Thailand	0.30	0.30	0.30
Turkey	20.40	20.40	20.40
United States	2774.60	2816.70	2850.00
Totals	5866.72	6796.98	8240.00

Source: www.demon.co.uk/geosci/wrtab.html, April 17, 2000.

Table 5.8 *Existing tidal plants*

Site	Installed capacity (Mwe)	Date in service
La Rance, France	240.0	1966
Kislaya Guba, Russia	0.4	1968
Jiangxia, China	3.2	1980
Annapolis, Canada	17.8	1984
Various in China	1.8	

Source: WEC (1993).

Although entirely predictable, tidal power is intermittent, yielding one or two pulses of energy per tide. The tides occur every 12 hours and 25 minutes and thus move in and out of phase with peak loads. The result is a low annual plant capacity factor in the range of 25% to 35 % (WEC, 1993), with no particular correspondence to peak loads. Pumped storage is possible, but capital costs are high and throughput efficiency is in the range of 75%. Tidal energy technology is relatively mature, and dramatic cost reductions are not expected.

Wave energy: Ocean waves are created by the interaction of the wind with the sea surface. The winds blow most strongly in the latitudes between 40° and 60°, and the wave energy potential is highest in this region, with additional potential where regular tradewinds prevail (around 30°). Coasts exposed to prevailing winds with long fetches like those of the United Kingdom, the west coast of the United States, and the south coast of New Zealand have the greatest wave energy density. The wave power dissipated on such coastlines has been estimated to be in excess of 2 TW. Estimates of the development potential by 2020 range as high as 12 TWh/yr (WEC, 1993).

Many different wave-energy devices have been developed at the prototype scale. Some are buoyant structures that are moored at or near the sea surface; some are hinged structures that follow the contours of the waves; some are flexible bag devices that inflate with air with the surge of the waves; others enclose an oscillating water column that acts like a piston to pump air; and some are shaped channels that increase wave amplitude to drive a pump or a fill a land-based reservoir. Prototypes of such devices have been deployed around the world – Japan, Russia, China, Sweden, Denmark, the United States, Taiwan, the United Kingdom, Canada, Ireland, Portugal, and Norway. The technology is relatively immature and should continue to decrease in delivered energy costs, which generally exceed $0.10/kWh today. Given the intermittent nature of the resource, wave energy systems generally cannot be credited with firm capacity and operate at capacity factors below 40% (WEC, 1993).

Ocean thermal energy conversion (OTEC): The difference in temperature between solar-heated surface waters and cooler waters at 1000 m can exceed 20 °C in the tropics and subtropics. This temperature difference can be used to produce power through a thermodynamic cycle similar to that of a conventional generator, where the heat of the surface water is used instead of fossil fuel combustion, and the cooler deep water is used to condense the working fluid after evaporation. The relatively small temperature difference between the surface and deep water implies a maximum theoretical efficiency of only 6.8% at 20°C and 9% at 27°C. In practice, throughput efficiencies are closer to

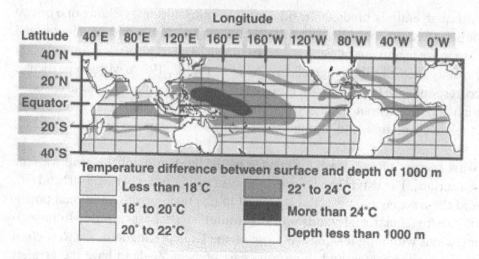

Figure 5.5   Ocean thermal energy conversion resource
(Source: www.nrel.gov/otec/what.html).

3%–4% (WEC, 1993). OTEC is a firm source of power that can also provide additional benefits in the form of mariculture and desalinated water supplies. Current costs are in the range of $0.12/kWh (US). OTEC costs are inherently high and are not expected to decline substantially because of the low efficiency associated with the relatively small temperature differences found in the ocean and the consequent large volumes of water that must be handled to produce power.

OTEC is the largest of the ocean energy resources. The area for possible exploitation is 60 million km². However, as shown in Figure 5.5, most of this area is offshore potential and inaccessible for power production. Estimates of the potential developable by 2020 have ranged as high as 168 TWh/yr (WEC, 1993), but are likely to be much smaller.

5.2.4.7 Total renewable resources

Table 5.9 summarizes the above renewable resources discussion by placing the potential estimates for each form of renewable energy in similar units. Table 5.9 reinforces the well-known fact that total renewable energy resources are formidable, especially for solar and biomass. It also shows a relatively small potential for geothermal and ocean resources to address global climate change.

The long-term technical potential of renewable energy, 450×10^3TWh$_e$/yr, is massive. This number is more than 10 times global primary energy consumption in 1996 (EIA, 1999, p. 142). The economic potential by the year 2025, 14 to 24 10³TWh$_e$/yr, is impressive as well. The high end of this range is about one-

Table 5.9 *Summary of renewable energy resources (10³ TWh_e/year)*

Resource	1990 consumption	2025 economic potential	Long-term technical potential	Annual total resource flow
Hydro	2	4–6	>14	>40
Geothermal	<1	<1	>2	>85
Wind	<1	1	>14	>21 000
Ocean	—	<1	>2	>32
Solar	—	2	>280	>300 000
Biomass	6	8–15	>140	
Total	8	14–24	>450	>300 000

Source: converted from IPCC (1996).

third of the projected global primary energy consumption in 2020 (EIA, 1999a, p. 142).

Clearly, renewable energy has vast potential. What Table 5.9 does not show is that the resources are dispersed and economic access to them can be constrained. The factors that constrain such access will be discussed in the next section.

5.3 Market Issues

There are a number of factors that will have to be overcome before renewables can contribute significantly to worldwide reductions in greenhouse gas emissions. For this discussion, we classify these factors into the following overlapping categories: economic, technical, institutional, and environmental.

5.3.1 Economic factors

5.3.1.1 Capital cost

Currently, a major barrier to the widespread use of renewable energy is the cost of the technology. In general, the cost is high relative to the cost of fossil fuel technologies because most renewable resources are not concentrated. Collection of these distributed, low-energy-density resources can require large amounts of collector material. For photovoltaics and solar thermal, this means large solar arrays or expensive tracking concentrator systems; for biomass, it means planting and harvesting large amounts of plant material; for wind, it means large areas swept by turbine blades. Exceptions include geothermal, where hydrothermal resources are concentrated by geologic forces, and hydro, where water flows have been concentrated in rivers.

The capital cost per unit output of many renewable energy systems is not as sensitive to system size as that of many conventional technologies. This ability to build small renewable energy plants at approximately the same cost-per-unit output as that of a large plant is generally considered an advantage in that it opens up distributed generation and other markets. On the other hand, some investors might consider this modularity to be a disadvantage in that significant economies of scale cannot be realized with larger projects. However, economies of scale are possible. Wind, photovoltaics, and small biopower systems can be largely produced in central manufacturing plants where economies of scale do exist. Economies of scale in manufacturing will occur at high levels of production. These levels have yet to be fully realized.

Intermittent renewable electric technologies can also have additional capital cost penalties. Additional capital costs are sometimes necessary to compensate for their intermittency, either by buying storage or a backup generator, or building some form of hybrid system using fossil fuels. Without on-site storage, intermittent renewable technologies like photovoltaics and wind produce electricity less than half the time. Thus, any dedicated transmission lines used to deliver their power to markets will be filled less than half the time, significantly increasing transmission costs. Even the cost of transmitting by non-dedicated lines can be assumed to be higher if firm transmission capacity is to be guaranteed for when the intermittent renewable energy source is available. These transmission costs will be somewhat mitigated as forecasting of wind availability improves since transmission capacity would then only need to be reserved when it is known with high confidence that the intermittent renewable energy source will be available.

As discussed in prior sections, the costs of renewable energy technologies are expected to decrease in the future both because of R&D efforts and because of industrial learning and economies of scale in manufacturing. While this is generally considered beneficial, past expectations of future improvements has led investors and consumers to delay their investments in the hope that they will be able to purchase or manufacture the new technology for less.

With the exception of biomass technologies, most renewable energy technologies have no fuel costs and relatively low O&M costs. Thus investors in renewable energy technologies reduce their exposure to future fuel price increases. However, the sword cuts both ways: owners of renewable energy systems cannot benefit if fuel costs decrease. Since the mid-1980s, fuel prices have been generally declining (at least in real terms when general inflation is subtracted out), and renewables have been at a disadvantage. However, recent year 2000 increases in oil and natural gas prices are reminding investors that price stability can be advantageous, and the risk reduction benefit of renewables is again a consideration.

5.3.1.2 Financing

The high first cost of renewable energy technologies demands that financing be readily available. However, there are certain unique attributes of renewables that frequently preclude the availability of financing. Primary among these is the uncertainty associated with the performance and costs of renewable energy systems. While most types of renewable energy systems have been available for a decade or two, their use is not widespread and financiers are often not familiar with them. Another problem common to any rapidly evolving technology is that the performance of last year's system is not necessarily reflective of this year's, yet it is all the financier has to refer to. For example, the capacity factor or annual generation of wind turbines has been continuously increasing due to improvements in equipment reliability, improved maintenance, and higher towers; yet most financial analyses use output data from older systems in the field.

Financing is also more difficult to obtain for smaller systems. The due diligence required for a smaller system can approach that required for a larger one. Similarly, all fixed costs must be spread over the smaller investment of a smaller system. While some renewable projects can be quite large, e.g., a 50 MW wind farm, they are small relative to fossil and nuclear projects of hundreds of megawatts. The most obvious way to overcome this obstacle is to try to group several renewable energy investments into a single financial package to spread the fixed costs of placing the financing over a larger base. This works best when the projects are relatively homogenous so that due diligence on one project supports the claims of all the projects.

5.3.1.3 Restructured energy markets

As the energy sector is increasingly privatized around the world, more and more decisions are being made almost exclusively on the basis of profit potential, as opposed to ensuring universal service in a sustainable manner. The implications for renewables are several-fold. First, the cost of capital in these markets is increasing to reflect corporations' hurdle rates for new projects and greater risks in privatized markets that do not have guaranteed returns. These higher money costs are especially detrimental to renewables, which are capital intensive, but have low operating costs. Secondly, developers can usually earn more money through larger projects. Limited resources at any one site frequently prevent renewable energy projects from being developed at the hundreds-of-megawatts sizes common for many fossil projects that can also benefit from economies of scale. Finally, the energy companies are not as concerned with the long-term sustainability and social issues that promoted R&D on renewables and deployment programs in regulated utilities. Witness the demise of many of the R&D and "demand-side management" programs at US utilities so common in the 1980s.

5.3.1.4 Lock in

The prevailing view of markets is that a free market will select the most economic technology or fuel. However, the most economic technology at the margin for the next incremental investment is not necessarily the same as the most economic investment from a long-term perspective. For plants with large infrastructure requirements, an investment at the margin can be much reduced from the full investment cost.

As an example, compare wind against coal for power generation. In a number of ways, coal is locked in, while wind is locked out. Coal is not locked in simply because it is in much greater use for power generation than is wind. Rather, it is locked in because the following infrastructure costs have been paid, and stakeholder groups exist that have vested interests in the continued use of coal:

- Coal-fired power plants already exist. The plant costs are sunk. To displace coal, wind must be compared to the O&M costs for coal, a much tougher economic test than comparison with the full capital and O&M costs of a planned coal-fired power plant.
- Coal mines already exist. If they did not, it would certainly be difficult to open a strip mine today. The support that coal mining has today from regions of the country that are highly dependent on coal-mining revenues would be non-existent.
- Rail transportation already exists. Railroads would find it difficult, if not impossible, to establish rights-of-way, centrally located rail yards, and acceptance of the noise and emissions if they were just now being introduced to allow coal-fired electricity generation.
- Regulations for the environmental control of coal plants have been hammered out over the last three decades. There would be no existing US plants exempted from the 1990 Clean Air Act Amendment provisions for SO_2 emission reductions. All plants would have to meet the New Source Performance Standards required of new plants.
- The inability of coal and nuclear generators to react instantaneously to changes in load has produced a generation mix of base, intermediate, and peak load (hydro and gas turbines) generators that provide system reliability in the face of changing loads.

Wind energy is still developing its infrastructure. Its visual and avian impacts have not been fully assimilated by the public and environmental regulation. There is no large stakeholder community of landowners and farmers currently deriving revenues from wind systems on their lands. Utilities have not adapted their generation mix and their planning procedures to the intermittent availability of wind resources.

Thus, when wind and coal are compared for the next increment of investment, the costs of the next increment of coal do not include the full cost of infrastructure development, while those of wind do. Coal is locked in; wind is not.

To partially circumvent this lock in, different forms of incentives have been put in place. Most are targeted at reducing the cost of renewable energy. These include investment tax credits, low-interest loans, and loan guarantees. Others promote more direct investment like the renewable portfolio standard proposed in the United States, the Non-Fossil Fuel Obligation in the United Kingdom, and the feed-in laws in Germany. These and other policies will be described further in a later section of this chapter.

5.3.2 Technical factors

While there are many technical factors such as system reliability, controls, and materials that impact renewables and other energy systems, the one technical problem that is somewhat unique to some renewable electric technologies is the intermittent availability of the renewable resource.

5.3.2.1 Intermittency

Two of the more promising renewable energy resources, solar and wind, are intermittent in their availability. Other renewable energy resources like hydro and biomass are not intermittent, but can have seasonal variations in resource availability. The intermittency of solar and wind raises issues related to utility system reliability and the capacity value of these renewable electric systems.

System reliability. Today's electric power grids are often vast networks of interconnected utilities and loads. The voltage and frequency of the power provided by utilities within these networks have to be within certain limits to ensure the safety of personnel and equipment and to ensure undue burdens are not placed on neighboring utilities within the network. Large amounts of intermittent wind and solar technologies within an electric grid may introduce larger-than-normal fluctuations of voltage or frequency, modify tie-line flows, and increase regulating duties at conventional generators.

Several studies have shown that speed fluctuations of wind will not cause stability problems (Herrera *et al.*, 1985 and Chan *et al.*, 1983). The large turbine inertia and low mechanical stiffness between the turbine and the generator provide excellent transient stability properties. Furthermore, newer variable-speed wind turbines and photovoltaic systems using self-commutated inverters do not require additional reactive power as the old systems did, but can actually supply reactive power to the grid and alleviate potential voltage instabilities and losses in the transmission system.

However, rapid transients in renewable electric output can overwhelm the regulation ability of conventional generating units in the system to provide automatic generation control (AGC). In this case, a sudden decrease (increase) in power from a renewable system will cause the system frequency to drop (increase) until the AGC can increase (decrease) the output of conventional generators to match the load once again. Some studies have found that the amount of photovoltaics that can be integrated into a system is limited to less than 16% (Wan and Parson, 1993) of peak load. Other studies indicate that wind should be constrained to no more than 15% (EPRI, 1979).

However, the limits are not so much technical as they are economic. The impact of high penetrations of intermittent renewables on a grid can be mitigated by increasing the spinning and stand-by reserves, by including more quick-response gas turbines and hydroelectric plants, by better forecasting the availability of the renewable resource, and by spreading the intermittent generators out geographically to reduce transients. One utility has integrated distributed photovoltaics into one feeder circuit equal to 50% of that feeder's capacity (Wan and Parson, 1993). One study (Grubb, 1987) found that under favorable conditions, more than 50% of the United Kingdom's electricity demand could be supplied by wind without storage facilities. With the proper incentives for reducing greenhouse gases, such favorable conditions may be economic, and system reliability should not be a significant constraint to the use of intermittent renewable electric systems.

Capacity Value. Inasmuch as a solar system will be available at most 25%–30% of the time and a wind system no more than 40%–50%, utilities cannot be 100% certain that these technologies will be available at the time of peak loads. Therefore, they sometimes give no credit to intermittent renewables for capacity in either their dispatch planning or their capacity expansion planning.

System operators frequently schedule their dispatching 24 hours in advance. Not knowing that far in advance whether intermittent renewables will be available, they ensure that the load can be met by dispatchable generation.[3] When

---

[3] In actuality, a forced outage can preclude a dispatchable fossil generator from being available. However, the probability of a forced outage of a fossil generator is generally much lower than the probability that an intermittent renewable energy technology will not be available.

the load actually occurs, the intermittents, with their near-zero operating costs, are dispatched first if they are available. Thus, intermittents are the first to be dispatched, but are not included in the planned dispatch order, nor given any credit for their capacity.

Capacity credit is also an issue for intermittents when planning for capacity expansion for a utility system or grid. Generally, the processes and tools used for system-wide planning of future capacity are not designed to handle the intermittent nature of wind and solar. However, a reasonable work-around is commonly used in which the capacity value of wind is estimated and then inserted into the capacity expansion model. The capacity value of intermittents can be calculated using the same probability-based indices as currently used by the industry, such as expected load-carrying capability (ELCC, expressed as a percentage of the nameplate capacity). Probabilistic measures such as ELCC can account for the coincidence of intermittent renewable resources with peak loads, the ability of spinning and standby reserves to mitigate non-availability, the role of storage, and the level of renewable energy penetration within the system. Studies of actual systems show wind ELCCs as high as 80% (Smith and Ilyin, 1990) (100% would be the equivalent of a perfect dispatchable technology), but generally ranging no higher than the capacity factor of wind or about 30%–40%.

Restructuring of the electric sector should reduce the impact of intermittency both on dispatch and in capacity planning. In a fully restructured electric market with marginal-cost, real-time pricing, dispatchers will not be as concerned with ensuring capacity is available to meet peak loads. If capacity is limited, real-time prices will rise and demand will fall in real time. The market will ensure supply meets demand. In this case, capacity will be valued differently and intermittency will be less of an issue.

Nonetheless, in a restructured electric market, generation at times of peak demand will be rewarded by higher prices. Fortunately, there are many instances where the availability of solar and wind are positively correlated with loads. For example, many utilities experience their peak loads in the summer due to air-conditioning. Generally, the largest air-conditioning loads exist when the sun is shining, or when solar electricity is available. Similarly, in those northern regions where the wind resource is at its maximum in the winter, regional loads may also peak in the winter. This coincidence of intermittent renewable energy availability with loads increases the ELCC and the value of intermittent resources.

5.3.2.2 Resource assessment

It is difficult to move renewable energy forward in those countries where renewable resources have not been well characterized. While estimates have been made for almost all countries, they frequently are made without national surveys. For example, most wind resource assessments are based on data collected at national weather stations, which can seriously underestimate resources on ridges and other prominences. Similarly, insolation data is often based on simple sunshine weather data, which lacks quantification and distinctions between direct and indirect insolation. Efforts need to be made to improve wind and solar assessment techniques and to conduct comprehensive national surveys for biomass, geothermal, hydro, and other renewable energy resources (Renne and Pilasky, 1998).

5.3.2.3 Substitution for oil

A second major technical limitation of renewables is that few of the renewable energy technologies directly address the transportation market and the displacement of oil. Oil displacement is important not only because oil is a limited natural resource, but also because petroleum use is the largest source of anthropogenic carbon emissions in the world today. Liquid fuels in the form of ethanol and biodiesel can be directly produced from biomass, but the costs are high relative to today's gasoline prices. Indirect substitution for oil is also possible through renewable electric technologies with the generation used in electric vehicles. This route requires not only improvements in the cost competitiveness of renewable electric technologies, but also in electric vehicle technology.

5.3.2.4 Land requirements

Many renewable energy forms are perceived as requiring extraordinary amounts of land for the collector systems required with the low energy density of most renewables. Table 5.10 shows that the land requirements for renewables do not greatly exceed those of conventional fuels when upstream processing is taken into account (e.g. coal mining). To some extent the values in Table 5.10 may even exaggerate the land requirements of renewable energy. For example, while wind machines may need to be separated from each other by five to ten blade diameters to optimize performance, the land in between can still be used for farming, agriculture, and other uses. Similarly, photovoltaics may require significant space for the collectors, but the more promising distributed applications frequently have the collector panels on a roof. The largest land requirements are associated with biomass, which typically converts less than 1% of the incident sunlight to potential energy stored in the biomass material.

Table 5.10 *Approximate land requirements for power production (hectares per MW)*

Plant type	Area
Gas turbine	0.3–0.8
Coal steam	0.8–8.0
Nuclear	0.8–1.0
Hydropower	2.4–1000
Wind	0.4–1.7
Photovoltaics	3–7
Biomass	150–300
Geothermal	0.1–0.3
Solar thermal	1–4

Source: OTA (1995).

5.3.3 Institutional factors

The economic and technical limitations of renewables presented above are being addressed largely through R&D and learning through actual production. Overcoming institutional factors generally requires some form of policy or social movement.

Probably the more acute institutional factors limiting renewables are a lack of familiarity and acceptance of the technologies, a lack of standards for the technologies and their application, and the availability of capital to finance a significant switch in the world's energy sources.

5.3.3.1 Technology familiarity and acceptance

For renewables to succeed in significantly reducing worldwide carbon emissions, there are a host of stakeholders that must become more aware of the benefits and costs of the technologies and their resource availability. These stakeholders include energy users as well as major equipment suppliers, financiers, utilities, and policy makers. Although information programs and demonstration efforts exist, the most convincing information will be successful, profit-making investments throughout the energy community.

Liberalized markets for electricity in the developed countries are already increasing consumer and utility awareness of renewable electric technologies. Green power programs are allowing consumers to choose their source of energy. In the United States, more than half the residential electric customers polled have indicated a preference for renewable energy (Farhar, 1999). When given the opportunity to participate in green pricing programs promoted by utilities, in which customers pay extra to the utility for the purchase of renewables, the number of participants is closer to 2% (Swezey and Bird, 1999). Nonetheless, these sorts of programs are bringing renewable electric technologies to the public forefront and serving as demonstrations of the technical possibilities. Such technology demonstrations and the rapid dissemination of their success are critical to increased market deployment.

Similarly, corporations are becoming more attuned to the need for renewables and their benefits. Somewhat surprisingly, the largest purchases of green power in the United States have come not from the residential sector, but from corporations seeking identification with the green movement. For example, Toyota has pledged to buy only green power for several of its California facilities. Similarly, the United States Postal Service has entered into an agreement to purchase green power for more than 1000 California facilities. On the supply side, several major energy companies like British Petroleum and Shell have initiated new renewable energy departments with major budgets and ventures planned. More than 75 US utilities currently offer green-pricing programs through which ratepayers can voluntarily pay to have the utility generate or purchase renewable electricity.

Technology acceptance issues are not limited to familiarity and comfort with the technology. There are environmental and cultural issues as well. Environmental issues are addressed in more detail below. Cultural issues include a number of societal concerns. For example, geothermal power plants in the western United States have been halted, at least partially, because of Native American religious values associated with underground geothermal energy sources. Similarly, hydro projects have been halted because the reservoir areas inundated would destroy local communities and recreational areas.

5.3.3.2 Standards and institutionalization

Standards facilitate commercialization of new technology by reducing the consumer risk in the purchase of the technology. Standards can also be used to ensure the technology is compatible with existing systems, e.g., photovoltaic rooftop systems need to be compatible with local building codes. Standards can also help manufacturers limit the number of product lines. For example, interconnection with the utility meter can be facilitated by common standards to which manufacturers can design their photovoltaic products. Interconnection standards with utility grids are the subject of much research today in both developed and developing nations.

Probably even more critical than standards is the institutionalization of renewable energy. When renewable energy technologies become the norm, significant cost decreases will be attainable for both the physical system and, perhaps more importantly, the transaction costs. Today, the design of a passive solar home is an exceptional event requiring an enlightened architect, builder, and local-government inspector. Similarly, installing a photovoltaic system on one's home can require significant consultation with, and education of, local electrical inspectors unfamiliar with distributed photovoltaic system interconnections. Nor are utility planners generally equipped to evaluate the capacity value of an intermittent source of generation within their system. City officials need education and convincing that landfill gases can be used profitably to generate local power. Due diligence exercised by financiers on low-head hydro projects is reduced as they become familiar with the technology.

5.3.3.3 Finance

In the section on economic factors above, we discussed financial considerations for individual projects. However, if there is a shortage of capital overall, only those projects with the very highest promised returns will be undertaken. A massive transformation of the energy system from fossil fuels towards renewables and other non-carbon technologies could create such a shortage of capital. Worldwide capital investment in all energy development is about US $900 billion per year or 5% of the total world Gross National Product (GNP) (WEC, 1993). If 10% of this were invested in renewables for each of the next 20 years, renewables would displace only about 8% of projected (EIA, 2000) fossil energy use in 2020.[4] Thus, to reduce carbon emissions through major reductions in fossil fuels will require significant worldwide capital redirection towards renewables. This problem of capital resource limits is especially acute in developing countries that already spend about 25% of their public sector budgets on power development with an estimated need of US $100 billion required per year (WEC, 1993).

5.3.4 Environmental factors

One of the principal benefits of many renewable energy forms is that they emit fewer air pollutants than do fossil fuels. In particular, non-combusting renewables (i.e. all but biomass) do not generate the nitrogen oxides that are common to all combustion processes, nor sulfur dioxide, which is common in coal and oil and, to a lesser degree, in natural gas combustion. Again with the exception

---

[4] If for 20 years 10% of the $900 billion worldwide investment were redirected towards renewables at an average cost of US $1000 kW of renewable electric capacity (the typical cost of wind energy today), the in-place renewables capacity (ignoring retirements) would be about 1800 GWe in 2020. At a 35% capacity factor, this capacity could generate about 24% of projected worldwide electricity consumption in 2020 or displace about 8% of projected world fossil use (EIA, 2000).

of biomass and some geothermal plants, renewables do not release any significant amounts of carbon dioxide to the atmosphere. Even biomass produces little net (about 5% of power plant CO_2 emissions, Mann and Spath, 1997) carbon dioxide because the plants grown uptake essentially the same amount as is released through combustion and upstream processes. Finally, non-combustion renewables generally produce no solid or liquid wastes and require no water for cooling or other uses.

While these environmental benefits are well recognized, they frequently do not impact energy purchase decisions. Such decisions are generally made on the basis of the financial costs to the decision-maker, not the environmental costs to society as a whole. While there have been multiple attempts to quantify and internalize these environmental costs (OTA, 1994), the area is fraught with difficulties, first in estimating the cost-per-unit energy and second in imposing these costs on individual decision makers. One partially successful method is the imposition by government of caps and trading schemes, as exemplified by the SO_2 caps and allowance system established in the United States under the acid rain provisions of the 1990 Clean Air Act Amendments. Under this phased-in approach, US emissions of SO_2 will be cut by more than half by 2010. While imperfect (e.g. existing SO_2 emitters have been granted rights to continue to emit, albeit at lower levels), and capable of being improved, this approach could be a model for CO_2 cap and trade provisions.

While there are clearly environmental advantages to renewables, there are also environmental issues associated with most renewable energy forms. With wind nearing competitive status with fossil-generated electricity, its environmental impacts are today under heavy scrutiny. Most apparent is the simple visual impact of towers and blades reaching up to 100 meters above ground level. Secondly, birds are occasionally killed as they traverse through the area swept by the turbine blades. This problem is being partially remedied by solid towers that lack struts for birds to perch on, by siting wind farms away from avian migratory paths and away from raptor ranges, and different avian repellent concepts. Noise levels and electromagnetic interference associated with some of the early machines have also been largely mitigated.

Geothermal power plants can release carbon dioxide and hydrogen sulfide from the brine as well as trace amounts of other gases. In the less-efficient binary geothermal power plants, the brine is reinjected to the ground without such air releases. In the United States, resources adequate for geothermal power are frequently found in more pristine western areas where the visual impact of a plant and associated transmission lines are resisted.

Photovoltaics have little impact at the point of use, especially if mounted on existing roofs or other structures. However, the manufacture of some forms of photovoltaics does require the use of toxic materials in or as the photovoltaic material, such as cadmium telluride, gallium arsenide, and copper indium diselenide. Toxic gases, liquids, and solid compounds are also used in the manufacturing process, especially in thin-film manufacturing. In all cases, proper manufacturing procedures exist to control any environmental hazards. Cells made of toxic materials may require controlled disposal when they are retired. The use of batteries in conjunction with photovoltaics for storage is potentially the greatest health and safety issue associated with the distributed use of photovoltaics. Batteries can be hazardous during their use and disposal. The development of low-cost, non-toxic, rechargeable batteries is critical for greatly expanded distributed use of photovoltaics (International Development and Energy Associates, 1992).

While biomass feedstock production for energy use would have environmental impacts similar to that of food-crop production, the level of production needed to displace a substantial fraction of fossil fuels for climate protection could severely exacerbate the problems introduced by non-sustainable agricultural practices. Concerns over soil fertility and erosion, promotion of non-native species, wildlife habitats, water requirements and water quality impacts, pesticide and fertilizer use, and biodiversity will all need to be addressed. Environmental impacts from biomass feedstock transportation and power production include limited air and water emissions. At the point of combustion it has been estimated that NO_x, SO_x, and particulates are released at rates 1/5, 1/10, and 1/28 of the maximums allowed by the US New Source Performance Standards for fossil-fueled plants. Approximately 95% of the carbon emitted at the point of combustion is recycled through the feedstock/combustion system, leaving net emissions of only 5% (Mann and Spath, 1997).

Although it is a relatively mature technology, hydroelectricity has encountered a host of environmental objections in the last couple of decades. Large hydro projects usually require the flooding of a vast area for reservoir storage. The areas frequently are scenic areas created by the river flowing through them. The disruption of both human and wildlife habitats can be a major environmental drawback. In addition, biomass material in the area decomposes underwater to produce methane, a more potent greenhouse gas than carbon dioxide. Additional methane releases occur as biomass material is deposited by the river and decomposed in the reservoir. No less significant are the potential impacts on fish that can't swim upstream beyond the dams and erosion of downstream riverbanks. The effects are mitigated by fish ladders and more evenly controlled releases of water through the dam to limit erosion effects. The impacts of reservoirs are eliminated in "run-of-river" installations that employ little or no reservoir, but which are more vulnerable to variations in the hydrologic cycle.

5.4 Regional Status and Potential of Renewables to Address Climate Change

The prospects for renewable energy vary tremendously around the world. The sources of the variation are physical resources, energy requirements, economic capabilities, existing infrastructure, environmental concerns, and politics. As shown in Figures 5.1 through 5.5, renewable resources are not uniformly spread, presenting different opportunities in different countries. Similarly, fossil-fuel resources are not uniformly spread across countries. Countries with plentiful fossil-fuel supplies may be less inclined to use renewables. In particular, natural gas, which is difficult and costly to transport, especially to areas that cannot be reached by pipeline, will compete strongly with renewables in those regions where gas is readily available.

The types of renewable energy technology that are most competitive also vary by region, not only due to resource availability, but also due to existing infrastructure. The primary infrastructure difference is in the availability of an electric transmission grid. It has been estimated that as many as two billion people do not have access to electric power (Stone and Ullal, 1999). In these regions, the economics of photovoltaics, biomass gasification, wind, low-head hydro, and other distributed renewable electricity technologies are considerably improved relative to central generation sources. For these loads, the cost of power from a central source would be considerably increased by the cost of the required transmission and distribution system. There is much research currently under way to design and apply tools to better compare mini-grids or "village power" systems with central generation/transmission options and to design hybrid systems with intermittent renewables, diesel generators, and batteries to optimally meet local loads. Other infrastructure issues include the availability of a trained technical corps to install, operate, and maintain renewable energy systems; local manufacturing capabilities; local financial institutions with access to capital; and designing and implementing local electricity billing systems.

The regional and national opportunities for renewables are also a function of policies and environmental concerns. Renewables have an advantage in countries like Germany with its electricity feed-in laws and the United Kingdom with its Non-Fossil-Fuel Obligation.[5] Countries like China, with local air pollution problems caused by extensive use of coal, will also find renewables increasingly attractive as they become more economic. Several

---

[5] Germany's feed-in tariff requires utilities to purchase renewables at a fixed percentage of the price consumers pay for their electricity, and the United Kingdom's Non-Fossil-Fuel Obligations require regional electricity companies to issue 15-year contracts to the lowest bidders in each non-fossil technology type, e.g. wind, nuclear, landfill gas. (Moore and Ihle, 1999)

European countries including Finland, The Netherlands, Norway, and Sweden have already implemented some form of carbon tax that at least implicitly provide an advantage to renewables (United Nations, 2000).

Resources, infrastructure, economics, and policies vary significantly from one country to another and even within countries. Unfortunately, an in-depth look at the opportunities at the country level is beyond the scope of this chapter. In the paragraphs that follow, we summarize the opportunities for two categories of countries beginning with the OECD countries and followed by developing economies.

5.4.1 Organization for Economic Cooperation and Development (OECD)

There are currently 29 countries in the OECD. The majority are located in Europe with the three largest countries of North America included – the United States, Canada, and Mexico – as well as Japan, Korea, Australia, and New Zealand. These countries use just over half the world's primary energy consumption today. Many have substantial renewable energy resources, but such resources are not uniformly distributed. As shown in Figure 5.1, strong solar resources exist in portions of the United States, Mexico, Australia, and New Zealand. Figure 5.2 shows that many OECD countries have some wind resources, especially on mountain ridges, but wind resources are large only in the United States, Canada, New Zealand, and the coastal regions of Northern Europe. Significant geothermal power opportunities for OECD countries exist primarily in North America, Japan, New Zealand, and portions of the southern Mediterranean, as shown in Figure 5.5. Although Table 5.4 is not broken down exactly along OECD lines, the first four lines indicate that OECD countries have less than one-third of the biomass waste resources and less than one-fourth of the potential for dedicated energy crops worldwide.

Some of these resources are less expensive than others. Figure 5.6 shows a supply curve for the additional cost (cost over and above the cost of alternative or fossil fuels) of renewable energy in the European Union. This figure was compiled from individual country estimates, not all of which were constructed consistently with one another (van Beek and Benner, 1998). The curve shows that today's cost-effective renewable energy potential is clearly only a small portion of the overall energy consumption by the European Union of 55 EJ/yr (European Union, 2000). As the cost of renewable energy decreases and that of conventional fuels increases over time, the supply curve of Figure 5.6 will shift downward (the difference in the costs of renewables and conventional energy decreases). The amount supplied might also increase as improved renewable energy economics may yield additional applications not considered in Figure 5.6.

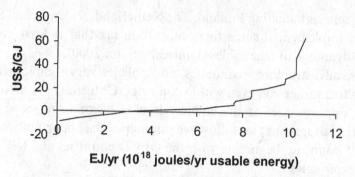

Figure 5.6   Renewable energy supply curve for the European Union (Source: developed from data in van Beek, 1998).

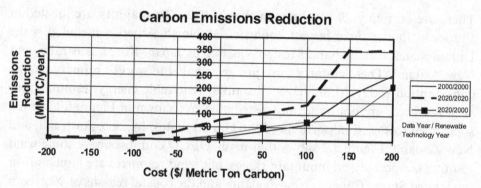

Figure 5.7   Renewable energy supply curve for the United States.

Figure 5.7 shows a supply curve for the potential of renewable energy in the United States to reduce carbon emissions. These curves differ from the European supply curve for renewable energy in that they are constrained, not only by the renewable energy available, but also by the amount of carbon-emitting, fossil-fuel technologies that could be displaced by renewables in each region of the United States. The solid curve in Figure 5.7 represents the potential displacements in 2020 given renewable energy technology and fossil-fuel costs today. The dashed curve represents the potential carbon displacements in 2020 assuming renewable energy technology costs decrease in accord with Table 5.1 (EPRI, 1997) and fossil fuel prices increase as estimated by the US Department of Energy (EIA, 1999b). For illustrative purposes only, the solid curve with box symbols assumes 2020 market conditions and fuel prices and assumes that no further improvements will occur in today's renewable energy technologies. Obviously, improvements to renewable energy technologies are a key element to low-cost carbon reductions in the next two decades. Fortunately,

Table 5.11 *Instruments used in OECD countries to promote renewables*

	Austria	Belgium	Denmark	Finland	France	Germany	Greece	Ireland	Italy	Japan	Luxembourg	Netherlands	Norway	Portugal	Spain	Sweden	U.K.	US
R&D	X	X	X			X		X	X	X	X	X				X		X
Tax incentives		X	X	X	X	X	X		X	X	X	X		X		X		X
Loan subsidies	X	X			X	X	X	X		X	X	X		X	X			
Capital subsidies	X	X	X			X	X	X	X			X		X	X	X		
Feed in tariffs	X	X	X			X	X	X	X		X	X		X	X	X		
Energy taxes	X											X						
Market liberalization						X		X						X			X	X
Information campaigns	X	X	X				X			X	X	X					X	X
Training					X		X	X				X	X					
Standardization			X		X			X	X	X		X	X					
Certification					X		X					X						

Source: van Beek and Banner (1998) and Goldstein *et al.* (1999).

one virtual certainty is that additional improvements in renewable energy technologies will continue, with the pace driven by the level of resources dedicated to the effort.

Many of the OECD countries are actively pursuing renewables. Table 5.11 presents a list of instruments used by many OECD countries to promote renewables. Almost all are engaged in some form of R&D to improve the basic cost and performance of renewable energy technologies. Similarly, all have adopted some form of policy to promote market deployment either through incentives, information, standardization, and/or certification.

In recent years, these policies have spurred considerable renewables development in the European Union, where renewable electric capacity grew by almost 50% in 1998. Much of this growth is in wind, with Germany increasing its capacity from 28 MW in 1990 to become the world leader in generation capacity at the end of 1998 with 2800 MW (Goldstein *et al.*, 1999). This growth in wind in the EU continued in 1999 at a 30% annual rate to reach a total of 8900 MW of wind in Europe (EWEA, 2000).

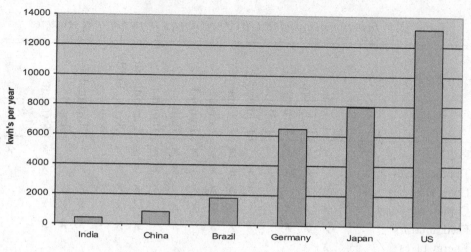

Figure 5.8   Electricity consumption per capita.

5.4.2 *Developing countries*

The economic situation and the characteristics of energy consumption in developing countries provide a distinctly different context for renewable energy development. The prospects for renewable energy are enhanced by a number of factors, including rapid population and energy demand growth, a shortage of electric generation capacity, and large numbers of people without access to the electric grid. On the other hand, renewable energy market development is often inhibited by less well-defined renewable resources, a lack of capital, a shortage of technical capabilities, a less structured legal system, and other infrastructural challenges. The removal of these kinds of barriers should open vast market opportunities to build renewable energy infrastructures before fossil fuels are as "locked in" as in the developed world.

5.4.2.1 *The potential market*

Two billion people in developing countries do not have access to electricity (Stone and Ullal, 1999, p. 19). This number is growing – the World Bank estimates that by 2050, seven billion people in developing countries will need electricity if universal service is to be achieved (Anderson, 1997, p. 192). Furthermore, per capita consumption of electricity in the developing world is very low, often less than 10%, that of the industrialized countries, as shown in Figure 5.8. This suggests that electrical demand growth could continue to rise for many decades before basic electrical technologies have saturated the market.

The demand for electricity is driven by another powerful force: economic development. Developing countries view electrification as a tool for poverty alleviation through economic development, much as industrialized countries did 50 years ago (UNDP/ESMAP, 1998, p. 10). They often carry out ambitious electrification programs in rural areas, because of the positive impacts electricity can have on socioeconomic development, especially among the poor. Electricity can provide jobs and improve the quality of life.

However, grid extension is expensive, running about $10000/km for a medium-voltage line (Anderson, 1997, p. 193). Electric customers in large areas of the developing world pay in excess of $0.25 to $0.35/kWh, which is the cost of energy delivered from a photovoltaic system (Anderson, 1997, p. 192). Renewable resources are often available locally and can obviate the need for much or all of the cost of grid extension.

5.4.2.2 Competition to renewables

Few developing countries have a full range of conventional fuel choices widely available at a reasonable cost. Natural gas, which is plentiful and inexpensive in many industrialized countries, is not available to most of the developing world because of the limited transmission and distribution infrastructure. Developing countries have most of the world's population and gas reserves, but with less than 10% of the world's gas pipeline system (Steinbauer *et al.*, 1998, pp. 260, 267), they account for less than 20% of global gas consumption (EIA, 1999a, p. 146).

Although lacking domestic fossil fuel resources, many developing countries are restricted from satisfying their energy demand from imports because of concerns about trade balance. Brazil has continued its ethanol program for many years, even though it requires substantial subsidies, because it displaced the need for importing oil (Keegan *et al.*, 1996).

The existing competition for renewables in rural areas is often candles, kerosene, or batteries, which can cost rural users between US $3.00 and $17.00 per month (Goldemberg and Mielnik 1998, pp. 3–14). A small photovoltaic system can be financed with monthly installments in this range.

Many developing countries have historically subsidized both fossil fuels and electricity, especially for residential applications. Subsidies were motivated by a desire to alleviate poverty or to develop the economy. Recently, developing countries have been making great progress in reducing these subsidies, sometimes replacing them with other programs to ensure that energy needs can be met. Fourteen of the largest developing countries reduced fossil-fuel subsidies by 45% between 1990 and 1996 (Reid and Goldemberg, 1997).

The opening up of the electric sector to competition in many developing economies will allow companies to offer renewable sources of energy as a competitor to large, centralized generation. In situations where renewables require less capital investment than line extension, such electric sector reform will stimulate renewable development (Kozloff, 1998, p. 1).

Much of the multinational aid and lending structure was established on the basis of large fossil-fuel power projects. Recently, however, many multilateral and bilateral donor organizations and financial institutions are encouraging renewable energy, including the World Bank, the Global Environment Facility, the InterAmerican Development Bank, the Asian Development Bank, and the US Agency for International Development (EIA, 1997, p. 134).

5.4.2.3 Barriers

Developing countries experience many of the same economic, technical, institutional, and environmental market issues as other countries with respect to renewable energy deployment. However, some unique conditions give rise to additional market issues and barriers.

Contracts can be difficult to enforce. Risk is reduced and private sector investment is facilitated when contracts are enforced. Electricity suppliers commonly use the contracts for the sale of electricity, known as power purchase agreements, as security for project loans. This approach, which is called project financing, is a preferred type for many renewable energy projects. Unfortunately, many developing countries have been unable to develop effective, low-cost enforcement of contracts, which makes project financing riskier and less feasible (Martinot, 1998, p. 910).

Financing can be expensive or inaccessible. Many developing countries experience high rates of inflation, which contributes to high nominal interest rates. Real interest rates are often much higher as well, exacerbating the problem. Most commercial lenders are unfamiliar with renewable energy technologies and lack experience with renewable energy projects. International financial institutions are often interested in renewable energy, but accessing this type of financing can be a lengthy and costly process. India addressed the financing problem in a creative fashion, and in doing so succeeded in developing nearly 1000 MW of wind capacity during the 1990s. A new institution, the India Renewable Energy Development Agency, was created to provide loans at reasonable rates (12% to 15%), for a large percentage of the project cost, at terms of up to 10 years (Jagadeesh, 2000). Unfortunately, few developing countries have this type of financial institution, and many lack secondary financial markets and other market infrastructure to handle capital flows (Northrup, 1997, pp. 17–18).

Fossil fuel subsidies continue to be substantial. In spite of recent progress in removing subsidies, a recent IEA study of eight of the largest developing countries confirmed that "pervasive under-pricing" of fossil fuels still exists, amounting to an average of 20% below market levels (IEA, 1999, p. 9). "In China, taxes and subsidies that discriminate against renewables in favor of fossil fuels are seen as the most important single constraint on the move towards healthy rural energy markets." (UNDP/ESMAP, 1998, p. 6). Ironically, the rural areas served by electrification programs, which is the sector for which renewable energy is most promising, is also the sector where fossil fuel subsidies are most persistent. These subsidies manifest themselves as electricity tariffs for grid power that send the wrong price signals, making off-grid renewable energy systems appear less competitive (Taylor, 1998).

Some electric sector reforms discourage investment in renewables. Regulations are typically written to address legitimate issues for conventional fossil fueled plants, but these regulations may impose barriers for smaller, renewable generators. Back-up power or "spinning reserve" requirements can add costs that make renewable projects unprofitable. The creation of spot markets, in which bulk power is available on very short notice for immediate delivery, discourages renewable energy technologies that are available intermittently. A spot market facilitates access to generation that can assure delivery of power during peak periods, and makes these types of plants more profitable and easier to finance. This tends to reward investment in fossil fuel plants rather than renewable energy projects (Kozloff, 1998, p. 1).

Information is more difficult to obtain. Countries that have had market economies for more than a century typically have public agencies that compile and disseminate information important to the private sector. Large, mature markets are served by private information providers as well. Developing countries have no such infrastructure. Many are just transitioning out of centralized systems in which there was no need for broad dissemination of information (Martinot, 1998, p. 909).

A lack of entrepreneurial skills, experience and spirit. Government-owned utilities and energy enterprises, which dominated the energy sector of many developing countries until recently, discouraged the development of small business. Decades of centrally planned economies did not nurture an entrepreneurial spirit or help develop the skills and experience for business.

Subsidies for renewable energy can inhibit market development. Subsidies are quite common for renewable energy development. There are often good reasons for these subsidies, such as "leveling the playing field" with conventional resources, or to enable the poor to enjoy the advantages of electricity. But subsidies are controversial, because they can smother innovation and prevent competition (UNDP/ESMAP, 1998, p. 17). Many countries provide "tied aid", in which funds can only be used to purchase goods or services from the donor country. Tied aid is recognized to have a chilling effect on commercial competition, and is not allowed under certain international agreements. Because renewable energy goods may not yet be considered "commercial", they can be excluded from these provisions (OTA, 1993, p. 47).

5.4.2.4 Renewable energy deployment

The numerous barriers to renewable energy deployment in developing countries have inhibited, but not prevented, market growth. In fact, some of the largest examples of renewable energy deployment are from the developing world. A few of these examples described below illustrate a variety of approaches, but, in each case, a proactive government effort was a key ingredient.

Wind energy in India. Windpower capacity in India was nearly 1000 MW in 1998 (Jagadeesh, 2000), which is about 10% of global wind capacity (Bijur, 1999, pp. 5–11). The development of this resource began with government-funded demonstrations in the 1980s. Financing from the Indian Renewable Energy Development Agency has played a key role, as have government subsidies (Keegan *et al.*, 1996). Since 1996, however, declining subsidies, economic factors, and problems arising from some poorly installed or maintained wind systems have dramatically reduced the growth in wind capacity (Jagadeesh, 2000, p. 160).

Ethanol in Brazil. The oil embargos of the 1970s spurred Brazil to initiate Proalcool, a program designed to substitute domestically produced ethanol from sugar cane for imported petroleum. In the early 1980s, the program grew rapidly and ethanol sales exceeded gasoline, largely due to government subsidies. Subsidies have declined since then, but ethanol maintains an important, if no longer dominant, share of the liquid fuel market (Keegan *et al.*, 1996).

Geothermal power in the Philippines. The capacity of geothermal energy grew from 3 MW in 1979 to 1455 MW in 1996, making the Philippines the world's second largest producer of geothermal electricity (Gazo, 1997). The government played a key role by establishing contracting mechanisms enabling private sector investment to occur (EIA, 1997, p. 132).

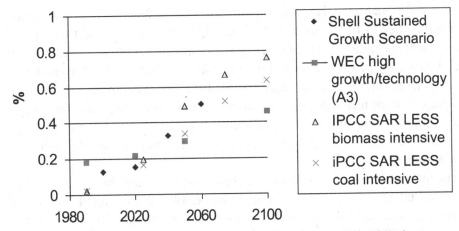

Figure 5.9   Scenarios of renewable energy contributions to worldwide primary energy (Sources: IPCC, 1996; Kassler, 1994; IIASA/WEC, 1998).

5.5 Scenarios for the Future

Renewable energy is frequently viewed as part of the future solution to reducing emissions of carbon dioxide from the combustion of fossil fuels. In fact, the United Nation's Intergovernmental Panel on Climate Change (IPCC), the International Institute for Applied Systems Analysis (IIASA) and the World Energy Council (WEC), as well as Shell Oil's renowned scenario developers have all developed scenarios that present futures for the 21st century in which much, if not most, world primary energy is derived from renewable energy (see Figure 5.9). Moreover, these scenarios are not considered to be outlying scenarios by their developers. The Shell "Sustained Growth Scenario" was essentially Shell's business-as-usual scenario when developed in the mid-1990s (Kassler, 1994). Perhaps more remarkable, the IPCC "Coal-Intensive Scenario" from the Second Assessment Report[6] (SAR) shows non-hydro renewables contributing 60% of all worldwide primary energy use by 2100 (IPCC, 1996).

Inasmuch as these scenarios also presume continued growth of world economies and of energy markets, an increase in the market share for renewables is an even larger increase in the absolute amount of energy provided. In the Shell "Sustained Growth" scenario, the contribution from non-hydro renewable energy by 2100 is more than 50% larger than today's total world energy use. Similarly, in the IPCC scenarios, the non-hydro renewables contribution is larger than today's total world primary energy consumption.

---

[6] The IPCC Third Assessment Report was undergoing review at the time this chapter was written.

Of course, such scenarios are fraught with uncertainties in their estimation and difficulties in their actual evolution. These include the role that climate change will play in market evolution, the price and availability of fossil fuels as the 21st century progresses, evolving technologies – both renewables and conventional, the extent of and access to renewable energy resources, and the infrastructure developments required to shift energy markets to renewable energy. The authors are usually careful to point out that scenarios are not forecasts, but planning tools that help to prepare us for change by outlining different possible futures.

The infrastructure required to attain these levels of renewable energy deployment will take many years to develop. For example, current world production of photovoltaic modules in 1999 is estimated to be about 170 MW annually (Energetics, 1999). Were this production increased by 25% for each of the next 30 years, annual production would increase to 137 GW by 2030, and total installations (assuming no retirements) from 2000 forward would be about 686 GW. At a 20% capacity factor, this world photovoltaic capacity would displace about 196 GW of coal capacity operating at a 70% capacity factor, or 12% of the world's coal capacity today. Thus, even with such aggressive growth, by 2030, the photovoltaic contribution to carbon emissions reductions is relatively modest. However, under such a growth scenario, by 2030 photovoltaics will begin making more substantial contributions, not only because the infrastructure will have developed, but also because industry will have learned how to bring the costs down. Based on historical cost improvements of 18% for each doubling in market installations (REPP, 2000), under the above scenario of 25% annual growth in production, costs will decrease by a factor of about 5 by 2030.[7] This would position photovoltaic technology for even faster market penetration and greater substitution for fossil fuels and their carbon emissions.

5.6 System-level Deployment

The levels of renewable energy use described in the above scenarios may require some major energy infrastructure changes. For example, the number of renewable energy technologies available today to directly displace petroleum use in the transportation sector is limited largely to ethanol and, perhaps, biodiesel. To meet the growing worldwide demand for transportation services in the second half of the 21st century, when world petroleum production will probably be in decline, will require other options. In the next few paragraphs, we

---

[7] The REPP reference to a progress ratio of 82% $(1-0.18)$ is based on growth in installations. The progress ratio is applied here to production because current installations are not well quantified. Mathematically this is justified if the production and installations are in a steady-state growth pattern.

briefly introduce what some of these major market shifts might entail for all energy sectors. We introduce market concepts in which alternative transportation forms reduce the need for liquid fuels, hydrogen becomes the common energy carrier, electric vehicles represent a means to use renewable energy more extensively, and superconducting transmission and storage are used to address the intermittency of leading non-hydro renewable energy technologies.

The scope of this chapter does not allow us to treat any of these topics in the detail they deserve and have received in multiple books and articles in the open literature. We focus here on the role that renewables might play in their realization, how they might provide larger markets for renewables, and some of the principal considerations in their development.

5.6.1 Hydrogen as a major energy carrier

Because it must be produced from other energy sources, hydrogen, as a fuel, is generally classified as an "energy carrier", not as an energy source. It is not a renewable fuel, but can be made using electricity from renewable energy sources or by using other energy feedstocks including natural gas, coal, biomass, and methanol.

The principal advantage of hydrogen is that it can be used as a fuel with no emissions at the point of use other than water vapor. Hydrogen can be directly combusted to produce heat and electricity or used at even higher efficiencies to produce electricity directly in fuel cells, a technology that is beginning to move into the marketplace. The major obstacles to the use of hydrogen as an energy carrier today are its cost of production, lack of an inexpensive storage technology, and the cost of fuel cells.

While there are no carbon emissions associated with the use of hydrogen, carbon is emitted in the steam reforming of methane, the most economic production method available today. Fortunately, there are ways to separate out the carbon emissions from the reforming process and to sequester that carbon. Research is under way to improve these processes and reduce their costs. If successful, such research could allow continued use of fossil fuels in a greenhouse-constrained world in centralized applications where the carbon is more easily separated and sequestered (Ogden, 1999).

Research is also under way to improve the production of hydrogen with renewable energy. The most cost-effective route today is to use electricity from renewable energy technologies to power an electrolyzer that separates water into oxygen and hydrogen. Polymer electrolyte membrane separators now operate routinely above 85% efficiency (Greenwinds, 2000), but they have relatively high capital costs approaching \$1000/kW of input electricity. However, the future is expected to be different both because R&D is bringing down the costs of these technologies and because higher value is being placed on reductions in carbon and local air emissions.

Not only is research reducing the cost of electrolyzers and electricity production from renewables, but novel uses of renewables to produce hydrogen are also under investigation. These include mutant algal strains that produce hydrogen at higher rates than natural strains, photocatalytic water-splitting systems using non-toxic semiconductors, and photoelectrochemical light-harvesting systems that can split water molecules and are stable in a water/electrolyte solution. Biomass feedstock alternatives for hydrogen production are also under investigation. Biomass pyrolysis produces a bio-oil that, unlike petroleum, contains many highly reactive, oxygenated components derived mainly from constitutive carbohydrates and lignin. These components can be thermally cracked and the product's steam reformed at 750° to 850°C with Ni-based catalysts, high heat transfer rates, and appropriate reactor configurations to produce hydrogen with minimal char deposits (Padro, 1998).

This research is under way at least partially because hydrogen can greatly increase the opportunities to use renewable energy. The advantages hydrogen presents to renewable energy are twofold. First, it presents a means to convert electricity from renewable energy systems to a hydrogen fuel that can be used for transportation. Thus, renewable energy contributions in this sector would not be limited to only biomass-derived fuels, but could be made by all renewable electric technologies. Of course, this would require improvements in hydrogen storage, delivery systems, and other infrastructures. Secondly, hydrogen can be used as a storage medium allowing electricity produced at off-peak load times by intermittent renewable energy systems to be used later for on-peak loads. This would eliminate reliability and dispatchability constraints on the use of intermittent renewable electric sources like wind and solar. More importantly, it would allow these intermittent renewables to receive firm capacity payments for the electricity they provide, greatly enhancing their economic competitiveness. These two enhancements would provide a route by which renewables could greatly increase their contribution to the reduction of carbon emissions worldwide.

Transition to a worldwide hydrogen energy market could be a massive enterprise. However, with renewables it could be accomplished in stages. Initially, hydrogen and fuel cells could be used to simply firm renewable electricity, increasing the capacity value of intermittent renewable sources. In this case, the hydrogen would be consumed by a stationary fuel cell producing power at the time of peak loads at the same location where it is produced by an electrolyzer, eliminating any need for pipelines. Subsequently, additional hydrogen could be produced for transportation use in local fuel cell vehicles, delaying the need for extensive pipelines. Inasmuch as renewables are widely distributed, a large portion of the transportation market might be served in this decentralized fashion.

While a hydrogen-dominated energy market could produce some unique benefits for renewable energy, it might also afford an opportunity to separate and sequester carbon from fossil fuels, allowing their continued use in a climate-concerned world. We can be assured that technology developments will determine the future routes followed. We can also be fairly confident that a hydrogen economy will not evolve under the market conditions of today. Natural gas currently offers nearly all of the benefits of hydrogen except complete carbon reduction, yet costs less than one-fourth of that of hydrogen from electrolysis.

5.6.2 Electric vehicles

Another route by which the contribution of renewable energy might be greatly expanded to further reduce carbon emissions would be through the successful development of cost-competitive electric vehicles (EVs). As with the hydrogen economy described above, the use of EVs is not tied only to renewables; the electricity to charge the batteries of EVs can come from any generation source. But EVs would provide some extra advantages to intermittent renewables. As with hydrogen, EVs afford renewable electric technologies the opportunity to serve the transportation sector as well as other end-use sectors.

Secondly, EVs afford renewable electric technologies a storage medium, reducing the impact of intermittent generation from renewables. There are several infrastructure routes that might be followed. They vary primarily in terms of who owns and controls the recharging of the batteries used in the EVs. If consumers own their own batteries and charge them with no real-time input as to when electricity is most available and least expensive, then EVs will provide no real advantage to renewable electric technologies other than increasing the overall load and allowing for more electric capacity growth, which renewables could provide. However, if electric system operators have some control over when the batteries are charged, either directly or through real-time price signals to consumers, then there is a limited opportunity to charge the batteries when the intermittent renewable energy sources are available.

However, most vehicle owners will insist that their batteries be charged some time during the course of the night so as to be available for use on the following day. This limits the length of time one can wait for the intermittent renewable energy source to be available. If all owners waited to just before they needed the battery in the morning, a new peak load would be created.

Greater use of intermittent renewables might be achievable if a central repository owned the batteries, and vehicle owners simply exchanged batteries as required, similar to filling one's car at a gas station today. This scheme would provide more opportunities for charging the extra batteries when the intermittent renewables are available. Furthermore, it might provide a source of central storage to the electric system (i.e., the batteries not in use in a vehicle could be tied to the grid) that could reduce the need for peak generation and help smooth out transients introduced by intermittent renewable electric sources.

Regardless of the infrastructure developed, EVs, together with renewable electric technologies, could allow the displacement of much of the petroleum currently used for transportation and the carbon emissions associated with its use. Their use will depend primarily on resolving issues associated with EV batteries – cost, weight, life, performance, and driving range (National Laboratory Directors, 1997).

5.6.3 Superconductivity

Superconductivity is the ability of certain materials to conduct electrical current with no resistance and extremely low losses. Recent developments allow superconductivity to be maintained at relatively high temperatures using liquid nitrogen. Electric power applications include wires, motors, generators, fault current controllers, transformers, and superconducting magnetic storage.

Superconducting components could provide several advantages to renewable energy. With superconducting wires, it should be possible to transmit power for long distances with few energy losses. Cross-continent lines and even intercontinental lines could ensure that intermittent renewable electric resources spread across the continent(s) are always available to meet a region's peak load. Superconducting lines might be another route for transmitting hydropower from huge resources in the unpopulated areas of Africa and Siberia to loads thousands of kilometers away. Furthermore, superconducting magnetic energy storage might allow the off-peak generation by intermittent renewables to be stored and used on-peak. Intermittent renewable energy generators could then be credited with firm capacity.

More R&D is needed to make superconducting transmission lines and magnetic storage a physical reality, especially to make it a cost-effective option. If it does materialize as a significant factor in a carbon-constrained electric system, it will give an advantage not only to renewables, but also to conventional generation technologies through more efficient generators, transformers, and transmission lines (DOE, 2000).

We have briefly discussed three areas of innovation that might lead to greater use of renewable energy. There are others, probably many that we haven't thought of at this time. The areas discussed above are not mutually exclusive, i.e., we could enjoy the benefits of both superconductivity and electric vehicles. There may even be synergisms between them.

5.7 Policy Requirements

Renewable energy is an important approach to reducing greenhouse gas emissions. Most renewable energy technologies produce no net greenhouse gas emissions. Some renewable energy technologies are well established and have been cost competitive for years. Many renewable energy technologies are still evolving and should be even more attractive in the near future.

The long-term potential of renewable energy is vast – annual production could some day exceed the total amount of global energy consumed today. IPCC figures show a potential for renewable energy 21 times as great as global energy use now.

Deployment of renewable energy thus far, with the exception of large hydropower, has been quite modest. Photovoltaic production, for example, is such a small industry that it could experience 25% annual growth for 30 years and still amount to only 12% of today's coal capacity. Many other renewable energy technologies are even less developed.

Growth in renewable energy continues to be inhibited by a number of factors. The higher capital cost and other factors make financing difficult. The small project size translates into lower levels of absolute profits and high proportions of transaction costs, which discourages private developers. Technical factors having to do with the value of intermittent resources and their integration with an electric system also work against renewable energy. Institutional barriers may be the most imposing of all.

Barriers could be overcome more easily if the true value of renewable energy in terms of environmental benefits, advantages to local economies, and national security advantages could be quantified and credited to renewables in all investment decisions. Such a blanket internalization of currently external costs and benefits is conceivable using a number of means, including tax credits, subsidies, and other forms of transfer payments. Each of these would require substantial intervention by government. Even if government was willing, it is not possible to accurately estimate the full value of externalities – and the results can be extremely sensitive to the estimate. If the externalities are underestimated, the benefit value may inappropriately exclude significant amounts of renewables from the market and increase social costs. If they are overestimated, society is disadvantaged by bringing on more costly renewables at unjustified levels.

The environmental benefits of renewable energy have spurred many attempts to encourage development using a plethora of different instruments, such as those shown by Table 5.11 for the OECD countries. Governments have sponsored direct research and development. Many forms of direct financial incentives have been tried. Voluntary efforts have been encouraged. Demonstrations and training programs have been undertaken to ensure the public is aware of the advantages. More recently, distinct targets for market penetration levels have been implemented, such as the British Non-Fossil-Fuel Obligation, or the Renewable Portfolio Standard considered by the US government and implemented by many states.

Table 5.11 would suggest that a wide range of strong sincere efforts to encourage renewables is under way. However, it is instructive to consider whether the resources being committed to the development of renewables are commensurate with the role renewables might be called on to play in addressing global climate change, much less with the multiple environmental, economic, and security benefits that renewables provide. Figure 5.10 shows the recent expenditures of OECD governments on renewable energy programs as a percentage of GDP. These are certainly well below the levels one would expect for a priority effort.

What more can be done? First, an international effort should be conducted to identify the full potential of renewables in the long term under a mandate to reduce carbon emissions to sustainable climate levels. We have made a brief attempt here to give some insight into this problem, but we had to rely on existing inconsistent sources that don't cover the entire spectrum of possibilities, either regionally, technologically, or over time. Such an analysis should include scenarios that consider alternative climate change solutions other than renewables, that compute the cost of energy imputed by the scenario, and that quantify the uncertainties as far as possible.

Second, renewable energy R&D efforts should be expanded and coordinated more on a worldwide basis. As shown by Figure 5.7 of the supply curve for carbon reductions in the US through renewables, the cost to reduce carbon through renewables will depend strongly on research success in the future. There is every reason to believe that, with a continuing coordinated effort, the R&D successes of the past will continue into the future.

Third, R&D should be expended on supporting technologies consistent with the findings of the analysis recommended in step 1 above. This could include

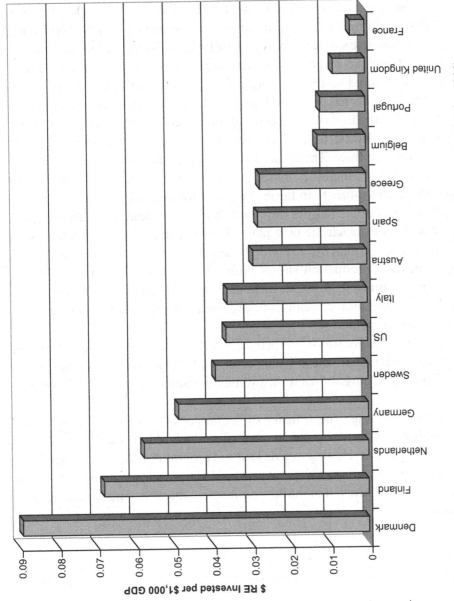

Figure 5.10   OECD renewable energy R&D expenditures (Source: Goldstein, 1999).

hydrogen production and infrastructure technologies, electric vehicles, super-conductivity, storage, and hybrid configurations.

Fourth, policies to address local air pollution and other environmental considerations should be structured so as to provide as much incentive to renewables and energy efficiency as possible. For example, the US SO_2 allowance system developed under the Clean Air Act Amendments of 1990 has been a significant economic success in reducing sulfur emissions with minimal market disruption. However, to gain political acceptance, the distribution of allowances was structured so that existing polluters received the initial free allowances.[8] Another alternative would have been to provide allowances annually to the electric sector based on the electricity produced, not the fuel input. In this way renewable electric technologies would also receive allowances for the electricity they produce. These could be sold by the renewable technologies to the fossil-powered generators. Thus, the renewables would receive the benefit of the allowances rather than the existing fossil-based polluters.

Fifth, identify opportunities to provide limited subsidies to renewables that will allow them to reach the next plateau in their development. Technologies close to market competitiveness are excellent candidates, especially if an increase in their production shows promise of industrial learning and cost reductions. As much as possible, these subsidy programs should be long-term with defined expiration conditions to provide as much certainty as possible to the investment community.

Sixth, establish programs to encourage voluntary purchases of renewable energy by government, the private sector, and by individuals. These "green power" programs have shown some promising results and appear to have great potential.

Seventh, encourage the transfer of technology to the developing world. Industrialized countries should increase efforts to build the markets for renewable energy in the developing world. Tied-aid programs, which encourage hardware sales but discourage competition, should be transformed into technology cooperation initiatives that build commercial markets and trigger private investment.

---

[8] The Clean Air Act Amendments of 1990 did provide for a set-aside of 300 000 allowances for utilities that adopted renewable energy technologies. However, less than 15 000 such allowances were claimed because:
1. They expired in 1990.
2. They could not be claimed by non-utility generators.
3. Only utilities adopting certain "integrated resource planning" procedures could claim them.
4. One allowance (one ton of SO_2) was allowed for each 500 MWh of renewable electricity provided – one-sixth the amount of SO_2 emitted from 500 MWh by those coal-fired generators controlled under the first phase of the SO_2 allowance program.

References

Anderson, D. (1997). Renewable energy technology and policy for development. Socolow, R., ed., *Annual Review of Energy and the Environment*, **22**. Palo Alto, CA: Annual Reviews, Inc.

Bijur, P. (June 1999). Energy Supply Technologies, Chapter 5 of *Powerful Partnerships: The Federal Role in International Cooperation on Energy Innovation*. Washington, DC: The President's Committee of Advisors on Science and Technology.

Cassedy, E. S. (2000). *Prospects for Sustainable Energy*. University Press, Cambridge, U.K.

Chan, S., Cheng, S., and Curtice, D. (October 1983). *Methods for Wind Turbine Dynamic Analysis*. EPRI El-3259, Electric Power Research Institute, Palo Alto, CA.

DOE. (2000). Web site: www.eren.doe.gov/superconductivity/pdfs/potential_of_supercon.pdf, June.

Energetics. (September 1999). *US Photovoltaics Industry, PV Technology Roadmap Workshop*. Columbia, MD.

EPRI (Electric Power Research Institute). (January 1979). *Requirements Assessment of Wind Power Plants in Electric Utility Systems*. Volume 2, EPRI ER-978, Palo Alto, CA.

EPRI (Electric Power Research Institute and US DOE). (December 1997). *Technology Characterizations of Renewable Electric Technologies*. EPRI TR-109496.

EIA (Energy Information Administration). (March 1997). *Renewable Energy Annual, 1996*. US Department of Energy, Washington, DC. National Energy Information Center.

EIA (Energy Information Administration). (1999a). *International Energy Outlook 1999*. US Department of Energy, Washington, DC. National Energy Information Center.

EIA (Energy Information Administration). (December 1999b). *Annual Energy Outlook 2000*. DOE/EIA- 0383(00).

EIA (Energy Information Administration). (June 2000). Web page: www.eia.doe.gov/oiaf/ieo/index.html, June.

European Union (July 2000). Web page: europa.eu.int/comm/eurostat/Public/datashop/print-catalogue/EN?catalogue = Eurostat&collection = 09-Database%20Information&product = SIRENE-EN

EWEA (European Wind Energy Association). (June 2000). Web page: www.ewea.org/y2k.htm.

Farhar, B. C. (July 1999). *Willingness to Pay for Electricity from Renewable Resources: A Review of Utility Market Research*. NREL/TP-550–26148, National Renewable Energy Laboratory, Golden, CO.

Gazo, F. (1997). Full steam ahead (a historical review of geothermal power development in the Philippines). *Proceedings of Geothermal Resources Council, Vol. 21, September/October 1997*. Republic of the Philippines: National Power Corporation.

Goldemberg, J. and Mielnik, O. (1998). Renewable energy and environmental concerns and actions in Latin America. Campbell-Howe, R., ed., *Proceedings of the 1998 Annual Conference of the American Solar Energy Society, June 1998, Albuquerque, New Mexico*. Boulder, CO: American Solar Energy Society.

Goldstein, L., Mortensen, J., and Trickett, D. (May 1999). *Grid-Connected*

Renewable-Electric Policies in the European Union. NREL/TP-620–26247, National Renewable Energy Laboratory, Golden, CO.

Greenwinds. (June 2000). Web site: www.pege.org/greenwinds/newpage/electrolyzer.htm.

Grubb, J. J. (1987). *The Integration and Analysis of Intermittent Source on Electricity Supply Systems*. Ph.D. Thesis, King's College, University of Cambridge.

Herrera, J. E., Reddoch, T. W. *et al.* (October 1985). *A Method for Determining How to Operate and Control Wind Turbine Arrays in Utility Systems*. Draft Report. Solar Energy Research Institute, Golden, CO.

IEA (International Energy Agency). (1997). *Key Issues in Developing Renewables*. Paris, France: Organization for Economic Cooperation and Development.

IEA (International Energy Agency). (1998). *World Energy Outlook*. Paris, France: Organization for Economic Cooperation and Development.

IEA (International Energy Agency). (1999). *World Energy Outlook: 1999 Insights*. Paris, France: Organization for Economic Cooperation and Development.

IIASA (International Institute for Applied Systems Analysis). (1981). *Energy in a Finite World, A Global System Analysis*. Cambridge, MA.

IIASA/WEC, 1998, *Global Energy Perspectives*, edited by N. Nakicenovic, A. Grubler, and A. McDonald, Cambridge University Press, Cambridge, UK.

INEEL (Idaho National Engineering and Environmental Laboratory). (2000). Web site: www.ineel.gov/national/hydropower/state/stateres.htm.

International Development and Energy Associates, Inc. (December 1992). *Technical and Commercial Assessment of Amorphous Silicon and Energy Conversion Devices Including Photovoltaic Technology*. Report No. 92–04 prepared for the Office of Energy and Infrastructure, US Agency for International Development.

IPCC (Intergovernmental Panel on Climate Change). (1996). *Climate Change 1995: Impacts, Adaptations and Mitigation of Climate Change: Scientific-Technical Analyses*. Contribution of Working Group II to the Second Assessment Report, Cambridge University Press.

Jagadeesh, A. (March 2000). Wind energy development in Tamil Nadu and Andhra Prodesh, India: institutional dynamics and barriers. *Energy Policy*, **28**, No. 3.

Johansson, T. B., Kelly, H., Reddy, A. K. N., and Williams, R. H. (1993). *Renewable Energy, Sources for Fuels and Electricity*, Island Press, Washington, D.C.

Kassler, P. (November 1994). *Energy for Development*. Shell Selected Paper, based on a presentation to the 11th Offshore Northern Seas Conference, Stavanger, August 1994.

Keegan, P., Price, B., Hazard, C., Stillman, C., and Mock, G. (1996). Developing country GHG mitigation: experience with energy efficiency and renewable energy policies and programs. *Climate Change Analysis Workshop: June 1996, Springfield, VA*. The International Institute for Energy Conservation.

Kozloff, K. (April 1998). Electricity sector reform in developing countries. *Research Report*. Washington, DC: Renewable Energy Policy Project.

Mann, M. K. and Spath, P. L. (December 1997). *Life Cycle Assessment of a Biomass Gasification Combined-Cycle System*. NREL/TP-430–23076, National Renewable Energy Laboratory, Golden, CO.

Martinot, E. (September 1998). Energy efficiency and renewable energy in Russia. *Energy Policy*, **26**, No. 11.

Moore, C. and Ihle, J. (October 1999). *Renewable Energy Policy Outside the United States*. Issue Brief No. 14. Renewable Energy Policy Project, Washington, DC.

National Laboratory Directors. (October 1997). *Technology Opportunities to Reduce*

US Greenhouse Gas Emissions. Oak Ridge National Laboratory, Oak Ridge, TN.

Northrup, M. (January/February 1997). Selling solar: financing household solar energy in the developing world. *Solar Today*, **11**, No. 1.

Ogden, J. M. (March 1999). Strategies for developing low emission hydrogen energy systems: implications of CO_2 sequestration. *Proceedings of the 10th National Hydrogen Association Meeting*, Arlington, VA.

OTA (Office of Technology Assessment). (August 1993). *Development Assistance, Export Promotion and Environmental Technology*. US Congress, Washington, DC: Background Paper.

OTA (Office of Technology Assessment). (September 1994). *Studies of the Environmental Costs of Electricity*. OTA-ETI-134, Washington, DC.

OTA. (September 1995). *Renewing Our Energy Future*. OTA-ETI-614 Office of Technology Assessment, Congress of the United States.

Padro, C. E. (August 1998). *The Road to the Hydrogen Future: Research and Development in the Hydrogen Program*. National Renewable Energy Laboratory, Presentation at the US DOE Hydrogen Program Review, April 28–30, 1998, Alexandria, VA.

Reid, W. and Goldemberg, J. (July 1997). Are developing countries already doing as much as industrialized countries to slow climate change? *Climate Notes*. New York: World Resources Institute.

Renne, D. S. and Pilasky, S. (February 1998). *Overview of the Quality and Completeness of Resource Assessment Data for the APEC Region*. APEC #98-RE-01.1, Prepared for the Asia-Pacific Economic Cooperation (APEC) by the National Renewable Energy Laboratory, Golden, CO.

REPP (Renewable Energy Policy Project). (June 2000). Web site: www.repp.org/articles/pvaction/index_pvactiona.html

Small Hydro Power, *Renewable Energy World*, July 1999.

Smith, D. R. and Ilyin, M. A. (1990). *Wind Energy Evaluation by PG&E*. Pacific Gas and Electric Research and Development, San Ramon, CA.

Solar Today, Nov/Dec 1999, p 19.

Steinbauer, T., Willke, T., and Shires, T. (1998). Natural gas pipelines: key infrastructure for world development. *Global Energy Sector: Concepts for a Sustainable Future*. US Agency for International Development. *Climate Change Initiative 1998–2002*. Washington, DC.

Stone, J. and Ullal, H. (1999). Electrifying rural India. *Solar Today*, **13**, No. 6, November/December, p. 19.

Swezey, B. and Bird, L. (August 1999). *Information Brief on Green Power Marketing*. NREL/TP-620–26901, National Renewable Energy Laboratory, Golden, CO.

Taylor, R. (July, 1998). *Lessons Learned from the NREL Village Power Program*. National Renewable Energy Laboratory. Golden, CO. Paper presented at the Second World Conference and Exhibition on Photovoltaic Solar Energy Conversion, Vienna, Austria.

Trainer, F. E. (1996). Book review: critical comments on renewable energy. *Energy*, **21**, No. 6, pp 511–17.

UNDP/ESMAP (Joint UNDP/World Bank Energy Sector Management Assistance Programme). (May 1998). *Rural Energy and Development Roundtable*. Washington, DC: The World Bank Report No. 202/98.

United Nations, (July 2000). Information Unit on Climate Change (IUCC), UNEP. Website: www.unfccc.de/resource/ccsites/senegal/fact/fs230.htm

van Beek, A., and Benner, J. H. (June 1998). *International Benchmark Study on Renewable Energy.* Ministerie van Economische Zaken.

Walsh, M. (1999). Oak Ridge National Lab. Unpublished biomass resource data. Personal e-mail communication, July 29.

Wan, Y. and Parson, B. (August 1993). *Factors Relevant to Utility Integration of Intermittent Renewable Technologies.* NREL/TP-463–4953, National Renewable Energy Laboratory, Golden, CO.

WEC (World Energy Council). (September 1993). *Renewable Energy Resources: Opportunities and Constraints 1990–2020.*

WEC (World Energy Council) (1998), 17th Congress. London, UK: World Energy Council, pp. 260, 267.

6

Carbonless Transportation and Energy Storage in Future Energy Systems

By 2050 world population is projected to stabilize near ten billion. Global economic development will outpace this growth, achieving present European per capita living standards by quintupling the size of the global economy – and increasing energy use, especially electricity, substantially. Even with aggressive efficiency improvements, global electricity use will at least triple to 30 trillion kWh/yr in 2050. Direct use of fuels, with greater potential for efficiency improvement, may be held to 80 trillion kWh (289 EJ) annually, 50% above present levels (IPCC, 1996). Sustaining energy use at these or higher rates, while simultaneously stabilizing atmospheric greenhouse gas levels, will require massive deployment of carbon-conscious energy systems for electricity generation and transportation by the mid 21st century. These systems will either involve a shift to non-fossil primary energy sources (such as solar, wind, biomass, nuclear, and hydroelectric) or continue to rely on fossil primary energy sources and sequester carbon emissions (Halmann and Steinberg, 1999). Both approaches share the need to convert, transmit, store and deliver energy to end-users through carbonless energy carriers.

6.1 Carbonless Energy Carriers

Electricity is the highest quality energy carrier, increasingly dominant throughout the world's energy infrastructure. Ultimately electricity use can expand to efficiently meet virtually all stationary energy applications, eliminating stationary end-use carbon emissions. This approach is unlikely to work in transportation, however, due to the high cost and low energy density of electricity storage. Chemical energy carriers, such as hydrogen, can more effectively serve transportation fuel and energy storage applications, offering much higher energy density at lower cost. Electrolytic hydrogen, extracted from steam with renewable energy, stored as a high pressure gas or cryogenic liquid, and reconverted to

181

electricity in fuel cells and or used to power hydrogen vehicles, will reduce emissions from both transportation and electric generation. Renewable resources and modular electrolytic technology also permit decentralized hydrogen production, circumventing distribution issues and barriers to market entry. In contrast, sequestration-based fossil-fueled systems must achieve economies of scale by relying on centralized production and hierarchical transmission and distribution of electricity, hydrogen fuel, and carbon (dioxide).

Renewable and fossil approaches may ultimately turn out to be complementary. Use of renewable sources would limit the sequestration burden to modest quantities using the most cost effective methods and reliable disposal sites. Previous analyses have concluded renewable electricity will be cost effective in combination with dispatchable carbonless energy sources (i.e. hydroelectric, fission, and biomass), to minimize energy storage (Union of Concerned Scientists, 1992; Kelly and Weinberg, 1993). Dispatchable carbonless sources only generate a fraction of current electricity, however, and are likely to be limited on the scale of burgeoning demand (Schipper and Meyers, 1992; Fetter, 1999). A future role for biomass, in particular, may be restricted due to competing uses for land, water, and perhaps other agricultural inputs (Smil, 1998). An alternative to expanding fission, hydropower, and biomass is to use modest fossil electric generation and carbon sequestration as a complement to wind and solar energy. An integrated hydrogen transportation sector complements renewable systems both by providing a large, but flexible, use for excess renewable electricity and by enabling dual-use of hydrogen fuel as utility energy storage and transportation.

This chapter surveys energy storage and hydrogen vehicle technologies, analyzing the integration of these technologies into increasingly renewable electricity and transportation sectors. The implications for greenhouse gas reduction strategies are examined using an aggressive efficiency scenario for the United States in 2020, the latest time horizon for which detailed sectoral projections have been made.

6.1.1 Conventional energy storage technologies

Low capital cost, but inefficient, gas-fired peaking plants are used to meet demand fluctuations in present utility systems. Demand fluctuations can also be met using energy storage to shift electric generation to more cost-effective times of day. Utility storage is employed today, in small amounts, using established principles of mechanical energy storage: elevated water or compressed air.

Hydroelectric pumped storage is the most widespread and mature technology, however the theoretical energy density of pumped hydro is quite low, requiring 3.7 *tonnes* (about 1000 gallons) of water traversing 100m of elevation to deliver 1 kWh. Pumped hydroelectric plants are consequently most viable on a large scale. The largest pumped hydro facility in the world today uses Lake Michigan and an artificial lake averaging 85m of elevation. It has a peak generating capacity of 2000 MW delivering up to 15 000 000 kWh over a period of about 12 hours, supplying the equivalent electric demand of about one to two million people. Roundtrip efficiencies approach 70%. At present 2% of electric demand is met by pumped hydro systems (Dowling, 1991).

Two disadvantages of hydroelectric energy storage appear in the context of future energy systems. The large scale nature of hydroelectric storage indicates that little if any cost saving will exist for electric distribution systems connected to a pumped hydroelectric facility. Finally, in the context of solar or wind intensive energy systems it seems unlikely that sufficient sites could be found in convenient locations (i.e. where natural formations provide low per kWh storage costs) to contribute more than a minor role in overall energy storage. River-fed reservoir hydroelectric capacity is probably best used to offset seasonal variations in solar or wind electric generation.

Compressed air energy storage (CAES) is also a reasonably mature approach, though only employed in a few sites worldwide. The energy density of CAES is about 50 times greater than hydroelectric storage. Air compressed to 100 atmospheres of pressure in a 20 gallon volume contains 1 kWh of energy. This energy density is still quite low, however: the same volume of compressed natural gas (CNG) contains nearly 100 *times* more energy. CAES is economic at larger scales (100–200 MW), relying on natural formations for low cost storage capacity, limiting widespread implementation. The compressed air can be run through turbines to generate peak electricity, although the heat of expansion must be supplied by thermal storage or fuel. For economic reasons, interest is greatest in applying CAES if additional fuel is burned with the precompressed air, enabling smaller turbines to match peaks in electric demand. In this application, however, the majority of the energy from a CAES system actually comes from fuel rather than compressed air (Gordon and Falcone, 1995). CAES systems are also not the most efficient method of gas-fired electric generation, and their greatest benefit is reducing the cost of generation capacity, rather than energy storage *per se*. CAES would have little to offer carbonless energy systems which do not rely much on fossil fuels, but do require renewable energy storage. Widespread application of CAES in a greenhouse gas context would require either carbon sequestration, or the use of carbonless fuel (i.e. hydrogen) produced elsewhere. The capacity for compressed air storage in CAES systems would be probably be more valuable as compressed *hydrogen* storage in the context of carbonless energy systems. Leakage has not been a difficulty when storing town gas (a mixture containing hydrogen) in underground caverns near Paris, France (Ingersoll, 1991).

6.1.2 *Advanced energy storage technologies*

Advanced energy storage technologies, in contrast to conventional energy storage, are characteristically modular, highly engineered systems without the scale and location constraints of pumped hydroelectric or compressed air storage. Approaches to energy storage include thermochemical (chemical couples), thermal (phase change materials), mechanical (flywheels), and electrochemical (batteries and electrolytic fuel production).

Thermochemical energy storage approaches capitalize on the high energy density of chemical energy storage and the use of low cost and abundant materials. Thermal energy can be stored in reversible chemical reactions (e.g. $2SO_3 \longleftrightarrow 2SO_2 + O_2$ or $CH_4 + H_2O \longleftrightarrow CO + 3H_2$) in which the reactants are transmitted though a "heat pipe" loop between thermal source and end-use over distances up to 100 miles (Vakil and Flock, 1978). For stationary applications, heat can be stored cheaply in the enthalpy of common materials (water, oil, or molten salts).

While thermochemical and thermal storage are expected to be low cost, the thermal energy stored is not as valuable as electric or fuel energy. The chief disadvantage of thermal energy storage *per se* is thermodynamic. Unlike electricity or fuels, thermal energy "leaks" continuously, and in proportion to the useful work which can be extracted (determined by the Carnot cycle). Today, large amounts of energy are used *as heat* for low temperature space and water heating, some of which could arguably be saved through judicious use of thermal energy storage. However, thermodynamics again present a disadvantage as future space and water heating needs could be supplied very efficiently using heat pumps. As a method of reducing carbon emissions, thermal energy storage is likely to be most useful at modulating solar power production to more effectively meet late afternoon peaks or nighttime electricity demands, using solar thermal electric plants (De Laquil *et al.*, 1990).

Electricity can be stored reasonably compactly and very efficiently as kinetic energy in flywheels. Flywheel energy storage is in the early stages of commercialization, and is targeted at uninterruptible power supplies (UPS), where the value of energy *reliability* far exceeds the value of energy. Flywheels can spin at very high velocities (10000–100000 rpm) in vacuum using magnetic bearings. They offer high efficiency (90%+) charging and discharging, low power related costs ($100/kW) and the prospects of very long equipment lifetime

(Post and Post, 1973). A 1 kWh flywheel module may weigh 10 kg and occupy 20 liters. Flywheel feasibility has advanced substantially with the advent of very strong and light carbon fibers and other composite materials. On the other hand, all of this specialized technology and materials (e.g. magnetic bearings to eliminate friction and provide rotor stability) can lead to high costs per unit of energy stored. Cost estimates are currently $100/kW and $600/kWh of storage capacity, although costs may fall to below $200/kWh (Post *et al.*, 1993) in mature mass production. Flywheels store relatively small amounts of electricity (1–300 kWh) and are probably best placed near end-users in the electricity system, easing the burden on distribution, providing peak power and reliability, and making future energy systems uninterruptible.

The chief alternative to flywheels is electrochemical energy storage. Batteries are heavier, and less efficient (70–80% turnaround efficiency), but more compact than flywheels. Batteries have lower capital costs ($100–$200/kWh), but also a much lower cycle life (1000s of cycles) placing in some doubt their role in bulk power storage. The availability of mineral resources for common battery materials (lead, nickel, cadmium etc.) is likely insufficient (Andersson and Rade, 1998) for globally significant amounts of energy storage (e.g. 24 hour storage, roughly 100 billion kWh would require 1–2 billion *tonnes* of battery materials) in future electricity systems. The most compelling energy application of batteries is efficient electrification of moderate range (100–200 miles) passenger vehicles, assuming battery mass and cycle life can be improved sufficiently.

Less well known than batteries is a closely related alternative: electrolytic fuel production. Electrolysis differs from battery storage in that the electrodes are not chemically changed during electrolysis and do not store energy as in batteries. Energy is instead stored in the chemical fuel produced. A number of electrolytic fuels have been proposed (e.g. lithium, aluminum, and zinc) whose technology is closely related to metal-air (oxygen) batteries. One advantage of electrolytic fuels is the decoupling of power (electrodes and electrolyte) and energy (fuel) functions which are combined in batteries. This reduces the capital cost of achieving high power or large storage capacity. Electrolytic fuels offer potentially rapid refueling and lower weight than conventional batteries, especially when using atmospheric oxygen as a reactant.

Hydrogen has been considered for decades as a universal electrolytic fuel and energy carrier (Cox and Williamson, 1977; Bockris, 1980; Winter and Nitsch, 1988; Ogden and Williams, 1989; Ogden and Nitsch, 1993). Historically, the feasibility of hydrogen has been limited by the fuel economy of passenger vehicles and the corresponding weight of onboard fuel storage systems to achieve good travel range. Recent advances in composite materials, as well as hybrid electric vehicles, have resolved these issues, enabling future hydrogen vehicles 2–3 times more fuel efficient than those envisioned 20 years ago. As a renewable energy carrier electrolytically produced from abundant water, hydrogen is capable of fueling all transportation sectors indefinitely. A spectrum of hydrogen storage methods allow hydrogen systems to be tailored to the economics of individual applications. Electrolytic hydrogen fuel is expected to have low capital costs of production (electrolysis), storage (compressed gas, cryogenic liquid, or chemical storage), and utilization (hybrid electric engines or fuel cells). Estimates in a utility context are $500–1000/kW and less than $5/kWh. The chief disadvantage is that the cumulative process efficiencies of each step in hydrogen systems lead to roundtrip efficiencies of 30–40%, roughly half that of more direct storage technologies. Decentralization may offset this to some extent, potentially making waste heat available for space and water heating.

It is clear from the above discussion that energy storage technologies are best suited to different roles. Flywheels can improve transmission and distribution reliability, storing and delivering electricity perhaps twice daily. At the opposite end of the temporal spectrum, buffered hydroelectric generation may be most useful in adjusting to seasonal electric supply and demand variations. Batteries, most useful contribution would be enabling high efficiency short range transportation. The most important role of thermal energy storage technology is probably allowing solar energy to contribute to nighttime electricity production. Electrolytic hydrogen can serve as a bulk energy storage *and* universal transportation fuel, even if somewhat energy intensively.

The usefulness of each technology will depend on how well these roles meet the needs of future electricity supply mixes which make increasing use of intermittent electric generation, as well as an evolving transportation sector potentially powered directly by electricity or indirectly through electrolytic fuels.

6.1.3 *Transmission technologies*

Transmission technology advances can also play a role in carbonless energy systems, potentially easing local renewable resource constraints by enabling solar and wind energy to be harnessed at greater distances from urban load centers. Transmission is expected to incrementally improve by going to higher voltages, with DC transmission replacing AC transmission lines for long distances. In the longer term high power and perhaps underground cryoresistive and/or superconducting transmission lines could ultimately allow for wholesale long distance electricity transmission, reducing energy storage needed for seasonal and day/night variations around the globe. Another futuristic option may be transmission of energy by relay satellite, similar to proposed satellite or lunar solar power (Hoffert, 1998). Capital costs of transmission are typically moderate relative to both electricity distribution and renewable electricity production. In future energy systems, however, full utilization of electric transmission capacity may become more difficult if large, but intermittent, solar or wind energy facilities are distant from population centers.

An alternative to electric transmission is hydrogen transmission by pipeline or cryogenic tanker, just as natural gas is transmitted today. Pipeline systems offer some buffer capability reducing their sensitivity to short-lived fluctuations in supply or demand. Hydrogen pipelines have been in operation for decades and it is possible for today's natural gas pipelines to transport hydrogen, albeit at reduced pressure and higher cost than today's natural gas (CRC Press, 1977; Bockris, 1980; Winter and Nitsch, 1988; Ogden and Williams, 1989; Ogden and Nitsch, 1993). For equal investment, new hydrogen pipelines also deliver substantially more energy than electric transmission lines, although the conversion losses (electric energy to hydrogen energy and back to electricity) may counter this advantage. The chief factors determining the efficacy of transmitting energy as hydrogen are the scale of energy demand necessary to justify pipelines, and the fraction of demand for transportation fuel vs electricity.

In the near future, energy storage and perhaps transmission improvements can improve electricity distribution and reliability as electricity markets become deregulated, but this will not impact overall *energy* use or emissions substantially. In the intermediate term, an increasing reliance upon intermittent (solar, wind), and/or less flexible (nuclear) electricity sources will at some point require significant *energy* storage, as distinct from *power* storage. This storage will be needed to match non-dispatchable electricity sources with varying electric demands (Iannucci *et al.*, 1998). Finally, energy storage as electrolytic fuels can extend the reach of carbonless energy sources to the transportation sector. This is especially important since the transportation sector is the highest value use of fossil fuels (Berry, 1996), the largest source of carbon emissions, and the least amenable to sequestration approaches.

6.2 Transition Paths Toward Carbonless Energy

Lawrence Livermore National Laboratory (LLNL) has developed a network optimization model (Figure 6.1) to examine these three stages of integrating renewables into utilities (reliability, intermittent intensive electric systems, and carbonless transportation). By constructing and analyzing model scenarios of future electricity and transportation systems attempts we quantify the

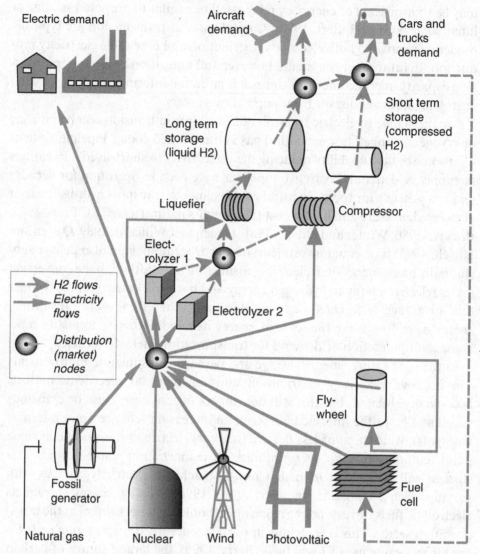

Figure 6.1: Schematic of a coupled utility electric generation and transportation system using nuclear, fossil, and renewable primary energy sources, with electricity and hydrogen as energy carriers.

characteristics of transition paths to carbonless utilities and transportation. The model calculates the economically optimum energy system to meet scenarios of electricity and/or transportation demands, arriving at the desired system structure in terms of energy supply sources, conversion technologies, and storage capacities. It simultaneously determines optimal operation of the system components, using patterns of electricity demand (Iannucci *et al.*,

1998), transportation fuel demand, energy available from solar and wind production, and capacities of long and short term hydrogen storage technologies. A more detailed model description and table of key assumptions used to generate these scenarios is given in Appendix 6.A.

Although there are many transition paths to a carbonless future, we wish to try to identify paths that are economically and strategically advantageous. Using the model discussed above, we have evaluated a broad range of possibilities, using various levels of nuclear, natural gas and renewable generation in combination with a transportation sector using natural gas and/or hydrogen in various amounts. In order to explore a representative example for which data was readily available, we chose to design scenarios based on US Energy Information Agency projections of electricity and transportation demands of the United States in 2020 (EIA, 2000). Under this scenario, within two decades, the United States will demand roughly 5 trillion kWh of electricity (about one third of world demand) and 4.6 trillion kWh of transportation fuel.

A subset of model results are shown as points in Figure 6.2. Each point corresponds to the projected cost of an electricity and transportation fuel system which can achieve the given level of overall carbon emissions from both sectors. These scenarios trace a transition which is efficient in terms of reducing carbon emissions for minimum cost, the *efficient frontier*. From an economic perspective an optimum emission reduction path should fall on this frontier since any other approach will cost more, have higher emissions, or both.

Figure 6.3 shows hourly variations in electricity demand and generation for a ten day period representative of selected cases along the efficient frontier. These cases illustrate the changes in generation patterns and structure along the efficient frontier as natural gas is displaced by renewables and hydrogen transportation is ultimately phased in.

6.2.1 High efficiency use of low carbon fuels

Our starting point is a scenario that efficiently uses natural gas and nuclear electric generation while transportation is fueled by natural gas (lower right point in Figure 6.2. In assessing different carbon reduction strategies, there is generally consensus that improved efficiency and fuel switching to natural gas are "no regrets" measures. To take this into maximum account, we have therefore chosen to measure scenarios using solar and wind electricity against a "no regrets" technically advanced carbon-conscious scenario fueled by natural gas.

Our 2020 US reference scenario relies essentially on very efficient use of natural gas. Hydroelectric and nuclear power (hereafter combined for

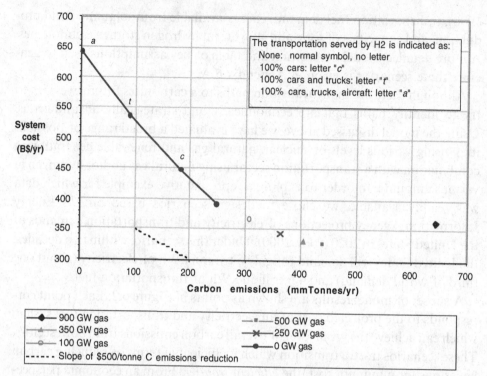

Figure 6.2   The efficient frontier for reducing US 2020 overall carbon emissions. Moving from right to left, natural gas generation capacity is reduced and replaced by renewable generation. Carbon emissions can be reduced to around 250 mmt (million metric tonnes) by displacing natural gas electric generation. Further carbon reductions are achieved by replacing natural gas fueled transportation with hydrogen-fueled transportation. Each case meets 5 trillion kWh of electricity demand and 4.6 trillion kWh of fuel demand and includes 100 GW of nuclear capacity in addition to the gas-fired generation capacities indicated.

simplicity and referred to as nuclear) provide nearly 20% of electricity and 10% of the generating capacity of a 1000 GW US electricity system. Natural gas fuels the remaining 80% of US electricity generation at 57% efficiency, in addition to the all the automobiles, trucks, and aircraft in the US transportation sector. The natural gas infrastructure is assumed to meet all these demands as efficiently as theoretically possible. Fuel cycle greenhouse gas emissions are neglected and natural gas leakage is *assumed* to be reducible to negligible levels. 250 million light-duty vehicles in the US are assumed to achieve an average fleet economy of 80 mpg through the use of lighter and more aerodynamic automobiles. Driving is assumed to rise only slightly to 14000 miles/yr. Natural gas at refueling stations is priced at $6.93–8.31/GJ (equivalent to $0.83–1.00/gallon of gasoline). Natural gas for utility electric generation is priced somewhat lower at $5.54/GJ.

Even with all the progress assumed in this scenario, US electricity and transportation fuel use is projected to produce 642 million metric tonnes of carbon emissions in 2020 and cost roughly $350 billion/yr. Rapid and full implementation of the "no regret" strategy, throughout the entire infrastructure, manages relatively moderate emission reductions relative to current levels (1000 mmtC/yr). Rising energy use driven by population and especially economic growth partially offset a near doubling of electric generation efficiency, a tripling of automobile fuel economy, and a near halving of the carbon intensity of fuels. Global emission targets consistent with stabilizing greenhouse gases at an equivalent doubling of carbon dioxide would limit worldwide gas fuel emissions to about 4000 mmtC/yr (Fetter, 1999). An eventual US target based on per capita share and adjusted to reflect transportation and utilities emissions alone would be about 100 mmtC/yr. Obviously, reducing carbon emissions to this level from the advanced fossil scenario (642 mmtC/yr) will ultimately require a massive shift from fossil energy to carbonless sources.

6.2.2 Displacing natural gas generation with renewables

To efficiently offset the greatest amount of carbon emissions, direct displacement of gas generation with wind and solar electricity is likely to be the best first step. The gas fired electric generation sector is projected to account for substantially more emissions (394 mmtC/yr) than a natural gas transportation system (248 mmtC/yr) which also has more complex fuel infrastructure issues than electric utilities. Employing carbonless sources in the utility sector also circumvents the energy penalties of converting wind or solar electricity to transportation fuel.

This forms the basis of a "utilities first" approach, in which fossil electric generation is successively displaced by renewables and ultimately eliminated. Further emission reductions are accomplished by displacing natural gas transportation fuel with renewable hydrogen, first in automobiles, then freight trucks, and ultimately in aircraft. The corresponding sequence of scenarios is shown in Figure 6.2. The cost breakdowns (Figure 6.4) and energy flows (Figure 6.5) indicate the dramatic shifts in energy supply, storage, emissions, and marginal cost as carbon emissions are reduced, and ultimately eliminated, using the "utilities first" strategy.

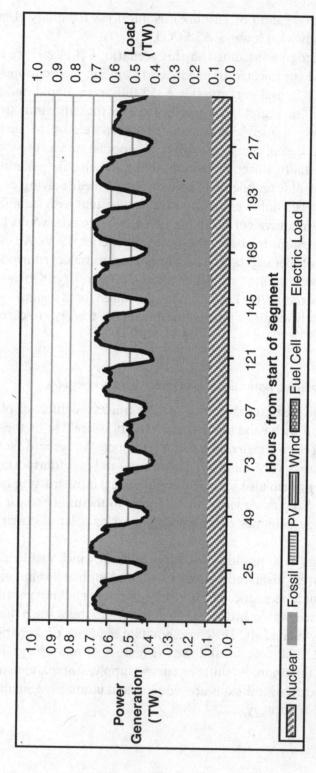

Figure 6.3 A series of dispatch diagrams from the model. These show the effect of changing the systems configuration along the efficient frontier. Each figure shows the dispatch of the electric generating devices over a ten day period starting at the 150th day of the year (end of April) serving the projected US electricity grid in 2020.

(a) Dispatch of an all gas and nuclear utility system. It simply follows the electric load. Note that in this scenario all transportation is fueled by natural gas (not shown).

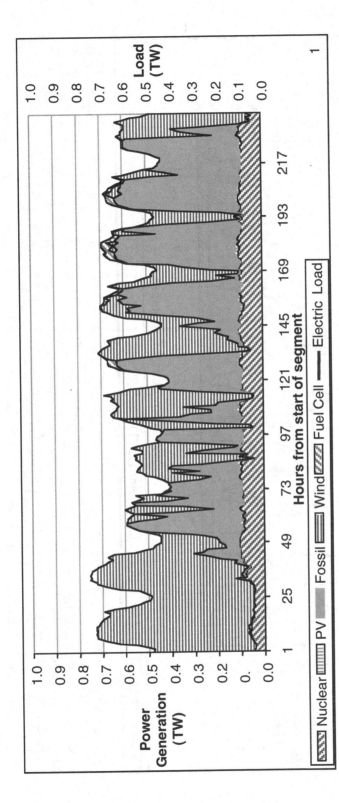

(b) Electric generation system with 0.5 TW of gas capacity blended with wind and a tiny amount of solar generation. This is the least cost system of all the model scenarios. The fuel cell is used occasionally to meet the peaks in the electric load. The wind generation often exceeds the electric demand by a small amount to provide energy needed to run the fuel cell.

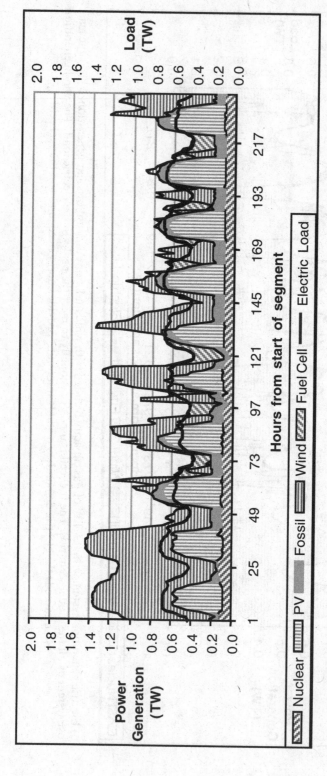

(c) System with 0.1 TW of gas capacity. The fuel cell is called upon more often to serve larger portions of the load on windless evenings and mornings. Excess wind and some weekend solar energy produce the hydrogen needed for the fuel cell. The wind capacity has reached its practical limit of 1.0 TW capacity. Although there is 1.0 TW of wind capacity in this system, it is not always fully dispatched. For example, at around hours 10 and 30 only about 0.8 TW is dispatched even though at these hours there is fully 1.0 TW available. This is because it is not economic to add enough electrolysis, compression, and liquefaction to accept the full output of the wind, PV, and nuclear simultaneously since such high outputs only occur occasionally during the year. At hour 121 the fuel cell displaces the gas generator. This is an imperfection in the algorithm's solution which occurs a few instances per year.

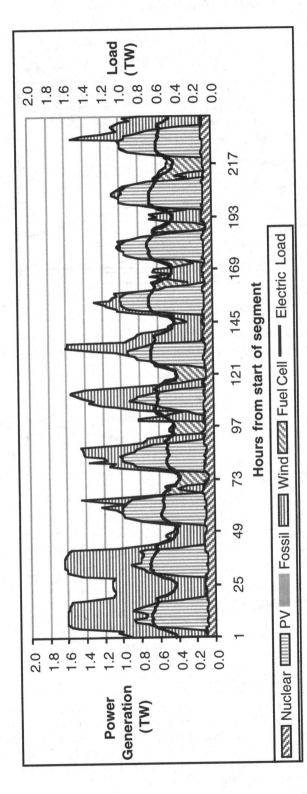

(d) Carbonless energy system serving electric demands and fueling hydrogen cars. Since the wind capacity had reached its limit in the previous case, all additional energy to make energy for the fuel cell and the cars must come from PV.

Figure 6.3 (*cont.*)

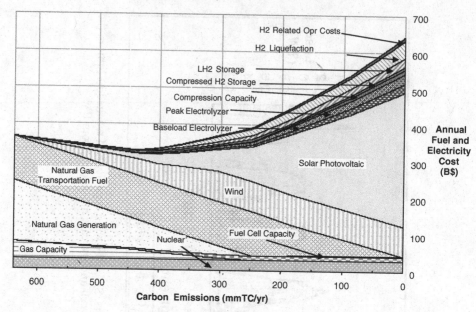

Figure 6.4   Cost breakdown for electricity and transportation systems using the "utilities first" approach to reducing carbon emissions along the efficient frontier in Figure 6.2.

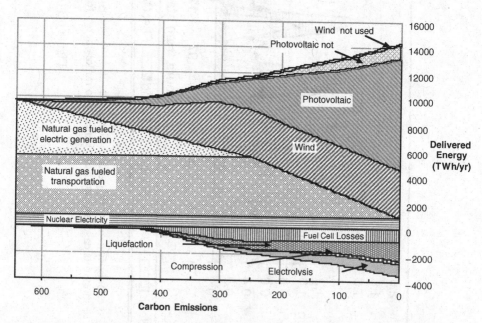

Figure 6.5   Energy flows for electricity and transportation systems using the "utilities first" approach to reducing carbon emissions along the efficient frontier in Figure 6.2. 5000 TWh of electricity and 4600 TWh of transportation fuel are delivered for end-uses. As carbon emissions are reduced, natural gas is displaced from electric generation, then from transportation. Losses are shown for hydrogen production, storage, and reconversion to electricity in fuel cells.

6.2.3 Integrating intermittent renewables

Initially, displacing natural gas generation with low cost wind energy is projected to reduce emissions and system costs (Figure 6.2, 500 GW scenario). Wind capacity can be fully absorbed by the electric grid, except during occasional nighttime hours of very low demand, and very high wind availability. Solar electricity will just begin to be cost effective, even though it is twice the cost of wind per kWh, augmenting generation during periods of higher (daylight) electric demand. Energy storage at this level of renewable penetration is minimal, accounting for roughly 2% of generation (similar to pumped hydro today). This storage essentially serves the same function as pumped hydro, reducing the cost of generating capacity. More than 95% of renewable generation will serve the grid directly, cost-effectively reducing total system emissions to two-thirds of the US 2020 reference case. Intermittent electricity, predominately wind, will account for 50% of the utility mix.

Later scenarios (Figure 6.2, 350 GW scenario) indicate integration issues increase the cost of emission reductions. Displacing more natural gas requires wind capacity levels near daytime peaks, as wind electricity costs much less than solar. At night wind capacity is high relative to demand, and must be stored, displacing natural gas only indirectly (and less efficiently) through fuel cells. As the mismatch between electric demand and wind generation grows, high cost solar electricity will cost-effectively contribute, flattening net electric demand patterns. This enables excess wind to be more cost-effectively used in electrolyzers, producing hydrogen. This hydrogen will be needed by fuel cells to meet peak demands in excess of the gas-fired capacity, accounting for roughly 5% of delivered electricity. Emissions are reduced to 404mmtC/yr at a marginal cost of $150/tonneC.

Reducing gas capacity further (Figure 6.2, 250 GW scenario) limits the ability of fossil generation to compensate for seasonal mismatches between demand and renewable (principally wind) supply. This will substantially increase solar generation, which matches seasonal demand patterns well. In winter, when solar is only partially available, and short term hydrogen storage is not large enough to cover periods between windy days, energy intensive liquid hydrogen will be used extensively. Fuel cell generation will grow to 10% of delivered electricity, being required every windless night and cloudy day. Carbon emissions will fall to 362 mmtC/yr at a marginal cost of approximately $300/tonneC.

Reducing emissions to 305 mmtC/yr (Figure 6.2, 100 GW scenario) will raise the wind capacity to the assumed maximum of 1.0 TW (Schipper and Meyers, 1992; Fetter, 1999). Additional displacement of natural gas will require increasingly costly solar capacity. Reduced gas generating capacity will also lead to greater fuel cell use (18% of delivered electricity). On days when wind and solar generation coincide, hydrogen will be sent to long term hydrogen storage since compressor capacity is exceeded. Liquid hydrogen flow will grow relative to capacity. Relying increasingly on high cost solar generation is the chief factor contributing to a marginal cost of $400/tonneC reducing emissions to 305 mmtC/yr.

Removing natural gas generation entirely from the utility sector (achieving 248 mmtC/yr), will increase solar generation to levels above peak daytime demands, leading to routine storage of excess solar electricity as compressed hydrogen to power late afternoon fuel cell generation. Fuel cell generation will peak at 20% of total electric demand. A totally carbonless utility system will require 7 trillion kWh of carbonless generation to reliably deliver 5 trillion kWh of end-use electricity, eliminating utility emissions at a marginal cost of $400/tonneC (Figure 6.2, 0 GW scenario). Further carbon reductions must then come from the transportation sector, essentially fueled by additional solar power.

6.3 Hydrogen Transportation Technology

US transportation fuel use is projected to reach nearly 4.6 trillion kWh, even if the Partnership for the Next-Generation of Vehicles (PNGV) succeeds in tripling automobile fleet fuel economy by 2020. Light-duty passenger vehicles will then account for approximately 25% of fuel use, aircraft for 35%, and heavy trucks for 40%. Fueling this demand with natural gas will produce direct carbon emissions of 248 mmtC/yr.

It will be easiest for hydrogen to displace natural gas in the light-duty vehicle fleet first. Passenger vehicles are idle 90%+ of the time, with fuel costs accounting for 5–10% of ownership costs. The development of hybrid electric cars and trucks and later fuel cell vehicles makes the prospect of achieving 80 mpg equivalent fuel economy over the entire vehicle fleet quite likely by 2020. This improved fuel economy is the single most important step in making hydrogen fueled vehicles viable, dramatically reducing refueling cost and the size, weight, and cost of onboard fuel storage.

Hydrogen vehicles are expected to require 5 gallons of gasoline equivalent (5 kg hydrogen) for a cruising range of about 400 miles. Hydrogen fuel can be stored onboard using lightweight composite pressure vessels, similar to natural gas vehicles. Other onboard storage approaches include hydrogen absorption and release from high surface area metal powders at moderate pressures, as well as cryogenic liquid hydrogen tanks operating near ambient pressures. Both approaches have been demonstrated in Germany for over a decade. As composite materials have improved in strength and cost, pressure vessel hydrogen storage is becoming more attractive. Another option is to insulate pressure vessels for cryogenic hydrogen service, using compressed hydrogen for routine refueling (perhaps at home or work), and cryogenic liquid hydrogen refueling for occasional long trips. Multi-layer insulation would be sufficient to store liquid hydrogen onboard vehicles even if parked for weeks at a time. Such a hydrogen storage vessel (Aceves *et al.*, 1998) is expected to be relatively compact (100 L), lightweight (100 kg), and low cost ($500–1000).

Hydrogen fueled freight-transportation differs substantially from light-duty passenger vehicles. A typical long-haul tractor trailer truck travels about 100 000 miles/yr, so fuel costs are a much higher proportion of total costs. This high sensitivity to fuel cost is the chief reason truck engines are so efficient today. In the future, fuel cost sensitivity will provide the incentive for compressed hydrogen storage onboard trucks. This can reduce hydrogen fuel cost substantially, as compressed hydrogen is less energy intensive to produce than cryogenic liquid hydrogen. 25–50 kg of hydrogen fuel onboard an 18 wheel tractor trailer would provide a range of 300 miles for a fuel economy equivalent to 6–12 mpg. Onboard storage of this hydrogen (at 5000 psi) will require a volume of about 250–500 gallons. A distinct advantage of hydrogen fueled trucks is the benefit to urban air pollution. Particulate, hydrocarbon, and carbon monoxide emissions would be eliminated, reducing, perhaps obviating, the need for onboard pollution control equipment. Hydrogen fueled tractor trailer trucks have been demonstrated in Japan.

Energy density considerations dictate that carbonless aircraft will have to be fueled with liquid hydrogen. From an emissions and energy perspective, cryogenic liquefaction of hydrogen fuel is not advantageous. Typically, liquefaction is very energy intensive, requiring up to 40% of the fuel energy in the hydrogen. It is also reasonably expensive, liquefaction plants are estimated to cost about $500/kW, while hydrogen compressors can be one-fifth this cost. The two principal advantages of liquid hydrogen are its low weight and low capital cost of large scale storage. The liquid hydrogen infrastructure at Kennedy Space Center is roughly one-tenth the scale required for a large civilian airport.

Liquid hydrogen has one-third the mass of jet fuel for equivalent energy, but nearly four times the volume. These characteristics can substantially reduce takeoff weight for cryogenically fueled aircraft, with the attendant advantages, at the cost of substantial changes in aircraft design. Calculations indicate (Winter and Nitsch, 1988) that hydrogen aircraft would likely use 10% less fuel for subsonic flight and 50% less fuel for supersonic flight than fossil-fueled counterparts. Given the uncertainties surrounding future air travel and to be conservative in comparisons, these potential advantages were neglected in the scenarios involving liquid hydrogen aircraft.

6.4 Displacing Natural Gas from Transportation

Eliminating carbon emissions from electric generation is only a partial step in the transition to carbonless energy systems. Further carbon reductions will require renewable fuel production, most likely hydrogen, given its universal applicability to all sectors of transportation. Fundamental energy balance indicates, however, that the overall costs of reducing emissions from transportation will likely be higher than reducing emissions from the utility sector. A kWh of renewable electricity delivered for end-use displaces at least 1.75 kWh of fossil fuel (natural gas). Alternatively that kWh of electricity would displace only 0.63 – 0.83 kWh of gas use though electrolytic hydrogen substitution in transportation. This basic factor is the underlying reason for the higher cost of carbon reduction in the transportation sector. Using solar electrolytic hydrogen, fueling 250 million passenger vehicles can reduce carbon emissions from 248 to 182 mmtC/yr at a cost of $850/tonneC. Displacing natural gas from freight transportation can reduce emissions to 90 mmtC/yr, increasing marginal cost only slightly to $950/tonneC, due to the lowered capacity factor of hydrogen production equipment driven by the solar patterns availability. Eliminating emissions from transportation completely requires displacing natural gas from aircraft with energy intensive liquid hydrogen, raising costs substantially to $1150/tonneC.

6.4.1 Joint carbon reduction from utilities and transportation

An alternative to the "utilities first" carbon reduction strategy just outlined is to undertake emission reductions from utilities and transportation sectors in tandem. Figure 6.6 shows model scenario results in addition to those in previous cases in which transportation and utility reductions are undertaken together. In the figure sets of points with the same fossil and nuclear capacity (100 GW nuclear in all cases) are shown together as a line. Points on the line indicate the cost and emissions with various hydrogen transportation options (e.g., none, cars only, cars and trucks, or cars, trucks and aircraft). The first point in each line is the case of no hydrogen transportation. These all fall on the efficient frontier. Interestingly, once low levels of fossil generation are achieved, scenarios that include some hydrogen transportation also fall on the efficient frontier of lowest marginal cost, or very close to it. Overall emissions and costs between scenarios are within 5% over this range of carbon emissions.

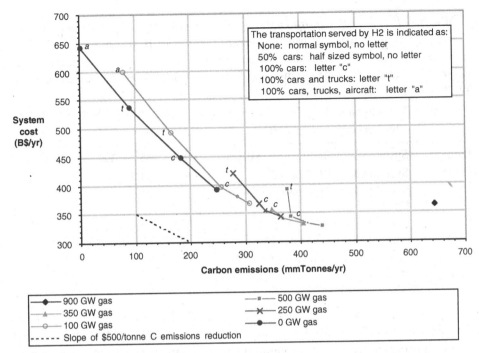

Figure 6.6   Additional scenarios showing the impact of hydrogen fueled transportation to electricity and transportation emissions and costs along the efficient frontier.

This suggests displacing natural gas from the entire (250 million) light-duty vehicle fleet will have costs comparable to reducing natural gas from utility generation when achieving emissions levels below 350 mmtC/yr.

The comparable costs of utility and transportation carbon reductions in this regime are principally driven by the projected higher efficiency of gas-fired combined cycle plants relative to peak power hydrogen fuel cells. Renewable hydrogen in utility energy storage can reduce more emissions by fueling light-duty vehicles than powering fuel cells. From the perspective of efficiently using natural gas, natural gas is more efficient than hydrogen at providing grid electricity while only equally efficient as a transportation fuel. In addition to improving carbon reduction efficiency, the flexibility of combined hydrogen storage and vehicle refueling can improve overall system capital and energy efficiency through higher capacity factors for renewable sources and fossil fuel capacity.

There is substantial value in maintaining approximately 20% of capacity as dispatchable technologies – costs begin to rise steeply as the last 100 GW of gas-fired capacity is eliminated from the utility system (Figure 6.6), especially in cases with hydrogen transportation. Retaining about 20% dispatchable

Table 6.1 *Break-even capital costs (at 5% discount rate) for notional 100% efficient flywheel storage in selected scenarios from Figure 6.6*

System description	Transportation demand served	None	None	None	Cars, trucks
	Fossil capacity (TW)	0.5	0.1	0.0	0.0
	Nuclear capacity (TW)	0.10	0.10	0.10	0.10
	Flywheel capacity (TWh)	10	5	5	5
Analysis results	System cost w/o fly wheel ($B/yr)	326.7	367.7	391.1	536.3
	System cost with flywheel (excluding cost of flywheel itself) ($B/yr)	309.7	341.7	361.9	455.8
	Annual cost savings ($B/yr)	17.1	26.0	29.2	80.5
	Annual cost savings per kWh storage	1.71	5.19	5.84	16.09
	Equivalent capital cost of savings, ($/kWh)	21.3	64.7	72.8	200.6

capacity minimizes the need for energy intensive storage. Strategically, this permits a broader array of complementary emission reduction technologies to be implemented earlier rather than later. Sequestration of modest gas-fired (potentially biomass) generation, in concert with transportation emission reductions from hydrogen fuel, can result in emissions levels (182mmtC/yr) well below those achievable by renewable utilities alone, and probably at lower overall costs than a strictly non-fossil approach to utilities and transportation.

6.5 Alternatives to Hydrogen Energy Storage

In the scenarios discussed earlier only cryogenic and compressed hydrogen energy storage were used, principally due to very low capital costs ($0.30/ kWh for cryogenic liquid hydrogen and $4.50/kWh compressed hydrogen), in spite of roundtrip efficiencies of only 30–40%. Figure 6.5 shows the significant energy losses (~20%) attributable to hydrogen storage and reconversion to electricity. Higher efficiency electricity storage alternatives (e.g. flywheels) could dramatically reduce these losses.

Additional scenarios (Table 6.1) exploring the potential of higher efficiency storage indicate that even 100% efficient electricity storage, an optimistic approximation of flywheels or batteries, would only become cost effective at less than $100/kWh as long as some dispatchable (e.g. natural gas) capacity remains in the electricity mix. Flywheels become more cost effective if fossil electric generation is eliminated entirely, and especially in cases with large solar electricity peaks and hydrogen transportation sectors. Flywheels, or any high efficiency electric storage, approach a value of about \$200/kWh of storage in this most favorable case. These cost levels may be ultimately achievable through higher performance, lower cost fiber-based composite materials. However, such advances in materials may also reduce the costs of composite pressure vessels used to store compressed hydrogen.

6.6 Hydrogen Vehicles as Buffer Energy Storage

In addition to the use of flywheels, energy efficiency can also be improved by using short-term compressed hydrogen storage over energy intensive storage as a cryogenic liquid. Additional short term storage capacity is potentially available within the transportation sector itself. All the scenarios modeled (Figure 6.6) in this chapter essentially take no credit for hydrogen storage onboard vehicles, using instead the conservative assumption that a future hydrogen transportation sector would have patterns of refueling identical to present fossil fueled transportation. This may not necessarily be the case. One intriguing possibility is the potential for energy consumers in a hydrogen transportation sector to adjust refueling and travel patterns, and perhaps end-use efficiency (speed), in response to near term and seasonal hydrogen supply levels. A responsive transportation sector could significantly impact renewable utilities. Fueling hydrogen transportation can account for 20–50% of electricity generation, and the hydrogen onboard vehicles at any given time is 2–4 *times* the amount in short term utility storage. If transportation demand turns out to be flexible, this could then buffer utility storage, improving hydrogen storage efficiency, perhaps ameliorating seasonal requirements for long-term cryogenic storage. Going one step further, some have proposed using fuel cell passenger vehicles *as* backup utility storage, noting that the generation capacity of parked vehicles in the United States would be on the order of *20 times* the projected US generating capacity.

6.7 Strategies for Reducing Carbon Emissions

In conclusion it appears that after end-use efficiency improvements are made, a robust carbon reduction strategy should prepare to follow a dual sector approach, making use of wind, solar, and a small but vital amount of dispatchable electric generation. While hydrogen storage and fuel cell systems can enable carbonless electric generation, hydrogen fueled transportation, especially light-duty vehicles, can also contribute effectively to carbon reduction, especially if consumers choose to impact emissions through more flexible

transportation decisions. Hydrogen transportation complements a modest
sequestration effort, should it be warranted, by essentially shifting transportation emissions to large scale utility generation facilities. A strategy composed
of these elements can cost-effectively eliminate the bulk of carbon emissions
from these sectors, while commercializing the technologies necessary to ultimately eliminate carbon emissions. Energy storage as both cryogenic liquid and
compressed hydrogen are likely to be cost effective, even more so in tandem
with hydrogen vehicles. Very high efficiency utility power storage technologies
(e.g. flywheels) will need to be very low cost (less than $100/kWh) in order to
provide substantial benefit, although the their value is greater in solar dominated energy systems.

Looking beyond the largely linear combination of transportation and utility
sectors explored in these analyses, there may be further synergies to be gained
by carefully integrating energy technology and use patterns, developing a more
complex and interdependent relationship between electricity and hydrogen fuel
production. Excess heat from gas fired generation and perhaps hydrogen fuel
cells could improve the efficiency of steam electrolysis. Pure oxygen, the
byproduct of electrolytic hydrogen, would also enable improved electric generation with natural gas, in fuel cells, if not in combined cycle plants. Even
further ahead, the functions of fuel cell, electrolyzer, and hydrogen compressor may be combined in a single electrochemical device, improving thermal
integration and electric network topology possibilities substantially. The key
technology throughout is that which links electricity and transportation fuel,
efficient hydrogen production by electrolysis.

References

Aceves, S. M. and Berry, G. D., 1998. Onboard storage alternatives for hydrogen
vehicles. *Energy and Fuels*, **12**, No. 1, 49–55.
Andersson, B. and Rade, I., 1998. *Large Scale Electric Vehicle Battery Systems:
Long-Term Material Constraints.* Fourth International Battery Recycling
Congress (IBRC). Hamburg, Germany.
Berry, G., 1996. *Hydrogen as a Transportation Fuel: Costs and Benefits.* Lawrence
Livermore National Laboratory report UCRL-ID-123465.
Bockris, J., 1980. *Energy Options: Real Economics and the Solar-Hydrogen System.*
Australia and New Zealand Book Company, Sydney.
Cox, K. E. and Williamson, K. D., 1977. *Hydrogen: Its Technology and Implications.*
Volumes 1–5. Cleveland Ohio.
De Laquil III, P., Lessley, R. L., Braun, G. W., and Skinrood, A. C., 1990. Solar central
receiver technology improvements over ten years. In *Solar Thermal Technology –
Research Development and Applications.* Hemisphere Publishing Corporation.
Dowling, J., 1991. Hydroelectricity, in *The Energy Sourcebook: a Guide to
Technology, Resources, and Policy*, R. Howes and A. Fainberg (eds.). American
Institute of Physics, New York.

EIA (Energy Information Agency). 2000. *Annual Energy Outlook 2000.* Washington DC.

Fetter, S., 1999. *Climate Change and the Transformation of World Energy Supply.* Center for International Security and Cooperation, Stanford University, Stanford CA.

Gordon, S. P. and Falcone, P. K., 1995. *The Emerging Roles of Energy Storage in a Competitive Power Market: Summary of a DOE Workshop.* Sandia National Laboratories, SAND95–8247.

Halmann, M. and Steinberg, M., 1999. *Greenhouse Gas Carbon Dioxide Mitigation Science and Technology*, Lewis Publishers, Boca Raton.

Hoffert, M., 1998. *Technologies for a Greenhouse Planet*, in S. Hassol and J. Katzenberger (eds.). *Elements of Change*, Aspen Global Change Institute, Aspen CO 81611.

Hogan, W. W. and Weyant, J. P. 1982. Combined energy models, in *Advances in the Economics of Energy and Resources*, **4**, 117–50, JAI Press Inc.

Iannucci, J., Meyer, J. E., Horgan, S. A., and Schoenung, S. M., 1998. *Coupling Renewables via Hydrogen into Utilities: Temporal and Spatial Issues, and Technology Opportunities.* Distributed Utility Associates, Livermore, CA.

Ingersoll, J. G., 1991. Energy storage systems, in *The Energy Sourcebook: a Guide to Technology, Resources, and Policy*, R. Howes and A. Fainberg (eds.), American Institute of Physics, New York.

IPCC (Intergovernmental Panel on Climate Change), 1996. *Technologies, Policies and Measures for Mitigating Climate Change.*

Kelly, H. and Weinberg, C. J., 1993. Utility strategies for using renewables, Chapter 23 in *Renewable Energy: Sources for Fuels and Electricity*, T. B. Johansson *et al.*, eds., Island Press, Washington D.C.

Lamont, A., 1994. *User's guide to the META•Net Economic Modeling System; version 1.2*, Lawrence Livermore National Laboratory, UCRL-ID-122511, 1994.

Ogden, J. and Nitsch, J., 1993. Solar hydrogen, in *Renewable Energy: Sources for Fuels and Electricity*, T. B. Johansson *et al.*, eds., Chapter 22, Island Press, Washington D.C.

Ogden, J. and Williams, R., 1989. *Solar Hydrogen: Moving Beyond Fossil Fuels World*, Resources Institute, Washington D.C.

Post, R. F., Fowler, T. K., and Post, S. F., 1993. A high efficiency electromechanical battery. *Proceedings of the IEEE*, **81**, No. 3, 462–74.

Post, R. F. and Post S. F., 1973. Flywheels, *Scientific American*, **229** No. 6, 17–23.

Schipper, L. and Meyers, S., 1992. *Energy Efficiency and Human Activity: Past Trends and Future Prospects.* Cambridge University Press.

Smil, V., 1998. Impacts of energy use on global biospheric cycle. In S. Hassol and J. Katzenberger (eds.), *Elements of Change*, Aspen Global Change Institute, Aspen CO 81611.

Union of Concerned Scientists, 1992. *America's Energy Choices: Investing in a Strong Economy and a Clean Environment*, Cambridge MA.

Vakil, H. B. and Flock, J. W., 1978. *Closed Loop Chemical Energy Systems for Energy Storage and Transmission (Chemical Heat Pipe)*, Final Report. Power Systems Laboratory, Corporate Research and Development, General Electric Company Schenectady, New York. Performed for US Department of Energy, Division of Energy Storage Systems.

Winter, C. J. and Nitsch, J., 1988. *Hydrogen as an Energy Carrier.* Springer Verlag Berlin.

Appendix 6.A Economic and technical assumptions used in model scenarios (2020 United States electricity and transportation fuel sectors).

Energy demand

Annual end-use electric demand	5.066 trillion kWh
Annual transportation fuel demand	4.589 trillion kWh
Light duty vehicles	1.254 trillion kWh
Freight transport	1.697 trillion kWh
Aircraft	1.638 trillion kWh

Primary energy supply

Transportation delivered natural gas price	$6.93–8.31/GJ
Utility natural gas price	$5.54/GJ
Reference case annual gas consumption	43 EJ/yr
Annual solar capacity factor	29.5%
Annual wind capacity factor	43.3%
Maximum wind capacity	1.0 TW
Nuclear/hydro capacity	0.1 TW

Economic and environmental

Discount rate	5%
Natural gas emission factor (combustion)	54g C/kWht
Fuel cycle emissions (production, compression, leakage)	neglected

Electric generation technology assumptions

Nuclear Fission	$2200/kW
Life	40 years
Operating cost	1 cent/kWh
Combined cycle plant	$600/kW
Life	30 years
Non-fuel operating cost	0.5 cents/kWh
Efficiency	57%
Wind	$655/kW
Life	15 years
Solar photovoltaic	$1500/kWp
Life	30 years
Fuel cells (peak)	$200/kW
Life	20 years
Efficiency (LHV hydrogen)	50%
Electric transmission and distribution	$200/kW
Life	30 years

Hydrogen production and storage technology assumptions

Steam electrolysis (baseload)	$500/kW
Life	20 years
Efficiency (LHV)	91%
Steam electrolysis (peak)	$250/kW
Life	20 years
Efficiency (LHV)	83%
Hydrogen compression	$100/kW
Life	15 years
Efficiency (LHV)	91%
Maintenance	0.1 cent/kWh
Compressed hydrogen storage (5000 psi)	$4.50/kWh
Life	20 years
Hydrogen liquefier	$500/kW
Life	20 years
Efficiency	71%
Maintenance	0.1 cents/kWh
Liquid hydrogen storage (large scale)	$0.30/kWh
Life	20 years

Appendix 6.B *Overview of the model system*

To conduct these studies we modeled an energy system potentially drawing on
a range of resources, both conventional and renewable, and supplying electricity and transportation fuels. The model can be set up to represent an entirely
conventional system, a system that relies entirely on renewable energy, or any
mixture. When modeling a renewable system, the model includes a fuel cell to
cover peak electric demands. In these analyses the system was always
configured to reliably serve the electric load.

The technological parameters (efficiencies and costs) represent a system that
might be developed some decades in the future (see Table 6.2 for cost and performance assumptions). We have used optimistic estimates of the costs and
efficiencies of various technologies including natural gas technologies, since
this should give a more useful picture of the tradeoff between these technologies in a future context.

Figure 6.1 shows a system schematic. The primary resources are wind, solar,
nuclear, and a gas fired turbine. A fuel cell is provided with enough capacity to
cover the electric demand if the wind and solar are not sufficient. This is sized
to just cover the maximum shortfall, not the maximum electric demand. A
flywheel storage option is also included in some of the runs. This can take
electricity from the grid and return it later.

The hydrogen transportation fuel sector is modeled directly within the model. These analyses evaluate various scenarios on the fraction of transportation fuel that is met by hydrogen. It is assumed that the balance of the transportation fuel is provided by natural gas. The costs and emissions from the natural gas fueled transportation are included in final cost and emissions results.

Both compressed hydrogen and liquid hydrogen storage have been included, along with the compression and liquefaction capacity needed to provide their outputs. For given capacities of long and short-term storage, the model determines the rates of filling and discharge for each of the storage devices each hour to minimize the overall cost. The model determines the required level of long-term storage capacity. Short-term storage capacity is varied based on the assumptions about wind and solar capacities and the transportation fuel demand.

Analysis method

Each scenario analyzed in this chapter required the optimization of both the system structure and annual operations under a set of constraints. The system structure is defined by the capacities of each of the components (e.g. generators, electrolyzers, storage, compressors, etc.). The system operations specify exactly how each component will be dispatched hour-by-hour over the year. This includes the allocation of electric demands to the generators each hour and the allocation of hydrogen production between the two electrolyzers. The storage devices must also allocate their purchases of hydrogen over time so as to meet their demands at the lowest cost.

Several constraints must be observed in the solution: the total demands for electricity and hydrogen must be met, and in most cases the maximum capacities of specific generating technologies were constrained. Once the capacity of a given generating unit is set, its production each hour is constrained. For the dispatchable technologies (the natural gas turbine and the nuclear generator) the constraint is constant each hour. For the renewable technologies the constraint each hour is a function of the capacity of the generator and the resource availability in each hour.

This requires a bi-level optimization approach: the structure of the system must be optimized and the operation of the system, given the structure, must be optimized. We have extended the META•Net economic modeling system (Lamont, 1994) to make these analyses. META•Net is a software system that allows the user to structure and solve models of economic systems. The system is modeled as a network of nodes representing end-use demands, conversion processes (such as generation or storage), markets, and resources. The markets represent the points in the system where a total demand (e.g. for electricity) will be allocated among a set of suppliers. META•Net finds a set of allocations each hour that is an economic equilibrium – all the demands are met and each market is in equilibrium.

META•Net finds the equilibrium solution through a series of iterations. Each iteration consists of a down pass and an up pass. On the down pass, the end-use demand nodes (here they are electric demand and demands for hydrogen transportation fuels) pass their demands down to the next nodes. When a node receives a demand it determines how to supply that demand. Market nodes sum up the total demand each period and then allocate that demand to all of its suppliers based on the prices that the suppliers require in that period. Conversion nodes determine the amount of each input they require in order to produce the demand. A conversion node may have several inputs. For example, a hydrogen compressor produces compressed hydrogen from uncompressed hydrogen using both hydrogen and electricity. They then pass the demands for the required inputs down the network to the nodes that provide those inputs.

Eventually the demands are passed down to the resource nodes. At this point the up-pass starts. The resource nodes determine the price (marginal cost) that must be charged in order to meet the demand. They then pass this price back up the network. When a conversion node receives the prices for its inputs, it computes the price required for its output including any capital or other operating costs. All nodes set prices to meet a target rate of return. These prices eventually are passed up to the end-use demand nodes. In general META•Net allows for price sensitive demands, however in this case the demands are fixed each hour. Through a series of iterations the allocations at the market nodes are adjusted until the total quantity demanded each hour is in equilibrium with the prices charged each hour. Further, the allocations within the market node are made until the prices (i.e. marginal costs) from the suppliers are equalized.

The economic equilibrium solution is equivalent to a cost minimization solution. Hogan and Weyant (1982) provide the formal proof of the equivalence between the equilibrium and cost optimization solutions for these sorts of network models. Intuitively the equivalence can be seen as follows: each of the production nodes determine their marginal cost of production each hour given the demands for output that they see. The market nodes allocate demand to each of the suppliers such that the marginal costs of production between the competing suppliers are equalized. The storage nodes are somewhat like market nodes, in that they have a total demand for releases – or "production" – over time. They attempt to minimize costs by purchasing this total amount at the hours when costs are minimum. The algorithm for the storage node also equalizes the marginal costs of purchases over time periods. In fact, correctly modeling the optimal operation of storage devices is very computation intensive. The algorithm used in this version of META•Net is an approximation to the true optimum. Through side calculations we find that the resulting system cost is within a percent or two of the true minimum cost.

Constraints

The renewable technologies have constraints in each hour, reflecting the hourly availability of the resource and the capacity that has been set. Each of the dispatchable technologies also has constraints reflecting the actual capacity available. These constraints are enforced by computing shadow prices whenever the demand to a node exceeds its constraint. The market allocations are actually based on these shadow prices. The shadow prices are adjusted until the constraints are just met.

Capacity modeling

The discussion above describes the optimization of operation and dispatch for each hour of the year. The model also optimizes the capacities of the components. The capacities are adjusted each iteration, but once they are set, they are constant for the entire year. Essentially, the capacities are adjusted until the marginal value of capacity is equal to the marginal cost of additional capacity. Note that we can also place maximum constraints on the capacity for any node.

7

What Nuclear Power Can Accomplish to Reduce CO$_2$ Emissions

7.1 Overview

7.1.1 General considerations

In the US we now emit 11% more CO$_2$ than in 1990; and at Kyoto we promised to reduce CO$_2$ emissions to 8% below 1990 levels in ten years for a decrease of 19% below today's levels. If all the electricity now generated by nuclear power were to be generated by coal, CO$_2$ emissions would increase by another 8%, making it more difficult to meet our commitment if we abandon nuclear power. About 30 years ago Dr. Glenn Seaborg, then Chairman of the US Atomic Energy Commission (AEC), testified to the Joint Committee of Atomic Energy of the US Congress (JCAE) that nuclear power would be comparatively benign environmentally (in particular, not producing appreciable CO$_2$) and also would produce electricity at a modest cost (Seaborg, 1968). This optimism was nationwide and worldwide. Since that time opposition to nuclear power has arisen, and nuclear power at the present moment is not being considered by most governments in the world as an option to meet energy and environmental aims and desires. Our purpose is to show ways in which nuclear power could help the world, and in particular the US, to meet commitments made at Kyoto. Our purpose is also to examine the causes of the changes in the fortunes of nuclear energy and to discuss the extent to which nuclear power can provide electricity worldwide safely and economically. After examining qualitatively possible scenarios for re-establishing equilibrium and ultimately to reverse the present trend in the use of nuclear energy, we end by showing examples of quantitative model results showing how nuclear energy can impact global climate change.

211

Table 7.1   *Cost of nuclear energy in 1971 (Benedict, 1971)*

Description	Coal	Nuclear
Unit investment cost of plant, dollars/kW	$202	$255
Annual capital charge rate per year	0.13	0.13
kilowatt-hours generated per year per kW capacity	5256	5256
Heat rate, million Btu/kWh	0.009	0.0104
Cost of heat from fuel, cents/million Btu	45	18
Cost of electricity, mills/kWh		
Plant investment	5.00	6.31
Operation and maintenance	0.30	0.38
Fuel	4.05	1.87
Total cost	9.35	8.56

7.1.2 Nuclear electricity has been cheap

We will firstly explain that nuclear energy was in the past very competitive with fossil energy sources and presumably could be again. This position is not a matter of optimism brought on by believing results from a model, but is one of accepting historical fact. Twenty-five years ago, Maine Yankee nuclear power plant had just been completed for a *total* cost of $180 million, or $200 per kWe of installed capacity. The Connecticut Yankee nuclear power plant was producing electricity at 0.55 cents per kWeh bussbar cost (i.e., generation cost at the plant boundary, before costs related to transmission and distribution are added), some part of which was needed to pay for the $55 million mortgage. The production cost (primarily fuel) was perhaps only 0.4 cents per kWeh. As reproduced in Table 7.1, thirty years ago, Benedict (1971) estimated average operating costs that were a little lower than this value and capital costs that were about 25% higher than for Maine Yankee. Taking no credit for learning, we could do as well if we could return to the optimism and procedures of thirty years ago. Allowing for inflation, the production cost could be less than 1 cent per kWeh, and, by keeping construction times down, the capital cost could be less than 2 cents per kWeh.

Yet the average operating cost of nuclear plants in the US today is 1.9 cents/kWeh (McCoy, 1998) and for a well-operated plant is still 1.4 (South Texas) to 1.5 (Seabrook) and 1.7 (Palo Verde) cents per kWeh. The construction cost of a new GE reactor is $1690 per kWe being built in Taiwan in about four years, leading to a charge for the capital of about 4 cents per kWeh. These costs are still very high and could be more if construction takes longer than four years.

7.1.3   Reasons for the cost increases

As noted in the last section, nuclear electricity *has been* competitive with electricity generated from other technologies. What has changed? Can it be changed back? Can it be partially changed back? Can nuclear energy be put on a new economic track?

Nuclear advocates expected in 1973 that, as more nuclear power plants were built and operated, both the construction cost and the operating cost would follow the decreases predicted by a "learning curve" (Fisher and Pry, 1971). The reverse has been the case, however; the costs have followed a "forgetting curve" (Wilson, 1992).

Some people argue that the increased cost has been caused by the need for increased safety. The safety of nuclear power in 1973, however, was probably better than for other comparable industrial facilities, has been steadily improved since then, and new designs promise further improvements. It is important to realize that the safety improvements have mostly come from improved analysis – which is (in principle) cheap. A complete probabilistic safety/risk analysis (PRA) for a nuclear power plant costs about $3 million, yet such analysis can pinpoint simple ways of reducing safety concerns. For example, based on the first PRA for the Surry reactor the mere doubling of an isolation valve reduced the calculated frequency of core melting by a factor of 3 for a cost of only ~$50k.

We have seen *no* careful study of how much improvement in safety margins has increased cost. Indeed, in 1984 when the Energy Engineering Board of the US National Academy of Sciences proposed a study of the subject; it was opposed by the utility industry, perhaps for fear of adversely influencing prudency hearings that were in progress before public utility commissions. Public utility commissions of several states refuse to allow cost overruns for nuclear power plants to be included in the rate base, and thereby to be recovered; a suspicion arises that public knowledge of the reasons for cost overruns could affect these "prudency" hearings.

Various reasons for the increase in the cost of nuclear energy include the following:

- In 1970 manufacturers built turnkey plants or otherwise sold cheap reactors as loss leaders, but turnkey operations can only account for a small proportion of the capital cost.
- Construction costs generally have risen since 1970 even when corrected for inflation.
- It may be that in 1972 we had good management and good technical people; but why has management got worse when that has not been true for other technologies?

- Operating costs rose rapidly in the 1970s because the rate of expansion of nuclear energy exceeded the rate of training of good personnel.
- A sudden rise in costs came in the late 1970s after the accident at Three Mile Island Unit II.
- Although mandated retrofits have been blamed for cost increases, this applies to existing plants and not to new construction.

Most people seem to agree that the principal *present* limitation in nuclear power development is related to diminished public acceptance of the technology. Decreased confidence and increased risk aversion drives excessive regulation, and excessive regulation in turn increases the cost. As noted above, this increased cost often reaches a factor of three even after correction for inflation. It is highly likely that nuclear power plants are safer today than they were in 1972. It would be hard to argue, however, that the actual safety improvements that have occurred have been the cause of the threefold increase in cost. Most improvements have resulted from more careful thought, using such approaches as event-tree analysis, but without excessive hardware expense.

Many people have suggested that the problem is that the regulation is more than needed for adequate safety, and this over-regulation increases the cost (Towers Perrin, 1994). In particular, many claim that regulation is too prescriptive and not based upon performance. A few of the arguments related to over-prescriptive regulations are as follows:

- The response to many regulations is to increase staff. The staff numbers at the Dresden-II power plant went from 250 in 1975 to over 1300 today (Behnke, 1997). This increased staffing costs money – 0.8 cents per kWeh, and it is far from clear that adding personnel improves safety.
- Shut downs (always costly) for failure to meet technical specifications occur even when the technical specifications have little effect upon safety.
- Any delay in licensing can seriously increase the capital cost, as interest payments incurred during construction accrue.
- A demand for safety-grade equipment in parts of the plant that have little impact on safety are expensive.

The problem is not unique to the US. In the UK the Atomic Energy Authority had to spend a lot of money making the Thorpe plant for reprocessing spent nuclear fuel as earthquake proof as an operating reactor; yet the inventory of dangerous material in a processing plant is far less than in a reactor, and the danger of re-criticality is remote (Hill, 1997).

In another paper (Wilson, 1999) and in Congressional testimony (Wilson, 1998a,b) one of us addressed the problem of excessive regulation; reasons why it inevitably appears and what can be done to avoid the problem were

Table 7.2 *Uranium supplies (Benedict, 1971); resource data plotted on Fig. 7.1b*

Uranium price $/lb U$_3O_8$	Resource tonnes	Cost increase, mills/kWeh		Electricity generated GWe yr	
		LWR	Breeder	LWR	Breeder
8 (base)	594 000	0.0	0.0	3470	460 000
10	940 000	0.1	0.0	5500	720 000
15	1 450 000	0.4	0.0	8480	1 120 000
30	2 240 000	1.3	0.0	13 100	1 720 000
50	10 000 000	2.5	0.0	58 300	7 700 000
100	25 000 000	5.5	0.0	146 000	19 200 000

addressed. The Chairman of the Nuclear Regulatory Commission recently addressed this question (Jackson, 1998) and emphasized this area as a vital area of research and subsequent implementation. This issue is discussed further in Section 7.4.1.2.

7.1.4 Are uranium-fuel supplies sufficient?

Various opponents of nuclear power have argued that the uranium fuel supply is insufficient to make it worthwhile to face the problems (whatever they may be) with nuclear energy. We show here that this is false. Thirty years ago Benedict (1971) reported that we had 20 million tonnes of uranium at prices up to $100 per pound of U$_3O_8$ (Table 7.2); the higher prices only raised the operating cost in an LWR by 0.5 cents per kWeh, which, although an excessive increase a quarter of a century ago, would now be considered acceptable. The total quantity of uranium resources (column 2 of Table 7.2) does not seem to have changed; subsequent columns of Table 7.2 are merely physical calculations from the first two columns. Therefore, Table 7.2 is as accurate today as it was 30 years ago. The Uranium Institute reported in 1998 that we have about 18 million tonnes of uranium in ore, proven reserves, reasonably assured supplies and possible supplies at prices up to $200 per kgU, as is depicted for the variously defined categories categories in Fig. 7.1, which includes the early estimates reported by Benedict (1971). We can afford appreciably increasing the initial fuel cost without significantly increasing the bussbar cost of electricity. This would produce in a light-water-reactor (LWR) system about 4×10^{15} kWeh (4.6×10^5 GWeyr) of electricity, or enough for over a century at the postulated year 2030 demand of 2500 GWeyr/yr.

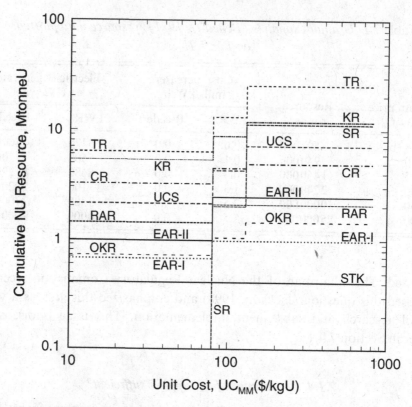

Figure 7.1a Uranium resources *versus* cost expressed in nine categories: STK = reported stocks; RAR = reasonably assured resources; EAR-I = estimated additional resources; OKR = other known resources; UCS = unconventional resources; EAR-II = estimated additional resources; SR = speculative resources (OECD, 1996).

A general rule about prices of any commercial fuels seems to have evolved. The time for depletion of reserves has stayed between 15 to 30 years for nearly a century! (Coal reserves are an exception.) If enough fuel exists for 30 years, little incentive can be generated for exploration, but, if the amount falls below 15 years, the profit motive ensures that exploration restarts. The present and anticipated use of nuclear power provides little incentive to explore uranium. Allowing for a price increase of 0.5 cents per kWeh, it appears we could have a future for nuclear power for 50 years without a breeder reactor (based either on a uranium or a thorium cycle), and possibly for many more years (Wilson, 2000). After perhaps half a century it would be wise to be ready to use alternative fission fuel cycles. The use of a thorium-based fuel cycle in light water reactors might postpone the need for a plutonium breeder reactor using fast neutrons. These alternative reactors and nuclear fuel cycles are discussed in 7.4.2.3. All in all, a factor of 1000 increase in effective fuel supply seems not

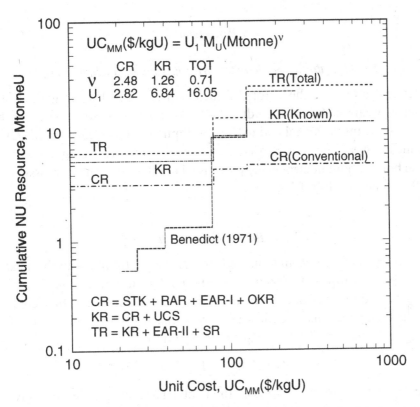

Figure 7.1b Uranium resources *versus* cost (OECD, Benedict) showing uranium resource cost models used for Conventional Resource (CR = STK + RAR + EAR-I + OKR), Known Resources (KR = CR + UCS), and Total Resources (TR = KR + EAR-II + SR) assumptions (OECD, 1996).

unreasonable. It would be impudent to project the existence of the human race beyond the 100 000 years implied by these factors.

7.1.5 Are public concerns real?

The public has expressed a number of concerns related to nuclear energy. These concerns are addressed in Section 7.3. We note here several general features of the public concerns that seem relevant to us. Often a concern is expressed that can be shown to be technically unwarranted or exaggerated. Yet the concern persists even after the truth is accepted; perceptions persist. This situation indicates to us that the statements of the anti-nuclear-energy opposition, while based on these perceptions, may not be a real indicator of the concern. Nevertheless, right or wrong, perceptions have consequences, and hence must be taken very seriously.

7.2 Background

7.2.1 How did we get here? A brief history of nuclear energy

Several descriptions of the history of nuclear energy clearly overlap the history of nuclear weapons (Goldschmidt, 1982). In Appendix A we show a chronology of important events related to the development of nuclear fission, starting with nuclear weaponry and proceeding to civilian nuclear energy. In addition, general books about nuclear energy discuss the history and the basic issues very clearly (Bodansky, 1997).

7.2.2 Basic physics of fission

An atomic nucleus of atomic mass A and charge Z (atomic number) can be thought of as composed of Z protons and $(A - Z)$ neutrons. These constituents are bound into the nucleus by the attractive nuclear forces. In accordance with Einstein's famous equation $(E = mc^2)$ this leads to a mass defect (or binding energy/c^2), ΔM, according to the equation:

$$M(A,Z) = Z\,M_p + (A - Z)\,M_n - \Delta M,$$

where M_n and M_p are the masses of the neutron and proton, respectively.

The mass defect per nucleon, $\Delta M/A$, is plotted in Fig. 7.2. The initial increase in $\Delta M/A$ as A increases is a result of nuclear forces, and the decreases of the per-nucleon mass defect as A becomes larger is a result of the repulsion of the Coulomb forces between the protons. From this curve it may be seen that if two light nuclei combine to form a heavier one (i.e., nuclear *fusion*), energy is released, but above the common element iron ($A = 56$) energy is released if a heavy nucleus is split into two lighter fractions (e.g., nuclear *fission*). It is also important that light nuclei usually have equal numbers of protons and neutrons, whereas heavy nuclei have a considerable neutron excess (uranium 238 has 92 protons and 146 neutrons). In this chapter we are concerned with fission. Heavy elements can be stable because the attractive nuclear forces are strong at small distances (10^{-13}cm), whereas the repulsive Coulomb forces act over much greater distances.

These basic features of nuclear masses were known at least as far back as the time of the early mass measurements (Aston, 1919), but it remained a puzzle how to unlock the energy available in the atom. The secret was discovered by Hahn and Strassman (1939), deliberately brought out of Nazi Germany by Lise Meitner, and theoretically described by Bohr and Wheeler (1939). When uranium is bombarded by slow neutrons (e.g., neutrons with energies close to

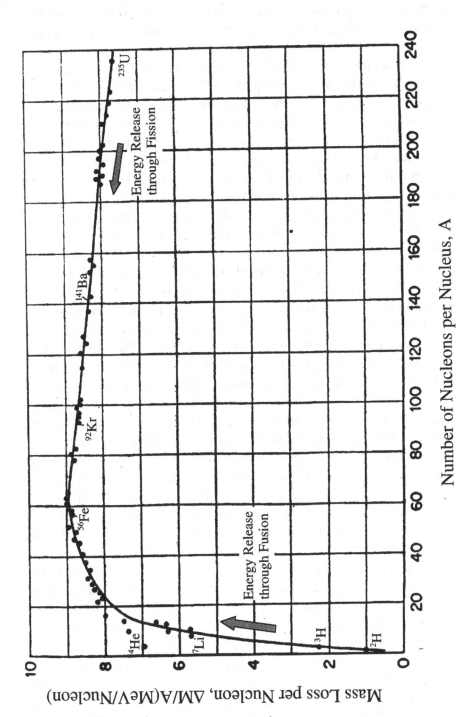

Figure 7.2   Binding energy (mass deficit) *versus* atomic mass showing fusion ($^2H + {^3H} \rightarrow {^4He} + n + 17.6$ MeV) and fission ($^{235}U +$ $n \rightarrow {^{141}Ba} + {^{92}Kr} + n + 205$ MeV) approaches to increased nuclear stability and attendant energy releases.

ambient thermal energy), this "thermal" neutron is absorbed into the nucleus and a compound nucleus is formed (in an excited state). If the compound nucleus has an atomic number divisible by four (an "even-even" nucleus), it will immediately split into two fragments of unequal mass. In the process, as first shown by Joliot, von Halban and Kowarski (1939), additional prompt free neutrons are produced (on average about 2.5 neutrons per fission). This release of extra (energetic) neutrons leads to the possibility of a chain reaction where at least one of the produced neutrons goes on to produce another fission, which in turn produces more neutrons. Approximately 81% of the energy produced – three million times as much per unit mass as in the burning of carbon – is in the kinetic energy of the fission fragments[1].

Of the naturally occurring elements, uranium of atomic mass 235 is unique in satisfying the requirement of Bohr and Wheeler (e.g., 236 is divisible by 4). Uranium 235 is only present to an extent of 0.71% in normal ores (99.3% is uranium 238 which is not fissionable by slow neutrons), so not all the nuclear energy in the uranium can be released. But other elements can be produced by capture of a neutron by the nucleus. In particular, when uranium 238 (^{238}U) captures a neutron it becomes uranium 239; successive radioactive (beta) decay leads to neptunium 239 (^{239}Np) and plutonium 239 (^{239}Pu, discovered by Seaborg and collaborators in 1940). A further neutron capture (by plutonium 239) leads to the excited nucleus plutonium 240 which, having an atomic number divisible by four, immediately splits. This sequence describes the process of breeding a "fissile" (meaning fissionable by slow neutrons) element plutonium 239 from the "fertile" element uranium 238, which is 141 times more plentiful than uranium 235. Further neutron capture leads to heavier nuclei which are unstable in nature and more easily fissioned, as is shown in Fig. 7.3a. In addition to uranium 235 being fissile, uranium 233 (also discovered by Seaborg) is fissile. While not present in nature, it can be bred from neutron capture on thorium 232 (^{232}Th), which is four times more plentiful than ^{238}U, as is shown in Fig. 7.3b.[2]

The important fissile elements for nuclear energy are ^{233}U, ^{235}U, ^{239}Pu and to a lesser extent the heavier transuranic elements (most plutonium isotopes, americium isotopes, curium isotopes, etc.). Fission can occur with either slow or fast neutrons. The probability of capture of a neutron by a nucleus, however,

---

[1] Typically, of the 210 MeV (3.2×10^{-11} J/fission, 19.3 TJ/mole, or 82.1 TJ/kg ^{235}U) released per fission, 81.0% appears as fission-product kinetic energy, 6.8% as fission-product decay, 3.9% as prompt gamma radiation, and the remaining 8.3% as neutrinos.

[2] ^{232}U is formed from the beta decay of ^{232}Pa, which in turn derives from (n, 2n) reactions on the fertile material ^{232}Th followed by a beta decay of the resulting ^{231}Th; the ^{232}U contributes significant alpha activity (heating), and from the 1.91-yr ^{228}Th daughter product, which leads to a number of short-lived daughters, some (^{212}Bi and ^{208}Tl) emitting very energetic gamma radiations.

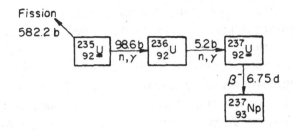

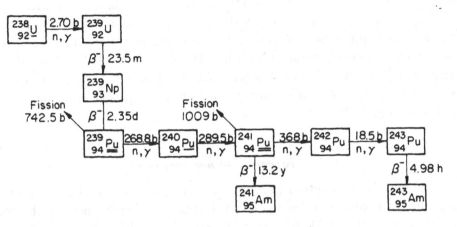

Figure 7.3a   Principal nuclear reactions and transitions in uranium, showing reaction
cross-sections (1 barn $= 10^{-28}$m^2) and transition half-lives.

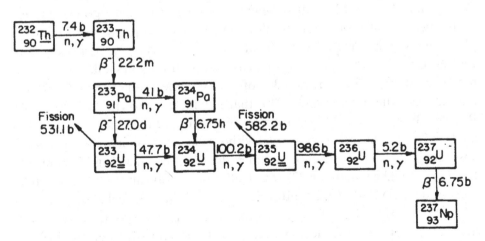

Figure 7.3b   Principal nuclear reactions and transitions in thorium, showing reaction
cross-sections (1 barn $= 10^{-28}$m^2) and transition half-lives.

is proportional to the time the neutron is within the range of the short-range nuclear forces (e.g., proportional to the inverse of the neutron speed, $1/v$) and, therefore, is greater for slow neutrons. It is, therefore, easier to make a chain reaction with slow neutrons, although it is well known that a chain reaction is possible with fast neutrons (as in a nuclear bomb). Nuclear reactors based on fissions from these slow or thermal neutrons require a "moderator" to slow down the energetic neutrons released from fission through collisions with the nuclei of the moderator. A desirable moderator will have light nuclei for slowing down (more energy lost by the neutron on average per collision), as well as a low neutron capture cross section (e.g., reduced "parasitic" loss of the valuable neutron to the moderator nuclei). This choice of one of the three fissile elements (^{233}U, ^{235}U, ^{239}Pu) combined with the choice of moderator gives a matrix of possible nuclear-reactor configurations discussed in Section 7.2.3 and elaborated in Table 7.3.

7.2.3 *Taxonomy of nuclear reactors*

The balance of the neutron production (through fission) and consumption (e.g., all absorptions plus leakage from the reactor core) is crucial to the sustainment of a controlled chain reaction for times sufficient to allow the safe extraction of an economic amount of energy (measured usually in units of GWtd) from a given mass of fuel (heavy metal, HM). For the U-Pu fission cycle depicted in Fig. 7.3a, ^{235}U is consumed, plutonium isotopes increase, and fission products build up. Figure 7.4 illustrates these changing nuclide concentrations in a large pressurized-water-reactor (PWR, a kind of LWR; Benedict, Pigford, and Levi, 1981), with time expressed in terms of integrated thermal energy per unit of initial fuel, GWd/tonneHM. As time progresses, fissions in plutonium isotopes increase, and neutron absorptions in the increasing concentrations of fission products increase. Typically (Marshall, 1983), each fission produces 2.6 neutrons, 63% of which come from ^{235}U and 32% are created from the fissioning of ^{239}Pu that is building into the fuel (Fig. 7.4). Roughly 39% of these 2.6 neutrons go on to produce fissions in the fuel, and another 39% are absorbed in the fuel without causing fission; of these 58% are absorbed in ^{238}U, 15% are captured in ^{239}Pu, and the remaining 15% are captured in ^{235}U. The remaining 22% of "lost" neutrons are absorbed in structure, coolant, fuel cladding, and control rods (76% of the non-fuel absorptions, or 17% of the original 2.6 fission neutrons), with fission-product absorptions accounting for 24% of the remaining "lost" neutrons [5% of the original ($^{235}U + ^{239}Pu$) fission neutrons]. Typically, leakage out of the reactor core *per se* is much less than a percent of the fission neutrons produced.

Achieving the correct mixture of fuel and moderator to assure an exact balance between neutron production and consumption over reasonable times of steady-state fission power production would be very difficult. Control of this desired steady state would be impossible for either thermal- or fast-spectrum reactors were it not for the delay in a small fraction of the fission-neutron emissions from certain fission products having to undergo beta decay before shedding an excess neutron. This delay ranges from a fraction of a second to nearly a minute, with an average delay being in the range 10–20 seconds. While the percentage of all fission neutrons that are in this delayed-neutron class vary with fissile isotope (0.28% for ^{233}U, 0.64% for ^{235}U, 0.21% for ^{239}Pu), the time scale for control of the neutron population in a critical reactor is increased from microseconds to milliseconds, which is in the regime of most mechanical control systems (e.g., neutron-absorbing control rods moved into and out of the reactor core).

On the basis of the foregoing description, a nuclear fission reactor requires four primary systems to provide safe and economic thermal-power generation while simultaneously satisfying the constraints imposed by the above-described neutron balance:

- The fuel form or fuel matrix;
- Moderator (low atomic mass) material that efficiently (few collisions required) slows down energetic fission neutrons;
- Structure within/surrounding the core (including fuel cladding); and
- Heat transfer/transport medium (coolant).

Additionally, while not directly impacting the neutron balance and the physics of the core, the system that converts the thermal energy released (mainly as fission-product kinetic energy) from particle kinetic energy to electrical energy on the grid is an essential element of the overall reactor system. Integral to the nuclear power plant, but not considered here, are all operational and off-normal safety and control systems. Within the neutronic constraints and limitations described above can be fitted a wide range of material options for the fuel, moderator (if needed), structure, and coolant. Material systems that have been considered and implemented in the past are listed in Table 7.3. Included in this table are typical materials with high (thermal) neutron absorption cross section used to control the neutron population within the reactor core by insertion into the coolant, into the fuel, or into separately manipulated control rods. Hence, a rich array of material combinations is available to define the economics, safety, and sustainability of nuclear energy (Todreas, 1993).

By far the most common reactor system in use today around the world is based on a fuel composed of fissile uranium (or plutonium) oxide, with both the

Table 7.3 *Materials options matrix for nuclear fission energy*

Fuel[a]	Fuel forms[d]	Control rods	Moderator[e]	Structure[g]	Coolant[h]	Conversion[i]
Fissile	Metal alloy	Boron (^{10}B)	Beryllium	Steels	Helium	Steam turbine[k]
• ^{235}U	Oxides	Cadmium	Light water	• Ferritic	Carbon dioxide	Gas turbine[l]
• ^{239}Pu[b]	Carbides	Europium	Heavy water	• Austenitic	Light water	MHD[m]
• ^{233}U[c]	Nitrides	Gadolinium	Graphite	Silicon carbide	Heavy water	NPL[n]
Fertile[a]	Cermets		none[f]	Concrete	Sodium	Chemical[o]
• ^{238}U	Dispersions			Fuel cladding	Lead eutectic	
• ^{232}Th	Liquids			• Zirconium alloys	Other liquid metals	
	• Aqueous slurry			• Stainless steels	• Potassium	
	• Molten salt			• Graphite/carbides	• Mercury	
	Gaseous (UF$_6$)				• Tin	
					• Lead	
					Molten salts	
					• FLiBe	
					• KCl/LiCl	
					Organic liquids	

Notes:

[a] A small fraction of the fertile isotopes ^{238}U and ^{232}Th undergo fission upon absorption of an energetic neutron.

[b] Plus ^{241}Pu in a thermal neutron spectrum; all plutonium isotopes fission in an energetic neutron spectrum; ^{239}Pu formed upon neutron absorption in ^{238}U according to Fig. 7.3.

[c] Formed upon neutron absorption in ^{232}Th according to Fig. 7.3.

[d] Solid fuel metrics clad in zirconium-based alloy or stainless steel.

[e] For decreasing neutron energy from few MeV at birth to near "room temperature".

f For fast or epithermal spectrum reactors, or for thermal systems with ^{235}U enrichments about 20%.

g Cladded fuel-pin grids, in-core support structure, pressure vessel, coolant piping.

h Coolant and moderator functions can be combined in some cases (e.g., light water); other coolants that have been considered include heavy water (D_2O), liquid metals (Na, NaK, Pb), and pressurized gases (CO_2, He).

i Refers primarily to thermal-to-electric conversion; non-electric applications envisage the delivery of high-temperature (800–900°C) process heat by means of a liquid-metal (sodium, sodium-potassium eutectic, lead or lead-bismuth eutectic; the use of chemical storage and transfer of nuclear energy has also been suggested.

k Largely the Rankine thermodynamic cycle; steam generation either directly within the reactor core or through a secondary coolant represent further sub-options.

l Largely the Brayton thermodynamic cycle; as with the steam-based conversion, either direct drive or the use of a secondary working fluid define sub-options.

m Magnetohydrodynamic conversion.

n Nuclear pumped laser based on direct fission-product energy conversion.

o For example, (endothermic) methane reforming at reactor site; transport of the $CO/H_2/CO_2$ synthesis gas mixture to a utility/user site; followed by (exoergic) re-menthanation and use of released chemical energy, and either reinjection or recirculation of the substitute natural gas.

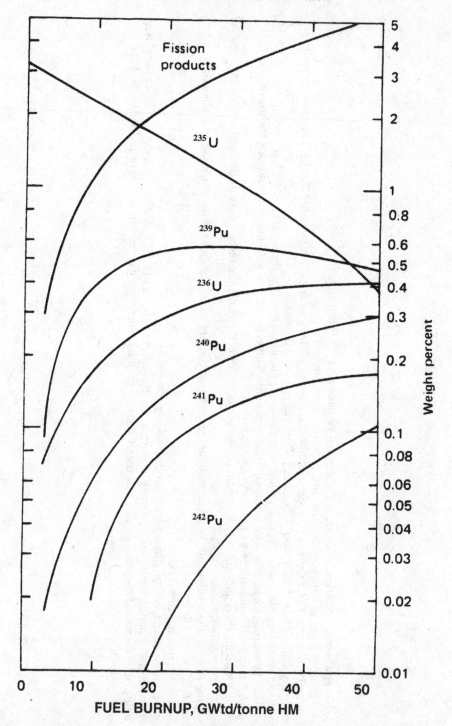

Figure 7.4 Change in fractional nuclide concentrations with burn-up for a 1060–MWe PWR, showing the depletion of the ^{235}U driver fuel and the buildup of both fission products and plutonium isotopes (Benedict *et al.*, 1981).

neutron-moderating and the cooling functions provided by ordinary (light) water, and with electricity generated by high-pressure steam driving a turbine-generator system. Approximately 76% of all commercial nuclear power plants (NPPs) are of this light-water reactor (LWR) kind [55% pressurized-water reactors (PWRs) and 21% boiling-water reactors (BWRs)]. Heavy-water-moderated/ light-water-cooled reactors operated on natural (unenriched in the ^{235}U isotope) uranium, and gas-cooled (helium)/graphite-moderated reactors represent important, but minor, contributors to the present mix of world nuclear power plants. Finally, a number of small reactors operated without strongly neutron-moderating (neutron thermalizing) materials and, therefore, sustained by fast or energetic neutrons, using liquid-metal (sodium) coolants, are being operated to investigate the physics and technology of such systems that might ultimately be needed to exploit the much larger ^{238}U fertile-fuel resource *via* the breeding of plutonium fissile fuel (Fig. 7.3a). Recently (Galperin *et al.*, 1997; Murogov *et al.*, 1995), interest is developing in exploiting the ^{232}Th–^{233}U breeding fuel cycle (Fig. 7.3b) in thermal-neutron LWRs; this interest is driven by desires to address both proliferation, resource, and waste concerns by building on the well-developed thermal-spectrum LWR technology. It has been claimed (West, 2000) that a fast-neutron reactor can also be used to produce ^{233}U uranium fuel that is as proliferation resistant as any LWR fuel.

As is seen from the chronology given in Appendix 7.A, the evolution of the somewhat hegemonic spectrum of commercial NPPs described above resulted largely from the needs and economics associated with a dual-use approach to military and peaceful applications of nuclear energy. Furthermore, the development and deployment of both nuclear power plants and the all-important "back-end" [e.g., used-fuel storage, reprocessing (if at all), and waste disposal] has been and continues to be subject to concerns and constraints related to the proliferation of nuclear-weapon materials to presently non-nuclear-weapon states. These concerns and constraints prevail even though nuclear-explosive materials have yet to be obtained for that dark purpose from the civilian nuclear fuel cycle (Meyer, 1984).

Appendix 7.B (Nuclear News, 1992) summarizes in table form contemporary reactor design and performance parameters for: pressurized-water reactors (PWR), boiling-water reactors (BWR), modular high-temperature gas-cooled reactors (MHTGR), Canadian heavy-water-moderated (D_2O) natural-uranium reactor (CANDU), and the advanced liquid-metal-cooled fast-spectrum reactor (ALMR). Prototypical diagrams for each of these four NPPs are given in Figs. 7.5–7.9, with narrative descriptions (Benedict *et al.*, 1981; Marshall, 1983; Cochran and Tsoufanidis, 1990) being given in the respective captions and associated endnotes. Similarly, Figs. 7.10–7.12 give "top-level" material flows fuel cycles for LWRs, MHTGRs, and LMFBRs,

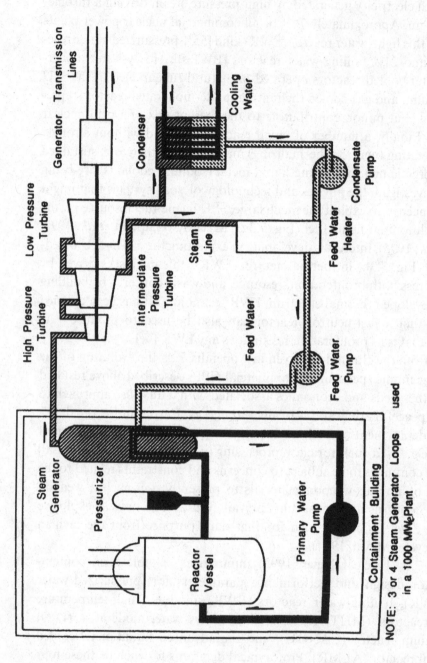

Figure 7.5   Schematic diagram of a pressurized-water reactor (PWR).[3]

---

[3] The PWR core is fueled with low-enriched (LEU, 3.5% ^{235}U) uranium-oxide (UO$_2$) pellets that are contained (clad) in zirconium-alloy tubes, through which the majority of the fission energy is conducted to the flowing pressurized-water coolant/moderator. The water is thereby heated as it passes through the reactor core and flows out of the core to a steam generator, where heat is exchanged to create steam in this secondary coolant loop; the steam is used to drive a turbine-generator unit to make electricity. The primary-coolant system (core, pressure vessel, primary-coolant pump, steam generators, etc.) are housed in a primary containment building; the fuel, cladding, pressure vessel, and containment building present sequential barriers to loss of radioactive materials to the environment.

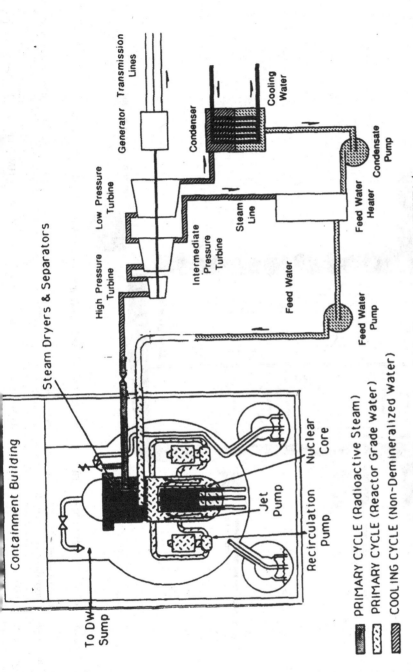

Figure 7.6   Schematic diagram of a boiling-water reactor (BWR).[4]

■ PRIMARY CYCLE (Radioactive Steam)
▨ PRIMARY CYCLE (Reactor Grade Water)
▨ COOLING CYCLE (Non-Demineralized Water)

4 Like the PWR, the BWR core is fueled with low-enriched (LEU, 3.5% ^{235}U) uranium oxide (UO$_2$) pellets that are contained (clad) in zirconium-alloy tubes, through which the majority of the fission energy is conducted to the flowing pressurized-water coolant/moderator. The water is thereby heated as it passes through the reactor core. Unlike the PWR, water boils as it passes upwards through the core and turns into steam within the primary pressure vessel. This steam is passed directly to the turbine-generator, condensed, and sent back to the reactor core by means of feedwater pumps. The primary-coolant system (core, pressure vessel) is housed in a primary containment building; the fuel, cladding, pressure vessel, and containment building present sequential barriers to loss of radioactive materials to the environment.

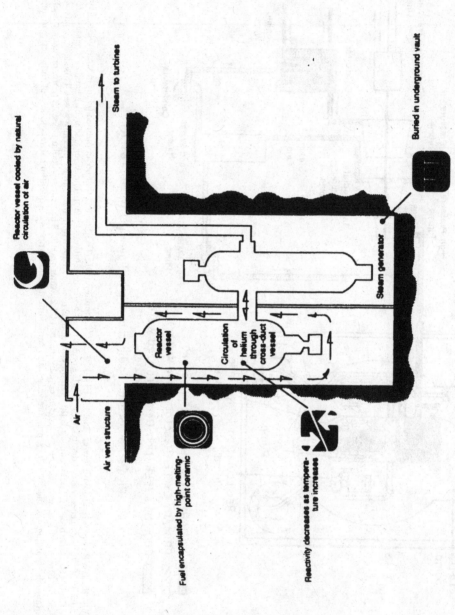

Reactor vessel cooled by natural circulation of air

Air vent structure

Air

Steam to turbines

Reactor vessel

Circulation of helium through cross-duct vessel

Steam generator

Buried in underground vault

Fuel encapsulated by high-melting-point ceramic

Reactivity decreases as temperature increases

Figure 7.7 Schematic diagram of a modular high-temperature gas-cooled reactor (MHTGR).[5]

[5] Although the HTGR has not been widely exploited commercially, its high level of inherent safety to coolant loss [large thermal mass, high-temperature (refractory) materials, and high-temperature helium, capable of direct cycle to a gas turbine], its high fuel efficiency when operated on a thorium fuel cycle, and its ability to produce process heat for non-electric application all combine to make the MHTGR an important concept in the long-term. Some of these benefits accrue at low power density and, hence, higher capital cost, which must be traded carefully with the very high fuel burn-up capability (>100 GWtd/tonneHM) and higher thermal-to-electric conversion efficiencies offered by this concept.

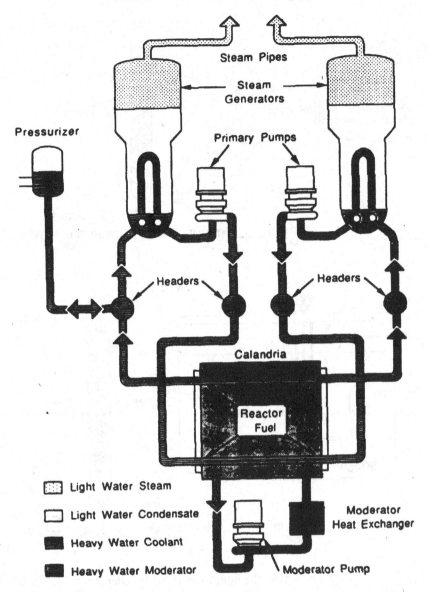

Figure 7.8 Schematic diagram of the Canadian deuterium natural uranium reactor (CANDU).[6]

---

[6] The CANDU uses natural (unenriched) uranium in oxide form (UO_2) as a fuel, and heavy water (D_2O) as a coolant. This combination of moderator (deuterium is a more efficient moderator than hydrogen when neutron-absorption properties are taken into account) and coolant improves the overall neutron balance to the extent that unenriched uranium can be used. The pressurized-water coolant and associated steam generators are essentially the same as that used in a PWR. The relatively low power density in the CANDU reactor core allows for refueling at full power ("on-line" refueling), thereby giving very high plant availability factors (>90%), and compensating for the (capital) cost penalties of operating a lower power-density system.

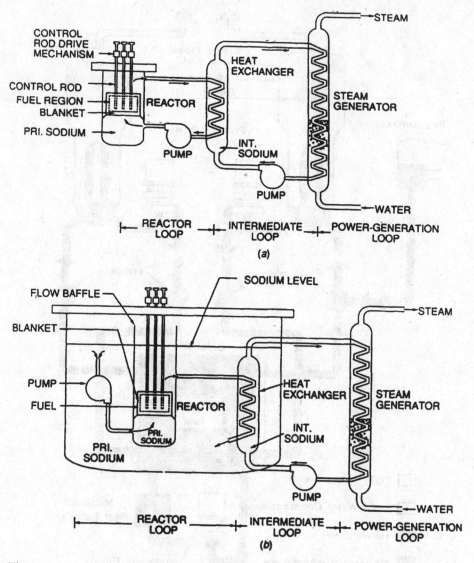

Figure 7.9   Schematic diagram of the liquid-metal fast breeder reactor (LMFBR)[7]

---

[7] Both a loop and a pool design for the LMFBR are shown. A moderator *per se* is not required for this fast-neutron sodium-cooled reactor; metallic, oxide or carbide fuels have been considered. The reactor is comprised of three main regions containing either stainless-steel-clad fissile (initially ^{235}U, later ^{239}Pu) or fertile (^{238}U) fuels – the core, the axial blanket, and the radial blanket. The uranium matrix in all three regions is comprised of depleted uranium from enrichment-plant tailings (i.e., ^{235}U content is below the 0.71% value found in natural uranium). The liquid-sodium coolant heats (and is activated by neutrons) as it passes through the core, and transfers this heat to a secondary coolant, which is not radioactive. It is then directed to a steam generator, where the steam so created is used to drive a turbine-generator unit.

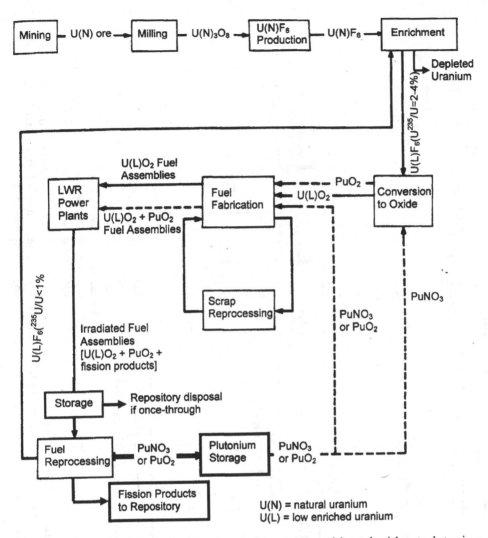

Figure 7.10   Schematic fuel-cycle diagram for LWRs with and without plutonium recycle. Parts of the fuel cycle where significant quantities of weapons usable nuclear materials may reside are indicated by the heavy lines (Willrich and Taylor, 1974), whereas dashed lines indicate process flows that are generally not yet implemented in the US.

along with indications of where in the fuel cycle elevated proliferation concerns may arise (Willrich and Taylor, 1974). For each of these three figures, operations conducted at distinct locations are designated by boxes, with the solid lines/arrows indicating inter-facility material flows; for situations where plutonium recycle is not occurring, dashed lines/arrows are used. Both the LWR and the MHTGR fuel cycles start with mining/milling/conversion (of oxide to

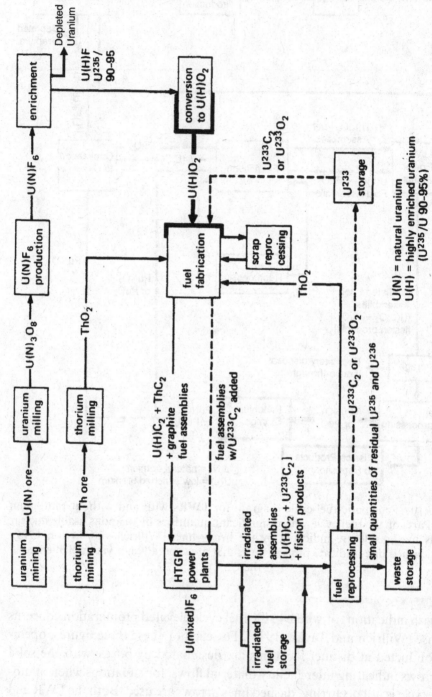

Figure 7.11 Schematic fuel-cycle diagram for MHTGR Th–U fuel cycle with and without ^{233}U recycle. Parts of the fuel cycle where significant quantities of weapons usable nuclear materials may reside are indicated by the heavy lines (Willrich and Taylor, 1974), whereas dashed lines indicate process flows that are not yet implemented.

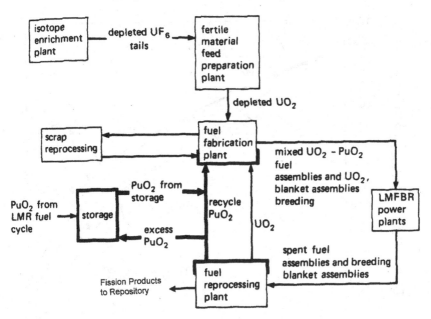

Figure 7.12   Schematic fuel-cycle diagram for LMFBR U-Pu fuel cycle based solely on the use of depleted uranium. Parts of the fuel cycle where significant quantities of weapons usable nuclear materials reside are indicated by the heavy lines (Willrich and Taylor, 1974). Generally, the pyro-chemical/electro-chemical (non-aqueous) processing of LMFBR irradiated fuel and blanket materials would occur integrally with each LMFBR power-plant operation.

gaseous UF_6)/enrichment operations, whereas the LMFBR fuel cycle depicted in Fig. 7.12 is assumed to be sustained on depleted uranium from the enrichment operations. A fuel cycle based on the use of thorium in LWRs would look similar to that for the MHTGR (Fig. 7.13), except that the ^{235}U enrichment required during the early phases of this Th-U/LWR fuel cycle would be below 20%.

7.2.4 Decay heat and fission-product radioactivity

The splitting of the nucleus leads to fission products that, according to both experiment and the Bohr–Wheeler theory (Bohr and Wheeler, 1939), are unequal in mass. These fission-product nuclei have neutron excesses (as the parent nuclei have) and are, therefore, unstable against beta decay. In general they will decay by a succession of beta-ray (i.e., energetic electrons) emissions until a more stable nucleus results. These decays result in two problems that are unique to nuclear fission energy. The radioactive decay produces heat (8% of the kinetic energy immediately after cessation of the nuclear chain reaction,

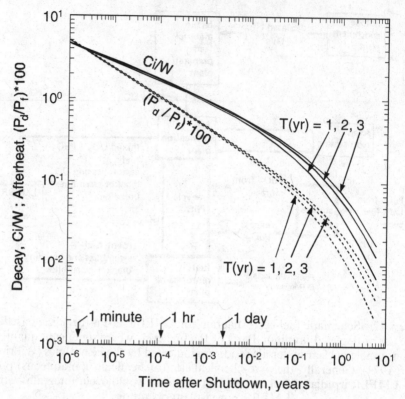

Figure 7.13 Fission-product radioactivity (solid lines) and nuclear decay heat per unit fission power (dashed lines) as a function of time after shutdown of the fission process for a range of full-power operating times.

~1% after one day, and ~0.2% after one month) and also various other radioactive emissions. The decay heat can produce an accident out of a malfunction of the coolant system in a non-passively cooled design, and the radioactivity, if released, can cause cancer and kill people. These two unique features cause considerable public concern. Figure 7.13 shows the nuclear decay heat as a percentage of full power as a function of time after the fission process has stopped, as well as the radioactivity as a function of time after cessation of the chain reaction. (Cohen, 1977; Benedict *et al.*, 1981; Bodansky, 1997). Both nuclear after-heat and radioactivity rates after cessation of fission fall roughly as $1/t^{1.2}$, as noted by Way and Wigner (1948), rather than exponentially because it is a sum of exponential decays.[8]

---

[8] Benedict, Pigford and Levi (1981, pp. 55–9) parametrize afterheat decays more precisely as $eV/s = 3.0/t^{1.9} + 11.7/t^{1.4}$, with this energy carried equally by gamma and beta rays at 25% each, and the rest carried off by neutrinos; also $Ci/W(\text{fission}) = 1.9/t^{0.2} + 1/(T+t)^{0.2}$, where T is the time at full power, and t is the time of cooling after reactor shutdown; decay heat as a fraction of full fission power is $0.0042(1/t^{0.2} + 1/(T+t)^{0.2}) + 0.0063(1/t^{0.4} + 1/(T+t)^{0.4})$.

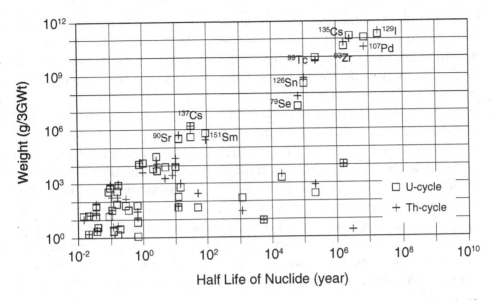

Figure 7.14   Distribution of specific fission yields and associated fission-product half-life at equilibrium for both the uranium and thorium fuel cycles (Takagi and Sekimoto, 1992).[9]

Figure 7.14 depicts key fission-product radionuclides in terms of masses in equilibrium per unit of nuclear power as a function of fission-product half-life. Only a handful of isotopes have a lifetime that is sufficiently long to present a long-term waste-disposal concern. It is important to realize that these calculations have no ambiguity. The only legitimate ambiguity is what are the consequences on the reactor core of this after-heating rate under conditions where the fission chain reaction has ceased, but the coolant flow is reduced or stopped. At small times after shut down (hours to few days), the important radionuclides from a public point of view are isotopes of radioactive iodine, because iodine concentrates in cows' milk and, after ingestion of the milk, in the human thyroid. In the time period from a year to 100 years the most important radionuclides are strontium 90 (a beta emitter that concentrates in the bone and can cause leukemia) and cesium 137, with half-lives of 28.1 and 30.0 yr, respectively. Both of these isotopes produce long-term consequences to health in the event of severe reactor accidents.

After 300 years the fission products have decayed to a level where they are overshadowed by the transuranic elements produced by (non-fissioning)

[9] These specific yields correspond to repository inventories that are in decay equilibrium with the corresponding reactor production rate. The main constituents of waste in the far future are the seven isotopes indicated in the upper right of this chart, which would be prime candidates for collection and transmutation (using either reactor or accelerator neutrons) to shorter-lived products to alleviate repository requirements.

neutron capture on the uranium/plutonium fuel. If these transuranics remain with the waste stream, the radioactivity level is about 10 times that of the original uranium ore. If plutonium and other transuranic elements are removed, the radioactivity of the waste stream is then less than the radioactivity of the original ore. Removal of plutonium in particular has the advantage that the waste disposal facility does not become a future "plutonium mine" for easily available bomb fuel after a few hundred years (Section 7.3.2.3). Removal of plutonium from the (still useful) uranium and the fission products (real waste) represents an inter-temporal trade-off related to human radiation exposure and proliferation risk about which clear resolution remains elusive. Additionally, the economics of plutonium recovery and recycle (to LWRs) is dictated largely by the price of uranium fuel (both for ore recovery and enrichment) and processing costs; generally, the economics based on present-day costs and technology are not favorable for plutonium recycle (i.e., reprocessing costs must be below ~1000 $/kgHM for breakeven when uranium costs are 100 $/kgU).

7.2.5 *Present status of nuclear power*

Nuclear energy (NE) was introduced over four decades ago into the commercial electricity market, and presently provides 18% of the global electrical energy supply and contributes ~8% of primary energy. A total generation capacity of 350 GWe is available in 442 nuclear power plants (NPPs) operated in 32 countries around the world (primarily in OECD countries, with a shift to the developing countries expected in the next century; Section 7.5). The mix of these reactors in 1993 was PWRs (55%, Fig. 7.5), BWRs (21%, Fig. 7.6, PWR/BWR = 2.7), GCRs (9%, Fig.7.7), PHWRs (7%, Fig. 7.8), water-cooled/graphite-moderated reactors (4.5%, all in the FSU), and other kinds of reactors (3.5%). Nuclear energy provides over 40% of the electricity in nine countries and 20–40% of the electricity supply in ten countries. If the 2130 TWeh of nuclear electricity generated in 1995 had been produced by fossil fuels, global carbon emissions would have been greater by 8%.

Nuclear energy was introduced rapidly in the 1960s and 1970s, with worldwide deployment rates of 13.5 GWe/yr in the 1970s and 14.6 GWe/yr in the 1980s; the average deployment rate in the US during this period was 5 GWe/yr (Diaz, 1998). No new orders for NPPs have been placed in the United States since 1978, leading to the capacity being pegged at 98.8 GWe. If nuclear energy is phased out in the US with no new plants being built, but with existing systems being operated for their presently licensed period, no nuclear power plant would be operating by the year 2030. On the other hand, if a 20-year license extension is approved, nuclear power generation would be extended to

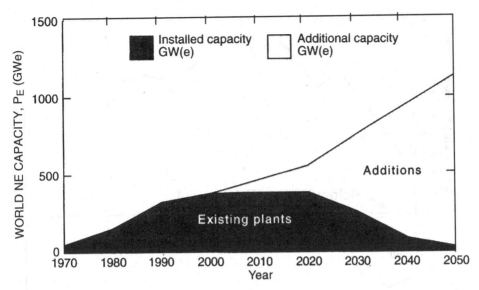

Figure 7.15   New NPP capacity additions required according to the medium variant case considered in the IAEA study (Wagner, 1997).

the year 2050. Figure 7.15 illustrates this phase-out scenario, as well as possible levels of nuclear energy capacity additions out to the year 2050 (Meneley, 1997).

After an increase in operating costs during the late 1970s, and a very rapid increase in costs following the Three Mile Island accident in 1978, the overall operational and economic performances of NPPs in the USA have improved since 1985, as is shown in Fig. 7.16 (Diaz, 1998). This economic performance, however, still has not returned to 1973 levels. The economic performance of nuclear power varies somewhat from country to country. A recent survey (OECD, 1998a) of 15 OECD countries (Belgium, Canada, Denmark, Finland, France, Hungary, Italy, Japan, the Republic of Korea, the Netherlands, Portugal, Spain, Turkey, the United Kingdom, and the United States) and five non-OECD countries (Brazil, China, India, Romania, and Russia) on the cost of electricity generation from nuclear, coal, gas, biomass, solar (photovoltaic), and wind is summarized in Table 7.4. Production-cost comparisons with fossil fuels are also included on Fig. 7.16, with nuclear energy production costs (unburdened of capital costs) being competitive. (Section 7.4 addresses country-specific technological responses to future NE developments.)

A decrease in collective radiation dose per unit of generated electrical energy has occurred (Lochard, 1997), and a dramatic decrease in radioactive liquid releases (without tritium) per unit of generated energy for NPPs is also reported. The number of accidents with significant radiological consequences

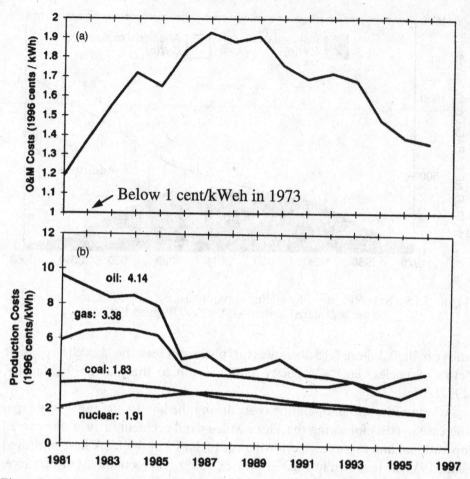

Figure 7.16   US NPP operations and maintenance costs (a) and electricity production
costs (b) over the past two decades (Diaz, 1998).

per unit of generated energy has also shown a dramatic decrease (Lochard, 1997).

7.2.6 Fuel supply

Following from the brief words in the overview (Section 7.1.4) we will show that fuel resources do *not* place a limitation on development of nuclear energy. The world uranium resources are summarized in Table 7.2 and Fig. 7.1 (Benedict, 1971; OECD, 1996, 1998a). Based on OECD estimates, Table 7.5 gives the energy equivalent of this resource. Proven reserves correspond to 1.5 MtonneU, reasonably assured reserves are 4.0 MtonneU, and total resources and reserves equal 18.5 MtonneU (excluding recoverable low-concentration

Table 7.4a *Economic data for electricity generation condensed from a range of sources taken from a recent NEI survey (1998)[a]*

	Capital cost ($/kWe)	O&M ($/kWey)	Fuel cycle (¢/kWh)	Total bussbar generation cost (¢/kWeh)
Nuclear	2110	49	0.65	3.2
Coal	1460	57		
Gas	810	24		

Note:
[a] All costs are expressed in 1 July 1996 dollars ($) or cents (¢) of the United States. Capital costs include contingency, interest during construction at 5% discount rate, and refurbishment and decommissioning (also discounted at 5%) when applicable. Fuel cycle and total generation costs are calculated at 5% discount rate for 40 years lifetime and 75% load factor.

Table 7.4b *Total bussbar generation cost (¢/kWeh)*

Year	1998	2030
Biomass	8	3.5
Solar PV	~30	$5.2(3.5+2)^b$ to $7(5+2)^b$
Wind	$8(6+2)^b$	$5(3+2)$

Note:
[b] The second number in the bracket is for the cost of backup electricity for these stochastic energy sources.

Table 7.5 *Global occurrences of natural uranium (Mtonne)[a]*

Category	Mtonne U	100 EJ[c]
Reasonably assured conventional reserves (<80 $/kgU)	4.0^b	324.1
Undiscovered conventional resources (80–130$/kgU)	2.8	226.9
Speculative conventional resources (>130$/kgU)	2.0	162.1
Speculative unconventional resources (130–600$/kgU)	9.7	786.0
Total reserves and resources	18.5	1500.0

Notes:
[a] Refer to Fig. 1a.
[b] 1.5 Mtonne (122 000 EJ) proven reserves.
[c] Based on complete fissioning of all uranium, of which 0.7205% is ^{235}U.

uranium in granite and seawater). Complete fissioning of all ^{235}U in the total reserves and resources (18.4 Mtonne) represents an energy resource of $\sim 10\,766$ EJ (1.5×10^6 EJ if all uranium were fissioned). Including the less-known thorium (^{233}U) resource would increase the 1.5×10^6 EJ resource by a factor of two to four. Compared to the fossil-energy resource listed in Table 7.6, the total uranium and fossil (including oil shale and clatherated methane) resources are comparable (e.g., 1.5×10^6 *versus* 1.0×10^6 EJ for total uranium *versus* total fossil), but the ratio of uranium fuel resource to fossil fuel resource without oil shale and clatherates (97 892 EJ) amounts to ~ 15.3.

The present world demands for total primary energy and for nuclear energy (350 GWe, $\sim 70\%$ plant availability, 35% thermal-to-electric conversion efficiency) amounts to 360 EJ/yr and 22 EJ/yr, respectively. At the present nuclear energy demand rate, most (90.7%) of the ^{235}U in the reasonably assured reserve category (4.0 MtonneU) would be consumed in a once-through (no plutonium recycle) fuel cycle by the year 2100. A linear increase of nuclear energy capacity to 2000 GWe would increase this 100-year uranium requirement by a factor of 2.7 (9.8 MtonneU), which amounts to approximately half of the ^{235}U in the total reserves and resource category (Table 7.2). These estimates are based on the total consumption of the ^{235}U resource; in actuality, only the fraction $(x_p/x_f)(x_f - x_t)/(x_p - x_t)$ is extracted in the enrichment process, which for x_p(product) $= 0.035$, x_t(tailings) $= 0.002$, and x_f(feed) $= 0.007205$ amounts to 77%. Advances that increase the efficiencies and decrease the cost of uranium enrichment (e.g., gaseous diffusion $\rightarrow$ centrifugation $\rightarrow$ laser isotope separation) can improve this situation first by extracting more of the ^{235}U isotope and second by removing ^{236}U buildup from recycled uranium fuel (Meneley, 1997). Generally, the long-term uranium requirement for a given demand scenario depends strongly on the efficiency with which the energy content in uranium (and thorium) is utilized, and the degree to which the world uranium resource is developed in the long run is dependent on the extent to which the four cardinal issues for nuclear energy are resolved (i.e., *safety, waste, proliferation*, and *cost*). Lastly, advances in removing economically even a small fraction of ~ 4000 MtonneU from seawater at cost 200–300 \$/kgU or less would indefinitely extend the resource limit for fission while possibly obviating the need for the fast-neutron, plutonium breeder.

Central to the sustainability of nuclear energy in the longer term (>100 years) is the efficient use of the uranium (and thorium) resource (Section 7.2.6). To illustrate quantitatively the present situation and the potential for waste, a simplified once-through uranium fuel-cycle for an LWR (OT/LWR) is considered, wherein uranium ore is mined at a rate F(tonneU/yr), enriched in ^{235}U content to a weight fraction x_p, and fissioned in an LWR of capacity P_E,

Table 7.6 *Summary of fossil fuel reserves and resources (EJ)*

Oil[a]		
• Identified reserves	6299	
• Undiscovered reserves	2626	
• Ultimate reserves		8925
Heavy and extra-heavy crude oil[b]		
• Identified resources	2719	
• Undiscovered resources	541	
• Total resources		3261
Bitumen,[b] original identified recoverable resources	2490	
Oil shale[b]		
• Measured resources	18741	
• Indicated resources	60559	
• Total resources		79300
Conventional gas[a]		
• Identified reserves	5239	
• Undiscovered reserves	4652	
• Total reserves		9890
Unconventional gas[c]		
• Coal-bed methane	9702	
• Fractured shale	17262	
• Tight-formation gas	7938	
• Clatherates	783216	
Remaining *in situ* after production	494	
Total less clatherates		35396
Coal[d]		
• Proven recoverable reserves	22941	
• Anthracite[d]	15217	
• Sub-bituminous	3942	
• Lignite	3782	
Additional recoverable reserves	14989	
Total reserves		37930
Total fossil fuel	960408	
• Less clatherates	177192	
• Less clatherates and oil shale	97892	

Notes:
[a] Masters *et al.* (1994).
[b] Masters, *et al.* (1987).
[c] Rogner (1996, 1997).
[d] WEC (1995).

thermal-to-electric conversion efficiency η_{TH}, and a once-through (no pluto-nium recovery/recycle) exposure of BU (GWtd/tonneHM). The rate of used-fuel generation is nominally 26 tonneHM/GWeyr. This spent material for a typical LWR contains $x_{Pu} \sim 0.009$ weight fraction plutonium, of which $\sim$73% is fissionable; for the present world nuclear energy capacity of 350 GWe, the plutonium generation rate amounts to $\sim$80 tonnePu/yr, without recycle as MOX (mixed plutonium-uranium oxide) back to the LWRs. The rate of poten-tial energy loss relative to P_E to both the repository and the depleted-uranium enrichment-plant tailings stream can also be estimated: $\sim$0.5% of the (electric) energy stored in the uranium resource is extracted by the OT/LWR fuel cycle. For the tailing and enrichment fractions $x_t = 0.002$ and $x_p = 0.035$, 6.4 units of natural uranium are needed to produce one unit of low-enriched uranium (LEU) feed to the reactor. The rate of uranium resource depletion amounts to 165.6 tonneU/GWeyr for these conditions, and at the present world nuclear energy capacity of $P_{Eo} = 350$ GWe, the cumulative uranium demand would be 5.9 MtonneU by the year 2100; if that capacity were to triple linearly to 1050 GWe by the year 2100, the uranium resource requirement would double in 2100 to 11.8 MtonneU. These uranium demands compare to 2.12 Mtonne in reserves, 4.51 MtonneU in known resources, and 15.5 MtonneU in total con-ventional resources (OECD, 1996) (Table 7.5; OECD, 1998a,b). Generally, fuel resources do not present an important limitation to the use of nuclear energy for most of the 21st century; public acceptance of this large, hazardous tech-nology does, however.

7.3 Public Acceptance

7.3.1 Public perceptions

Four cardinal technical issues are identified here that seem to mold public acceptance of nuclear energy and related technologies; *safety* (avoidance of both operational exposure and severe accidents); *waste* disposal (avoidance of long-term exposure); *proliferation* (of nuclear weapons); and *cost*. Before dis-cussing each in turn (Section 7.3.2), the essential elements of comprising/defining the issue of public acceptance *per se* are addressed in the following order: nuclear legacy, social concerns; and public opinion metrics (polls).

7.3.1.1 Nuclear legacy

As indicated in Appendix 7.A, the development of nuclear weapons and nuclear energy followed a dual and often symbiotic track that, in the public's mind, has welded a connection between these two activities. Little differentiation is made between nuclear-weapons waste and nuclear-energy wastes. Nuclear energy is a generator of materials that can be clandestinely diverted to the nuclear-weapons dark side. The true cost of nuclear energy is sometimes obfuscated by which side has actually paid the bill. Lastly, the ultimate public image of an accident involving a nuclear reactor is inexorably drawn to the nuclear dark side. Together, these connections define for nuclear energy a somewhat muddled and imprecise, if not incorrect, *nuclear legacy*. Two contrasting stereotypical views about the nuclear legacy can be identified as: (a) that of the scientific technologist; and (b) that of a skeptical environmentally concerned public. Both are perceptions and, as noted earlier, perceptions have consequences.

A starry-eyed nuclear-energy technologist notes the successes. Starting with a dream in 1945, nuclear power produced 1% of the US electricity within 30 years by 1975–80. This position is supported by:

- A record of occupational safety equal to any industry;
- No important release of pollutants to the environment;
- A safety record (in western reactors) that has resulted in no loss of life to the public;
- A plan (albeit stalled) for waste disposal that is superior to the disposal of any other waste in society;
- Technological and analytical advances for environment and safety that lead other technologies;
- An open scientific discussion superior to any other industry; and
- All of the above-listed accomplishments have been achieved while producing electricity at reasonable cost.

A starry-eyed environmentalist sees the same facts differently:

- The nuclear industry has been a driving force (or at least a cover) for proliferation of nuclear weapons;
- The nuclear industry has failed to be honest with the public and covered up malfunctions accidents and other wrong doing;
- Radiation, even at low levels, is uniquely dangerous, and no amount is or can ever be safe;
- The radiation hazards associated with nuclear waste will last millions of years, and we are laying up insurmountable problems for our grandchildren;
- When properly accounted, nuclear energy is expensive and is becoming more expensive; and
- The whole nuclear enterprise is dominated by an unacceptable elitism.

While the optimism of the technologist is and will remain the driving force behind this technology, the doubts of the environmentalist are the brakes that, once understood in a context that is broader than that defining either the technologist or environmentalist perspective, must be released before anyone can proceed. In the sections below we discuss these issues and perspectives in some detail; resolution of these issues is crucial to any sustainable advancement of this technology.

7.3.1.2 Social concerns about nuclear power

Kasperson (1993) has noted that "public response to nuclear energy is value-ladened and cultural in context; this condition has far-reaching implications for efforts to win greater acceptance of this technology".

More generally, the public acceptance issue is defined under a societal-cultural paradigm rather than in terms of a technological-economic one in which the expert operates. The introduction of nuclear energy as a limitless and nearly "cost-free" source of "labor-liberating" energy simultaneously with a war-making tool having nearly unlimited destructive power created a kind of public schizophrenia that combined total acceptance of the new source of energy and complete fear of its military dark side. This dichotomy of public acceptance and fear, that previously was kept at abeyance and separated, has dissolved as:

- The economic benefits originally promised never materialized;
- Safety issues when calculated became visible and caused concerns;
- Fear of the spread of nuclear weapons increased; and
- Public trust and credibility in any regulatory authorities and most government activities decreased.

The chronology laid out in Appendix 7.A lists a few of the defining events that energized the metamorphosis of nuclear energy from "the great hope" to "the great concern". Without public acceptance, this technology cannot advance, even if technical solutions to the four cardinal issues exist or emerge. While no single source of the public concern with nuclear energy can be identified, Kasperson associated the following societal vectors with this public concern:

- The nature of risk perception by the public;
- The aforementioned legacy of fear;
- The perception of benefit;
- Value conflicts and shifting cultural settings; and
- Diminished institutional credibility and trust.

A committee of the National Academy of Sciences (NAS, 1991) noted that "forces shaping public attitudes towards nuclear power are social-cultural in nature, and are not (directly) resolved within a technological-economic paradigm"; these forces are related to:

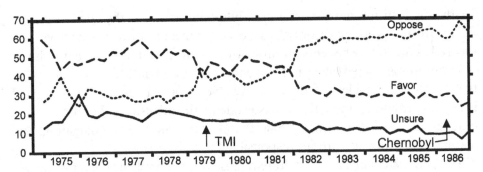

Figure 7.17   Public attitudes towards nuclear power in the US during early years (Kasperson, 1993).

- No need for new electricity generating capacity;
- An increased cost of nuclear energy;
- A lack of trust in industry and government advocates;
- Concerns over the effects of radiation at low doses;
- Concerns that means of disposal of high level waste do not exist; and
- Concerns that nuclear electric power will increase the proliferation of nuclear weapons.

An important feature is that the general public judge risk and hazard differently than experts. The public finds difficult the assessment of low-probability events and tends to over-estimate related impacts, particularly if such events have potentially catastrophic consequences and are well advertised. This tendency on the part of the public is reflected in the under-estimation of the risks associated with chronic diseases that occur with high frequency, but have few fatalities *per* event and are not persistently or sensationally advertised (Kasperson, 1993). Figure 7.17 summarizes public attitudes towards nuclear power in the US and the correlation with the TMI accident.

Most analysts agree that the fear of nuclear weapons and the connection with nuclear energy is deep seated, and this fear is sustained by novels and the cinema. While the public accepts higher risks associated with technologies that offer higher benefits, support for such technologies falls rapidly if the perceived benefits do not come to fruition; nuclear energy is in this position. The issue of public value conflict is complex, evolving, and culture dependent; typical examples of these conflicts are:

- The aforementioned connection between nuclear weapons and nuclear energy and the attempt to create a "civilian" control of nuclear energy under the hopes and claims of increased general welfare (e.g., limitless, cheap energy); and
- Value judgments driven by the moral responsibility to future generations in dealing with long-lived nuclear waste.

Lastly, the institutional issues of credibility and trust, as they determine the going level of acceptability or risk associated with a hazardous technology and the trade-off with the benefits promised by the implementation of that technology, are strongly dependent on the openness or closed nature of the society. In the US, the long neglect of the civilian nuclear waste issue, as well as related issues that reside on the nuclear waste side of the institutional house (among other events) have diminished public confidence in the rule-promulgating, regulating, governing institutions responsible for nuclear energy.

The parameters of a given culture represent a strong determinant of acceptance or rejection of large, potentially hazardous technologies like nuclear energy. Grid–group analyses of political cultures (Douglas, 1970) have been applied to technical and industrial societies (Thompson, Ellis and Wildavsky, 1990) and most recently and more specifically to nuclear energy by Rochlin (1993) and Rochlin and von Meier (1994). The degree to which individuals within a given social unit interact is "measured" by a group parameter; a free market is a social unit with weak group interactions, and communal collectives have strong group interactions within each respective social unit. The grid parameter "measures" the interaction between agents within social units; weak grid interactions correspond to weak constraints (egalitarian behaviors), and strong grid interactions tend towards bureaucratic, hierarchical systems. In short, the group parameter establishes the rules for group incorporation, and the grid parameter prescribes individual rules for members in that group. This grid–group structure is illustrated and further elaborated by Kasperson (1993), Rochlin (1993) and Rochlin and Suchard (1994), in matrix form in Fig. 7.18. The discrete (four) hierarchical social structures indicated on this figure should be viewed as a continuum. Rochlin suggested that only if the socio-cultural and socio-political attributes of a given technology are compatible with the social and political culture attributes of the system into which that technology is being introduced will that technology gain acceptance by the public of the host system.

Figure 7.18 (Rochlin, 1993; Rochlin and von Meier, 1994) gives a heuristic view of these constraints for nuclear energy and explains in part the success experienced in the Former Soviet Union (FSU), France and Japan. Since any given system (country in this case) temporally migrates in this grid–group space, the level of public acceptance of a given hazardous technology can shift, giving rise to "legacy stresses" (e.g., the in-place technology cannot readily undergo metamorphosis in response to cultural shifts, and in fact can determine the rates and directions a given system (country) drifts in 'grid–group space'.). Finally, social and political culture attributes should also play a role in determining the large-scale introduction of some of the "small-is-beautiful" renewable energy technologies [e.g., solar (PV), solar (H_2), wind, OTEC,

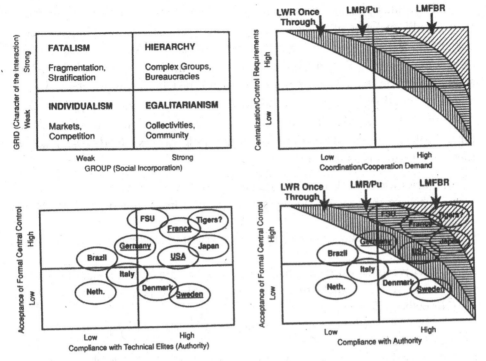

Figure 7.18   Application of group-grid analyses (Douglas, 1970; Thompson, 1990) to nuclear energy (Rochlin, 1994, 1993). The underlined countries were analyzed in Rochlin and Suchard (1994); Tigers refers to fast-growing Asian economies.

biomass, geothermal, etc.] (Johansson *et al.*, 1993), and boundaries similar to those drawn in Fig. 7.18 for nuclear energy could be established [most likely in the low–high regions (egalitarian) of grid–group space].

7.3.1.3 Public opinion polls

Most nuclear advocates have had the personal experience of discussing (arguing) the issues with anti-nuclear scientists. Often the advocate can conclusively disprove the contention of the anti-nuclear scientist. Nonetheless, the anti-nuclear scientist often remains anti-nuclear for reasons he does not or cannot articulate. Although this leads to a problem in interpreting public opinion polls, few alternatives are available. In addition, it is well known that public opinion polls can give apparently different results depending upon the precise formulation of the questions being asked. While the poll quoted below was commissioned by the Nuclear Energy Institute (NEI), which has an obviously pro-nuclear interest, that poll was actually performed by an independent polling organization (NEI, 1998).

While public opinion about commercial nuclear energy in the US seemed to be generally in favor of nuclear energy in its formative years, changes began about 1970. The changes during 1975–86 are illustrated in Fig. 7.17 (Kasperson, 1993). The marked change at the time of the Three Mile Island accident is evident and is more important than the Chernobyl accident. A more recent (1995–98) survey of (US) public attitudes towards nuclear power (NEI, 1998) indicates a significant shift towards a more positive disposition, at least for the college-educated group to which that poll was restricted; a summary of these poll results is as follows:

- 61% favor use of nuclear energy as one way to provide electricity (June 1998);
- 67% believe that nuclear energy should play an important future role (March 1997);
- 70% believe that nuclear energy will be important in meeting future energy needs (May 1997);
- 62% believe that the option to build more NPPs should be maintained (June 1998);
- 74% believe that manufacturers of US NPPs should play a leading world role (November 1995);
- 72% believe that NPPs should be considered for electricity generation "in your area" of the US (June 1998);
- 80% agree that it is a good idea to renew licenses to current NPPs that meet Federal safety standards (June 1998);
- 63% believe that nuclear energy is clean and reliable (November 1995); and
- 81% think we should reduce unneeded stockpiles of weapons plutonium by processing it into fuel and using it for electricity (February 1997).

This survey of 1000 adults has a 3% 'margin of error'. The most surprising finding of this survey is that those expressing favorable opinions thought (mistakenly) that the majority of the US public held negative views. While these more recent US trends signal a shift towards a more favorably disposed public, no survey of this kind can assure a strong reduction in the public concern over the four cardinal issues, or the occurrence of adverse events or forces (real or perceived) that can re-energize public concerns.

In summary, threading through the four cardinal issues (e.g., safety, waste, cost, proliferation) are concerns about health effects of low-level radiation, which impact safety, waste, and cost issues, and the generally low level of public trust in either governmental or industrial advocates of nuclear energy (NAS, 1991). This combination of the four cardinal issues interwoven with threads of public distrust of institutions and things radioactive in some countries is shrouding nuclear energy in a difficult-to-shed fabric. Nonetheless, these issues define the problems, solutions, and prospectus for nuclear energy. While economic and safety improvements *vis á vis* the "re-engineering" route are important, "re-culturation" of nuclear energy is sorely needed to make any headway on the all-determining issue of public acceptance.

7.3.2 Nuclear energy: four cardinal issues

As discussed above, the public concerns about nuclear energy can be condensed into four cardinal issues: safety, waste, proliferation, and cost. The following sections elaborate on each in turn.

7.3.2.1 Safety

It is important to recognize at the outset that a nuclear reactor *cannot* explode like a nuclear bomb in spite of the fact that a 1000-MWe LWR contains enough fissile matter to make 3–5 nuclear weapons. It is not easy to achieve a bomb explosion, so that when effort is made to avoid an explosion it will succeed. The bomb-crucial fissile material contained in the core of a nuclear reactor is surrounded and intermixed with so much material that interferes with the forming of the required bomb configuration, that it is not unlike gathering the fertilizer from many thousand acres of corn fields to fashion a chemical bomb of the size that destroyed the Federal Office Building in Oklahoma-City. Nonetheless, meetings of anti-nuclear groups often convene with a picture of the mushroom cloud that follows a nuclear explosion. Apart from this common misconception (which has not completely disappeared with time) there would be little public concern about the safety of a nuclear reactor if radionuclides were not released and people were not exposed to the radiation emitted from the radionuclides, either during a serious accident (from the view point of damage and loss of capital investment), during routine operating conditions, or from effluent resulting from a range of post-operational activities (e.g., reprocessing, transport, waste disposal, etc.). Even under accident conditions, the radioactivity contained in the core of an operating reactor will not be dispersed (and, therefore, people will not be exposed) unless the nuclear fuel melts or evaporates and releases the contained material to the accessible environment. Even if released from the fuel, the radionuclides must pass through a series of engineered barriers (fuel cladding, pressure vessel, containment building, etc.) before entering the environment. If released, however, these radionuclides have the potential of increasing the radiation dose for many people.

All modern reactors (the Russian RBMK reactors of the type operated at Chernobyl are an exception) are designed with many inherent safety features. For example, if a light-water-reactor overheats and the water evaporates, the disappearance of the moderator shuts down the nuclear chain reaction; this is a safety feature not possessed by the RBMKs. The RBMKs also have tonnes of a highly combustible moderator (e.g., carbon), and the fuse to ignite it (e.g., zirconium/water, should sufficiently hot conditions be achieved, which they were), and for a brief, ignominious period, Chernobyl was a fossil-fuel power plant out of control! Although an LWR shuts down if the water escapes following a loss of coolant pressure through (say) a pipe or vessel rupture, or the inadvertent opening of a value by a confused operator, the decay heat (initially 6–7% of the full thermal power of the full-up core, but decaying rapidly, as shown in Fig. 7.13) is still sufficient to cause melting if nothing, or the wrong thing, is done. Under these conditions radioactivity would be released. All western power reactors have a containment to retain the radioactivity in the event of core melting. At Three Mile Island in 1978 about one third of the core melted. Although many of the comparatively innocuous noble gases (argon, krypton, etc.) were released, the dangerous radioactivity was contained within the primary containment or auxiliary buildings. (Some of the radioactive iodine was absorbed by the water and pumped into the auxiliary building.) Much of the radioactive inventory actually remained within the damaged core/pressure vessel. It is estimated (for example) that only 15 Curies of radioactive iodine, which corresponds to less than one millionth of the inventory in Three Mile Island (TMI), was released to the environment.

Since 1974 a systematic procedure for estimating the probability and consequences of an accident has been applied, where all the barriers to fission-product release are broken (WASH 1400). The calculated probability is low ($<10^{-4}$ core melts per year and $<10^{-6}$ accidents with serious release of radioactivity per reactor year). Nonetheless this raises concern, since for $\sim$2000 reactors expected under a number of scenarios given in Fig. 7.19 (IAEA, 1997; OECD, 1998b; Krakowski, 1999), the resulting ($<$1/5yr) rate of core melts and ($<$1/500yr) serious reactor accidents would raise serious questions in the public mind (not to mention the corporate mind) of the social and economic viability of this technology.

Public concern about safety issues was driven by the demonstrated acts, omissions, and incompetence leading to the TMI accident. This concern developed in spite of the fact that no loss of public life or property resulted from the TMI accident; the utility was the main (financial) casualty. This concern exists also in spite of the fact that such an industrial accident would have caused little concern in other industries that do not release radioactive material in severe accidents. Indeed few people noticed in the US when railroad cars carrying flammable gases and toxic chemicals overturned a few weeks later in a Toronto (Canada) suburb; this accident led to the largest evacuation in peacetime North American history.

The acts and omissions of incompetence associated with the TMI loss were

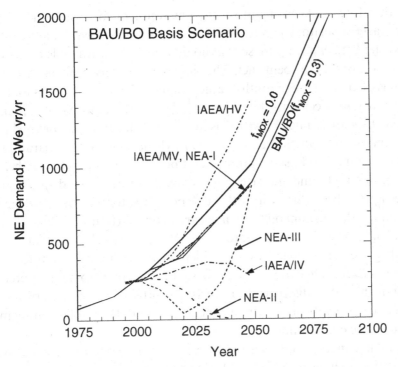

Figure 7.19   Comparison of world NE generation scenarios (IAEA, 1997; OECD, 1998b; Krakowski, 1999).

not limited to one area, but instead were threaded/shared/communicated throughout the responsible institutional system, as structured under the unusually diffuse paradigm that defines US nuclear energy (unlike those in France or Japan, shown in Fig. 7.18):

- The reactor manufacturer had not performed a rigorous safety analysis using the procedures developed by Rasmussen five years before;
- The utility had not performed a safety analysis; the operators did not understand the reactor and in particular the behavior of a boiling-water PWR; and
- The NRC panicked and did not understand when the danger of radioactivity release was over.

These public concerns were reinforced by the accident at the Chernobyl nuclear power plant even though the RBMK reactor was neither built nor operated according to standards of safety in Western industrialized countries.

7.3.2.2 Waste disposal

As shown in Figs. 7.13 and 7.14 (discussed in Section 7.2.4), nuclear waste remains hazardous for hundreds and thousands of years. Means must be found, therefore, to keep it out of the environment for a long time. But a nuclear waste repository is not a reactor or a bomb (Bowman and Venneri, 1996; Kastenberg *et al.*, 1996), so that no driving force is available to melt, disperse, or evaporate the spent fuel. The time scale for any problem with waste disposal is of the order of months, rather than on the scale of hours that characterize a reactor accident; this time scale leaves mankind plenty of time to react. Furthermore, "nuclear waste" is not some amorphous, ill-defined radioactive compost over which the technology has little control. By separating actinides and/or long-lived fission products (at both some cost and some benefit), these repository lifetime and size determinants can be directed to alternative destinies for the betterment of both the repository storage, for future generations, and for the sustainability of nuclear energy (Arthur and Wagner, 1996, 1998; Takagi, Takagi, and Sekimoto, 1998). While experts believe that both the probability and consequences of an accident in a waste repository are low, provided that appropriate compositions are sent to the repository, an appreciable segment of the public distrusts the experts. Uncertainties enhanced by the very long time scales involved also drive concerns of both the public and segments of the scientific community.

The public concern about waste is driven by the absence of a safe disposal repository for high-level waste (e.g., fission products and some activated structural materials). Concern over the interleaving of the proliferation issue (if plutonium is inadvertently and incorrectly treated as a waste product), the longevity of the waste, and the lack of closure on this issue are primary forces behind the waste concern, both within the US and abroad (Shlyakhter, Stadie, and Wilson, 1995). Generally, more can be done in building public trust with regard to this issue of nuclear-waste management, if feasible "above-ground" solutions having (public) acceptable time lines can be developed, while dispelling the "out-of-sight/out-of-mind" nature of the present approaches that have only the short view and nuclear phase out in mind.

7.3.2.3 *Proliferation of nuclear weapons*

The proliferation issue is entwined with the issues of cost and waste, and general resolution is made difficult by the historical connection between nuclear weapons and nuclear energy, and a lack of quantitative differentiation between the difficulty of building massively destructive weapons using source materials originating from the civil nuclear fuel cycle *versus* other sources of materials (including increasing potential for aggressions using chemical–biological materials). A large literature exists on proliferation (Willrich and Taylor, 1974; OTA, 1977; Meyer, 1984; Davis and Frankel, 1993; Gardner, 1994; Reiss, 1995) but there is comparatively little describing how to have a nuclear power program without compromising the demands of non-proliferation. One expert commentator argued that proliferation of nuclear weapons is the only aspect of nuclear energy where a technically literate person has reasonable criticism (Cottrell, 1981).

As is indicated in the chronology given in Appendix 7.A, the nuclear energy industry is an outgrowth that occurred in parallel with, and in some cases was partially supported by, national efforts to build nuclear arsenals. During the Cold War era and before threats of nuclear terrorism and significant sabotage related thereto, the enormity of the task of constructing nuclear arsenals, combined with the challenges and struggles of a fledging nuclear-power enterprise, reduced perceived connections between civilian nuclear power and the desire of the state to acquire nuclear weapons; the two were not strongly connected prior to about 1970. In fact, a common belief that to some extent prevails even today was that a nuclear weapon could not be constructed from the plutonium generated in high-burn-up commercial spent fuel (e.g., Sakharov, 1977). The connection between civilian nuclear power and nuclear-weapons proliferation was made visible to public concern only after the growth of terrorist activities in the early 1970s, although the world has lived with the spectre of nuclear-weapons proliferation since the first detonations in 1945, and proliferation control has been a key element of the socio-political consciousness ever since.

While a deceptively simple phrase, the definition of "nuclear-weapons proliferation" has been broadened (OTA, 1977) from one pertaining only to the acquisition of nuclear-explosive devices or weapons by countries not currently possessing them (NPT context) to encompassing any country that has acquired the capability to produce rapidly any nuclear-explosive device or weapon. Policy debates over ways to reduce the risk of proliferation center largely on three key issues (OTA, 1977):

- The likelihood, rate, and time delay of proliferation *via* various alternative routes (diversion from civilian nuclear power, clandestine indigenous facilities, and/or direct purchase or theft of key components);
- The nature and consequences of proliferation (regional *versus* global impacts on political or military stability); and
- The differing assessments of political and/or economic impacts of different policy options.

A system of safeguards and security has emerged from the chronology given in Appendix 7.A that applies to all "declared" nuclear facilities, and in particular those facilities that comprise and support nuclear energy. While the IAEA is responsible for setting standards of (nuclear) Materials Protection, Control, and Accounting (MPC&A) for each state operating such a facility, through negotiations with the state and facility, and for verifying that these standards are being met, the actual responsibility for assuring that a given facility is not being used in the proliferation of nuclear weapons rests with the state and not with the IAEA. Generally, each nuclear facility reports to the state, who in turn reports to the IAEA, and in cooperation with the state the IAEA verifies compliance with the MPC&A standards originally agreed between the state and the IAEA. Additionally, the IAEA engages in surveillance and accounting activities to reduce the cost and effort associated with direct verification actions. In short, a strong regime of nuclear-materials control, accounting, and physical protection surrounds all facilities required of nuclear energy, and this regime is one that exists under IAEA–state mutual (negotiated) agreement, state enforcement responsibility, and IAEA verification/ surveillance/accounting.

While the expertise required to operate a safe and economic nuclear-energy enterprise can easily be "diverted" to support activities that aid in the proliferation of nuclear weapons, by far the strongest connection between civilian nuclear power and proliferation is the potential for providing nuclear-explosive material to that dark activity. Correlations between states having research reactors, power reactors, and latent proliferation-related capabilities and states that actually have developed nuclear weapons is at best weak (Meyer, 1984). Nevertheless, the potential for a nuclear-power state to acquire nuclear weapons through the civilian nuclear fuel cycle, while small, is non-zero and must be accommodated. Figures 7.10–7.12 illustrate points within a number of nuclear fuel cycles that are most susceptible to the diversion of nuclear material (plutonium). The above-described safeguards and security umbrella extend to each facility that comprises the flow charts in Figs. 7.10–7.12. The nuclear-weapons value of plutonium in spent fuel can be significantly diminished by increasing the time during which power is generated from that fuel (e.g., increased exposure or burn-up, GWtd/tonneHM) (Beller and Krakowski, 1999). Once the used fuel and contained fission products are discharged from the reactor (typically after three years, with this power-production time increasing as more advanced fuels are developed), the radiation field created by the contained fission products makes a casual retrieval of weapons-usable from this used-fuel plutonium impossible. Should economics and politics allow reprocessing of the spent fuel to recover and recycle the unused plutonium and uranium fuels back to the reactor and enhance utilization of that resource, these points where diversion from the fuel cycle to clandestine use are few, contained, and a strong focus of MPC&A activities.

Generally, sources of nuclear-explosive materials that are simpler and more attractive than the forms derived from the civil fuel cycle are readily identified. The South Africans, for example, used the Becker nozzle to separate uranium isotopes (additional clandestine uranium enrichment using advanced (e.g., higher production rates) difficult-to-detect technologies, as well as smuggling and theft of nuclear-weapons materials and components represent other alternative sources). In spite of this situation, proliferation remains in the public mind one of the four cardinal issues for nuclear energy. In this regard, with the rise of terrorist acts since the 1970s, the end of political bi-polarity accompanying the end of the Cold War, the rise of multi-polarity and regional conflicts, the increased use of terrorist tactics having large-scale impacts against established governments, and the weakening of the powers of individual states as globalization of economies proceeds, guidance for the physical protection of nuclear materials has increasingly been strengthened as an important component of nuclear safeguards and security (Bunn, 1997; Kurihara, 1977).

7.3.2.4 The cost of nuclear electricity

It is likely that the majority of Americans will be in favor of the electricity generators that will produce the cheapest electricity. It is to be expected, therefore, that public concern about nuclear power will not be ameliorated unless nuclear energy is economically competitive with alternatives.

The cost issue revolves around the general trend of high capital costs of the technologies that are used to utilize the very abundant and cheap fuels (e.g., the progression fossil → fission → solar → fusion shows increasing capital costs to utilize cheaper and more abundant fuels). At present, no perceived urgency exists in the Western industrialized countries for new electric-generation capacity. Fossil fuels remain cheap and abundant, and, while the cost varies across regions (OECD, 1998a), nuclear energy is at present perceived to be more costly than alternatives. As indicated in Fig. 7.16 (Diaz, 1998) and Table 7.4 (OECD, 1998a), these perceptions often conflict with reality (here we assume that the perception of the expert is more real than the perception of a lay person).

7.3.3 Gaining (or restoring) public confidence

Abraham Lincoln wrote a letter to Alexander McClure that stated: "If you once forfeit the confidence of your fellow citizens, you will never regain their respect and esteem". A more modern writer (Slovic, 1993) has emphasized the need for *trust* in technologists and assessors of risk.

Great uncertainty characterizes means to (re)gain public trust. In developing a plan for the future of nuclear energy, a conflict arises between the technical/economic paradigm and the societal/cultural paradigm. Both paradigms have their place. The technologist offers his plan using the first (technical/economic) paradigm. The ultimate fate of that technology, however, is largely determined in the streets, the market place, and in the courts under the second paradigm. As noted previously, Kasperson (1993) argued that the evidence points to a public response to the nuclear technologies that is "value-laden and cultural in context". This condition has far-reaching implications for any effort to win increased societal acceptance of nuclear energy. The following multiple pathways to increased public acceptance of nuclear energy were suggested (Kasperson, 1993):

- Demonstrate a record of safe operation of present NPPs;
- Contain the potential for catastrophic risk;
- Continue to improve present NPPs;
- Develop new, reduced-risk and standardized NPPs;
- Separate nuclear energy from nuclear weapons (we would add in so far as possible);
- Re-discover the benefits of nuclear energy;
- Reduce the impacts of future oil price shocks;
- Increase energy security;
- Mitigate GHG emissions;
- Improve competitive prices;
- Maintain steady progress on waste management, leading to sustainable nuclear energy;
- Begin with specific waste facilities (repositories, MRSs);
- Close the fuel cycle;
- Plan, develop, implement no-actinide, minimum-(long-lived)-fission-product systems (Arthur, Cunningham, and Wagner, 1998; Takagi, Takagi, and Sekimoto, 1998);
- Create and implement fair, open, equitable institutions for the administration of nuclear energy; and
- To the extent necessary, break with the past (even if the technologist believes in that past).

Given the capacity to "re-engineer" and based on the previous discussions, the key to resolving satisfactorily the four cardinal issues to a degree needed to impact public acceptance is "socializing/culturating" the application of this technology (Kasperson, 1993); elements of this "socialization/culturation" include:

- Demonstrated public and occupational safety, even in the face of capital($)-but not necessarily human-effects-intensive events;
- Reduced and standardized nuclear power plants;
- Total separation of nuclear energy from nuclear weapons;
- Re-discovered benefits of nuclear energy either in competition or symbiosis with other renewable sources of energy;
- Total waste containment/control/management;
- Open/fair/equitable administrative institutions.

The detailed means by which resolution of nuclear energy's four cardinal issues are translated into positive forces for public acceptance is beyond the scope of this chapter. They must use tools available from both technology and engineering sciences and from the social, institutional and political sciences. The development of mutual understanding and appreciation of each science and more open, transparent and simultaneously efficient regulation is essential. Nuclear energy must maintain and increase the distance from nuclear weapons. The "reculturation and socialization" process must occur in unison with the "re-engineering" scenarios described in section 7.4.2.2.

7.4 Future Directions

7.4.1 Technological responses to a nuclear-energy future

7.4.1.1 Management responses

Desirable responses on the part of managers of both the development, implementation, and regulation of all systems required for safe and sustainable nuclear energy are largely captured by the multiple pathways towards public acceptance suggested by Kasperson (1993) and by Slovic (1993), as summarized in Section 7.3.3. From the perspective of government managers of things nuclear, of paramount importance are: (a) the need to maintain a long view with respect to an evolving nuclear infrastructure as related to safety, waste, and advanced (economic, future niche-filling) systems, including non-electric applications; and (b) maintenance, if not enlargement, of both functional and institutional barriers between nuclear weapons and nuclear energy, in spite of any economic incentive to the contrary (e.g., breeding of weapons-directed tritium in commercial nuclear power stations). The responses of industry managers to this issue focus largely on: (a) the demonstration of a record of safe operation of present nuclear facilities (both power plants *per se* and all supporting facilities); (b) continued improvement of safety and economic characteristics of existing nuclear power plants; (c) development of new, reduced-risk, standardized NPPs; (d) dealing with the often opposing rules and constraints of centralized *versus* distributed electricity generation in an energy market that increasingly is becoming global both in extent and interconnectedness. Lastly, from a regulatory viewpoint, management of the crucial rule-making/compliance functions must increasingly seek to achieve performance-based, less-adversarial operations while protecting the safety and long-term interests of the public. These responses of all management components (governmental, industrial, regulatory), depending on region and history, must break with past practices in a way that deals with the emerging nature of the risks associated with this large and potentially hazardous technology, as perceived by a range of publics, in a way that does not jeopardize basic trends towards increasing democratization and liberalization of governing systems that are to benefit from such technologies (Hiskes, 1998).

7.4.1.2 Regulatory issues

We address here the problem of regulation and the intricate and complex relationship between regulator and licensee. There is always a conflict in regulation. A regulator must ensure that "a nuclear power plant is operated without undue risk to the public". On the other hand, it has been said that "the power to regulate is the power to destroy". It is evident that safety is a large part of the cost of nuclear power. Without attention to safety a containment vessel, often one-third of the capital cost of a nuclear power plant, would not be necessary. But the important issue is for how much safety should one pay? If one demands too much (beyond reason, as defined by or in reference to other comparable risks), nuclear energy will inevitably be priced out of any market.

The first attempt to address "how much safety" for radiation exposure was begun by the first Nuclear Regulatory Commission (NRC) to take office some 14 years ago when astronaut William (Bill) Anders was chairman. After two years of public hearings started by the AEC the NRC set some radiation and safety guidelines in the rulemaking document RM-30–2 (NRC, 1975). The Commission ~proposed that expenditure on radiation exposure reduction should be made if it costs less than $1000 per person-rem (prem), now doubled to $2000 (Kress, 1994); this number is higher than anyone in the hearings had proposed. Somewhat later the National Council for Radiation Protection and Measurements (NCRP, 1990) suggested that the number be between $10 and $100 per prem for dental and medical exposures. A corollary was implied, but not explicitly stated. If a proposed dose-reducing action would cost more than this, it should *not* be done. The US EPA (1998) in their draft "Guidelines for Preparing Economic Analyses" suggest a number of about $5 million per life saved, and later use the $6.1 million for their arsenic "rule" (EPA, 2000), which corresponds closely (using a linear dose response relationship) to the number in the above rule of $2000 per person-rem, and should probably include the time spent talking about the particular issue.

Issues with a direct impact on severe accident probability are more difficult to address. In the 1980s the Advisory Committee on Reactor Safeguards (ACRS) made a study that led to the promulgation by the Commission in 1986 of a set of *safety goals* (Federal Register, 1986):

- "Individual members of the public should be provided with a level of protection from the consequences of nuclear power plant operation such that individuals bear no significant additional risk to life and health"; and
- "Societal risks to life and health from nuclear power plant operation should be comparable to or less than the risks of generating electricity by viable competing technologies and should not be a significant addition to other risks".

These somewhat vague goals were supplemented by the following *quantitative objectives*:

- "The risk to an average individual in the vicinity of a power plant of prompt fatalities that might result from reactor accidents should not exceed 0.1% of prompt fatality risks from the sum of prompt fatality risks from other accidents to which members of the US population are generally exposed"; and
- "The risk to the population in the area near a nuclear power plant of cancer fatalities that might result from nuclear power plant operation should not exceed 0.15 of the sum of cancer fatality risks resulting from all other causes."

These objectives are met for light water reactors (with containments) if a subsidiary objective is met: "The core melt frequency must be less than 1/10 000 per year". Although not stated, it was implied that steps to decrease core melt frequency still further were unwarranted and it was not worth the expense to undertake them. For simplicity we address only this "intermediate" safety goal here, but the same argument can be applied to the more fundamental safety goal.

A fundamental problem arises in implementing *goals* as opposed to issuing or following *regulations* – no definitive way of proceeding can be found. But studies can be made retrospectively to see whether they are met. An independent study (Tengs *et al.*, 1995) suggests that expenditures in the nuclear industry for *radwaste* have been over $1 million per person-rem – 1000 times the goal. It seems that either the regulations (in this case probably the technical specifications) are stricter than needed, that the industry is spending more than the regulations call for, or the total amount of money is so small it is not worth worrying about. The procedure does not, however suggest how they be relaxed or whether the cost decrease is large enough to be worth the bother.

Similarly, the ACRS has repeatedly stated that it is not sensible to regulate on the basis of a probabilistic risk assessment (PRA). But a PRA *can* be used to discuss retrospectively whether reactors that were designed and operate under existing regulations meet the goals. If they meet the goals, fine. If they do not, regulations might be tightened. On the other hand, if the safety goals are met with a large margin maybe the regulations can be relaxed. Indeed, the important parts of a PRA can now be put on a small PC or laptop so that the effect of any small change in procedures can be quickly calculated.

As difficult as it is to reduce the severity of a regulation, it is not easy to forgive a deliberate violation of regulations even when that violation does not result in any safety goal being exceeded. But again, immediate and rapid effort in this direction seems warranted. If a procedural violation has occurred, the NRC must of course act in some way because such violations can escalate. But a graded response seems sensible. The power plant might be shut down, as were the four power plants at Millstone and Connecticut in 1996, when a "whistle blower" pointed out that the technical specifications had been routinely violated. But a graded response would suggest that a restart be permitted (and replaced by a fine) as soon as it was determined that the procedural violation did not result in the safety goals being exceeded. With fast computers a PRA can be set up to do such an analysis within a week or two at most.

The US NRC was spawned in 1974–75 from the old Atomic Energy Commission to separate the promotional role of nuclear energy from the regulatory role. It was already geographically separated by putting the promotional arm of the AEC in Germantown, Maryland, and the regulatory arm of the AEC in Bethesda. But unlike the mandate given to the AEC by the Atomic Energy Act of 1945, the NRC has no mandate to keep power plants in operation – only to ensure that the power plants operate without undue risk to the public. It was left to ERDA and now the Department of Energy to promote nuclear energy and to provide the balance. It is important to realize that the utility companies cannot and will not by themselves perform this function of balance. The utility companies are under close local or regional control, and historically have shown extreme reluctance to challenge any regulatory body. A great unbalance in power exists. A regulator often has the ability to keep a power plant shut down for an extra day – an action which costs the utility company $1 million per day. A counterbalance to ensure that this power is used wisely and well cannot be found. The Nuclear Regulatory Commission has been sued in the courts (in what seems to be the preferred procedure in the US for obtaining balance) by one or another group opposed to nuclear power, but to the best of our knowledge has not been sued by utility companies. Any regulator will automatically adjust his strategy to minimize lawsuits – and probably that is easiest done by ensuring that the number of lawsuits from each side is equal. If no one is actively promoting nuclear energy, therefore, the regulation will inevitably become more strict and will force unnecessary price rises until price competition destroys the industry.

The cost of over-regulation at Millstone is huge and seems to have been deliberately understated in many reports so far. We take it here to be the busbar cost of replacement electricity of about $3 million a day or 2 billion dollars so far. The effect on public health is also huge. Supposing the replacement electricity to come from a mixture of fossil fuels and hydro power in the average proportions, each power plant replacement costs over 50 premature deaths a year from air pollution (Wilson and Spengler, 1996), or over 400 deaths so far.

In addition to the above we must also remember the cost that regulatory delay involves. The license hearing for the Maine Yankee construction permit was only six hours long. The operating license hearing a few years later was only two days long. More recently some hearings have lasted ten years. Temporary storage of high level nuclear waste and uranium isotope separation are inherently safer than operating a reactor – the latter much so. Yet the hearings for storing waste in the Goshute Indian reservation are optimistically estimated to last four years and plans for a uranium isotope separation facility in Florida were abandoned after seven years of no progress.

These procedures are largely under the control of NRC. Indications are that NRC may be changing. The chairperson of the NRC stated in October 1998, "Regulation, by its nature, is a burden, but that burden must be made clear, based on risk insights, with performance expectations clearly defined" (Jackson, 1998). However, she did not go so far as to discuss the ALARA and safety goals directly.

Two steps can help to regain a balance in regulation. The first is a procedure to decide to regulate nuclear power in a more efficient way (including deciding upon how much regulation is necessary). The second necessary step is to find a group which will play the active promotional role that is so necessary in the US system and those patterned after it. This second step could happen by the DOE returning to the political concept of 1973 when the AEC was broken up. Other mechanisms might be found which should constantly call the regulator to task when he takes actions that exceed his own goals. In the US that approach would have to involve lawsuits because that is where the action finally occurs in any subject.

7.4.1.3 Government responses

From both historical and future policy perspectives, the response of individual governments to the prospects and directions of a nuclear-energy future varies widely. At both cultural (Fig. 7.18), social, political, and economic levels, each country having or contemplating a nuclear energy alternative prioritizes and emphasizes differently the four cardinal issues facing civilian nuclear power (cost, safety, proliferation, waste). This section summarizes a sampling of governmental responses to the issues facing nuclear energy and the way in which local orientation to these issues shape the nuclear-energy futures.

Sweden. Sweden has adopted two approaches which seem to be unique. The 1980 referendum, which was widely considered a vote against nuclear energy, could in fact be called a reprieve. The following three alternatives were presented in a widely publicized referendum of 1980 (Lindstrom 1992).

- Immediate shut down or phase out;
- No new plants and phase out by the year 2010;
- Further expansion.

It is a well-known fact of political life that most people, when faced with three alternatives that they do not understand, choose the middle one. This propensity can be enhanced by making the extremes very unpalatable. An immediate shut down would have provided much disruption of the Swedish economy; whereas an expansion of nuclear energy seemed unnecessary in a country where enough power plants were operating or under construction to supply half of the electricity generation by nuclear fission. Despite every effort made to inform the electorate about different nuclear energy options, some confusion still exists among the public. One person remarked on Swedish TV on the day of the referendum "I still don't understand why we want nuclear power, when I've got electricity in my house already" (Price, 1990, p.72). The government declared that a necessary condition for continuation was to have a solution to the waste problem. Swedish scientists did not propose a simple solution; the comparatively expensive solution required encasing the waste in solid copper containers, which would not be eroded, and *then* putting them into an area of little ground water. In nearly twenty years since the nuclear referendum, Sweden in 1999 still has not shut down a nuclear power plant and may be the only country with a politically *accepted* solution to nuclear waste.

France. France has a fair quantity of hydropower in the south east, but has very little coal, oil or natural gas. After the "oil shock" of 1972/1973, France made a bold decision to base all the electricity expansion in France on nuclear power, and with nuclear reactors of a specific type (pressurized water reactor, PWR). These PWRs were made by a French company, Framatome, under license from the Westinghouse Corporation. Public opinion surveys similar to those carried out in the US show that the French public have the same perception of safety as the US public (Slovic, 1993). Nonetheless, nuclear power seems to be well accepted in France. Several possible reasons have been advanced for this situation:

- The French have a high degree of confidence in their professional engineers. Jealous Americans describe this as the dominance of the personnel from the Ecole Polytechnique (founded by Napoleon). This belief in professionalism is one of the characteristics that make France so different from other developed countries (Jasper, 1990; Price, 1990). Associated with this belief in professionalism is a pride in achievements in art and technology, particularly those of a spectacular nature.

- France has a centralized political structure in contra-distinction to the federal structure of Germany, or the political power of the states in the US (Nelkin, 1971, 1974). For a technology like nuclear energy that has national and even transnational ramifications, the centralized structure may be particularly appropriate.

- Opposition to nuclear power in the west has often been considered to be a left-wing phenomenon – a revolt of the people against oppressive industry or government. In France, the powerful French communist party supported nuclear energy because Moscow supported it. A small example of this was Nobel Laureate Frederic Joliot-Curie, a leading communist who was a member of the French resistance movement in WWII. Few opponents of the Government wanted to repudiate him.

- The press have generally been in favor of nuclear energy. It is not easy to assess the extent to which this media support reflects public attitudes and the extend to which it molds them. Generally, the press have been very sophisticated in discussions of nuclear energy.

- Those who live within 30 km of a nuclear power plant have especially reduced rates for electricity as a kind of compensation for being near an industrial facility. This rebate considerably reduces local opposition to nuclear power plants.

- France has a representative government rather than an American style democracy. The average French citizen expects their elected representatives to govern, and does not expect to be second-guessing them all the time by letters, faxes and referenda. This acceptance of rule leads to a historically close cooperation between government and industry.

In spite of the six reasons listed above, opposition to nuclear power is growing in France, and the concerns seem to echo those in the US.

Germany. Germany built several pressurized-water reactors. In East Germany the reactors were VVER 400s of Russian design (now shut down), and in the west PWRs similar to those sold by Westinghouse. In addition, an active research program in new reactor types was pursued, particularly the "pebble bed" (high-temperature, gas-cooled, HTGR) reactor. This activity in Germany slowed after the Three Mile Island accident and came to a halt after Chernobyl. The differences from the French situation are discussed by Nelkin (1983), who pointed out that the decentralized political structure allowed nuclear power to be used as a political weapon in "states-rights" issues. In addition the closeness of Germany to Chernobyl, and the relationship between nuclear power and nuclear weapons, make the adverse concerns about nuclear energy seem more important in Germany.

Former USSR. During the Communist regime in the USSR, nuclear power was closely coupled to the military industry. Even now, both are under the same industry in Russia (MINATOM). Nuclear power was centralized in the USSR. Abundant electricity was considered to be very important for the Soviet state, according to Lenin's dictum "Communism is Socialism plus electricity". After Chernobyl, hostility to nuclear power developed in many ways. The accident itself demonstrated to the governing elite that they had failed to manage adequately a modern technology; the governing technocratic elite immediately put expansion plans on hold. Hostility to nuclear power is particularly widespread in the Ukraine and in Byelorussia, with other regions suffering heavily from Chernobyl fall-out. It was said that the people of Byelorussia did not want any kind of nuclear power – even safe nuclear power (Price, 1990). The fact that the central government had failed to produce a safe system, coupled with the fact that the central government kept the details of the accident and the radioactive deposition secret from the people (Shlyakhter and Wilson, 1992), were used by the fledgling opponents of central rule to discredit the USSR. After voting for, and gaining, independence from the USSR in December 1991, the parliament of the Ukraine voted to halt all expansion of nuclear power and to shut down the Chernobyl nuclear power plant in December 1993, which had been cleaned up, the destroyed unit "entombed", and the remaining units restarted, after tremendous effort and work at a moderately high radiation exposure of the 600 000 clean-up workers or liquidators. Economic realities changed this anti-nuclear, Chernobyl-driven position; when faced with paying world market price for gas and oil from Russia, the Ukraine recommenced construction of the partially finished nuclear power plant in 1992 (although it was shut down nine years later), and rescinded the vote to shut down Chernobyl.

Nuclear power does not seem to be high in the concerns, either for or against, of either the Russian Government or its people. A division of opinion and position seems to exist, however; the ambitious nuclear expansion program was canceled after the Chernobyl accident, and only a part of it has recommenced.

Japan. After WWII Japan followed the lead of the US in many ways, including taking up baseball as a national sport. This parallelism also included the development of nuclear power. But the fact that Japan is an island with few indigenous fuel resources makes the incentive for nuclear energy far greater than in the US. Japan entered two world wars driven in part by a search for fuel supplies. In 1914, Japan joined England and France against Germany to gain control of the coal in the German concessions in Manchuria. After the US and Holland had imposed an oil embargo in 1941, and less than a one-month supply of oil was left, Japan bombed Pearl Harbor. It is easy and usual to stockpile many years supply of nuclear fuel, while it is difficult and uncommon to stockpile more than a few months of oil or gas. With nuclear power Japan can more easily sustain another oil embargo by an unfriendly nation.

In addition, decisions are made and changed slowly in Japan. Although the world supply of uranium seems adequate for more than 50 years, Japan, nevertheless, is pursuing a strong breeder-reactor program and is stockpiling separated plutonium for the purpose. Also, Japan is continuing its nuclear power expansion with six advanced boiling-water reactors (ABWRs) that are either built or under construction.

China. China has a larger energy intensity (ratio of energy use to gross domestic product, GDP) than western countries and this intensity is decreasing steadily. This decrease is in accord with various scenarios for development of Hafele *et al.* (1981). Present plans of the Chinese government call for most of their expansion of electricity supply to be fueled by an abundant supply of coal (with enormous increase in CO_2 emissions and potential for global warming (Fang *et al.*, 1998). At the present time nuclear powered electricity is perceived as too expensive to compete, even in favorable locations (such as the SE China coast), where the transportation route for fossil fuels is very long.

7.4.2 Technological responses

Concerns and potential problems related to a (global) nuclear energy phase-out scenario have been listed in Section 7.3. This section deals with a few "re-engineering" directions required primarily for a nuclear-energy growth scenario, albeit, some of the technologies described below would be necessary even for a nuclear-energy phase-out scenario.

Shlyakhter *et al.* (1995) and others have addressed the possibility of modifying the technology to meet perceived issues. In examining the technological responses to these issues noted in the following sections, the extent to which these responses (listed in Section 8.3.3. above) are met as elaborated.

7.4.2.1 General approach to a nuclear future

Todreas (1993) suggested a general paradigm for bridging to a nuclear-energy future based on a progression from safe and economic current reactors (LWRs) to evolutionary LWRs (ELWRs), and finally a movement into advanced nuclear systems that would address the longer-term problems of resource, waste, and sustainability. Building on that paradigm, Fig. 7.20 elaborates this progression, which might be comprised of the following essential elements:

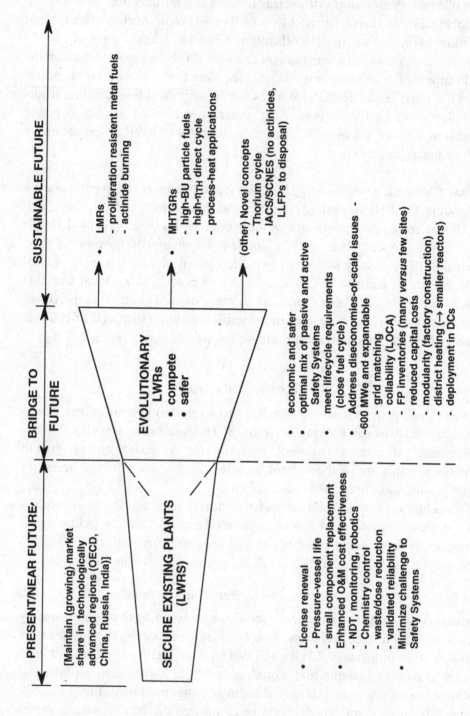

Figure 7.20 Possible (almost generic) growth scenario for nuclear energy (Todreas, 1993).

- Existing NPPs be secured through license renewal, continued reductions in O&M costs, and reduced demands on safety systems;
- (Evolutionary) LWRs continue to develop safer and more competitive systems;
- Substantial progress must be made on the technologies required to assure that prior to the year 2100:
- All but the operationally minimal stocks of separated plutonium be eliminated;
- All inventoried plutonium remains unseparated and isolated by a high radiation barrier;
- All waste directed to a repository should be free of both actinides and long-lived fission products to the maximum extent practicable (Takagi, Takagi, and Sekimoto, 1998); and
- The world NPPs operated with the minimum inventories of plutonium in all forms (Arthur, Cunningham, and Wagner, 1998).

A three-phase growth scenario for nuclear energy is suggested. In this paper Todreas does *not* presuppose (in the first instance) any basically new reactor types or basically new systems. The first three bullets (above) are already being considered in the US. Plutonium recycle in LWRs, as practiced in Europe, is also a consideration. It is unclear, however, how the cost can be reduced.

If further development is needed, a rich array of technical solutions to the four cardinal issues that define the public concern (e.g., safety, waste, proliferation, cost) can be identified. This richness of technical solution and innovation is reflected in the large combination of materials available to perform the basic and essential functions needed to generate thermal and/or electrical energy from nuclear fission, as discussed earlier in Section 7.2.3. Furthermore, nuclear reactors are "tunable" (Todreas, 1993) to create materials, to destroy materials, or to provide process or space heat, simultaneously with the generation of electrical power. The material and intellectual resources needed to "re-engineer fission reactors for safe, globally sustainable, proliferation-resistant, and cost-effective nuclear power" are not in short supply.

7.4.2.2 A three-phased approach

The following generic features of the plan suggested by Todreas are summarized below:

Phase I (Present/near-future):

- Secure existing NPPs (mainly LWRs) through license renewals (pressure-vessel life, small-component replacement);
- Increase O&M effectiveness and associated cost reductions (NDT/robotics/remote-monitoring, coolant chemistry control, waste and dose reduction, validated reliability);

- Reduce challenges to safety systems (optimized balance between passive control and operator intervention in matters of plant safety);
- Begin gaining control of the waste issue [initiate a system of international monitored retrievable surface storage (IMRSS) systems for used fuels in preparation for Phase-III activities];
- Begin reduction of separated, inadequately secured plutonium inventories; and
- We would add here (re)gain balance in the regulatory process (Wilson, 1999, and Section 7.4.1.2.).

Phase II (Bridge to the future):

- Continue development and deployment of evolutionary LWRs [economically competitive, safer, standardized, flexible capacities, simplified (fewer valves, fewer pumps, reduced piping, less HVAC ducting, reduced seismic building volume, less control cable, etc.)];
- Meet key life-cycle requirements (close the fuel cycle under conditions required prior to attaining sustainability, e.g., fissile-fuel breeding);
- Optimize balance between passive and active safety systems; and
- Address diseconomies-of-scale issues on a per-region/application basis [~600 MWe and expandable, grid matching, size *versus* configuration *versus* coolability, fission-product quantity *versus* number of sites, reduce (installed) capital costs, modularity (factory *versus* site fabrication, site capacity)].

Phase III (Sustainable future):

- Enter into technologies required for a competitively sustainable nuclear-energy future that includes (possibly) proliferation-resistant breeding of fissile fuels from the world uranium and thorium resources;
- Use non-electric applications (when competitive); and
- Either direct or support facilities that eliminate all actinides and long-lived fission products (LLFP) from passing through to the externalities that will define Phase III.

The evolutionary Phase II would continue the ongoing process of developing and improving the present LWRs, with particular emphasis on a wide range of areas, including:

- Safety: increase use of passive safety systems, leading to optimum use of passive and active systems to maximize safety;
- Core and fuel designs: improve fuel integrity under accident conditions and maximize energy derived from a given mass of fissile material;
- Containment: develop advanced systems to provide improved safety against potential releases of radioactive materials to the environment;
- Reliability and load-factor increases;
- Economics: lower both capital and O&M costs;

- Efficiency: increase net electrical output per unit of thermal power generated;
- Load following: improve capability to track electrical loads; and
- Construction times: reduce under support of improved designs (standardization) and licensing programs/procedures.

Phase II must also address the problems of waste and proliferation and, therefore, would emphasize both actinide and long-lived fission product (LLFP) control. It remains for the technologist to assure that resolutions of these two key issues are presented while assuring good progress on the remaining two (cost and safety).

7.4.2.3 Fuel chains and cycles

The "once through" (OT) LWR fuel chain depicted in Fig. 7.10 and elaborated in Fig. 7.21 can be expanded to include two recycle schemes:

- The "closed cycle" merely recycling plutonium; and
- The "self consistent nuclear energy system" (SCENES) (Sekimoto and Takagi, 1991; Takagi, Takagi, and Sekimoto, 1998; Takagi and Sekimoto, 1992).

The "closed cycle" based only on (thermal-neutron) LWRs are capable at most of only two or three recycles of plutonium as MOX before the nuclear reactivity decreases, caused by shifts in plutonium isotopics which make this fuel inefficient in a thermal-neutron spectrum; the harder spectra developed in a fast-neutron reactor such as an LMR (Fig. 7.12) is needed to fission most of the higher plutonium isotopes. Nevertheless, some decrease in plutonium inventories is possible if the MOX/LWR fuel cycle is used. For the "continued growth" (Variant-I) scenario considered in the recent NEA study (OECD, 1998b), the decrease in "non-reprocessed spent fuel" for the case where MOX fuels occupy a core volume fraction of 30% for all LWRs (Fig. 7.19: NEA-I variant) is of the order of 40–50% (Krakowski, 1999). For these NEA cases, tailings from the uranium enrichment plant are taken to be $x_t = 0.003$; decreasing this tailings fraction to $x_t = 0.0015$ decreases the cumulative natural uranium requirement in the year 2050 by 1.4 MtonneU (a 25% decrease over the OT/LWR case). Adopting the MOX/LWR fuel cycle for this variant would save an additional 0.6 MtonneU (a 10.7% decrease in cumulative natural uranium demand). Table 7.7 summarizes the time evolution of nuclear energy generation, carbon-dioxide impacts, spent-fuel accumulations, natural-uranium resource demands, as well as impacts on carbon-dioxide emissions, for the Variant-I case (OECD, 1998b) depicted in Fig. 7.19.

Although the "closed cycle" based on using MOX in a LWR degrades the attractiveness of the reactor-grade plutonium for use in nuclear weapons (Beller and Krakowski, 1999), it far from eliminates this material. Furthermore, the

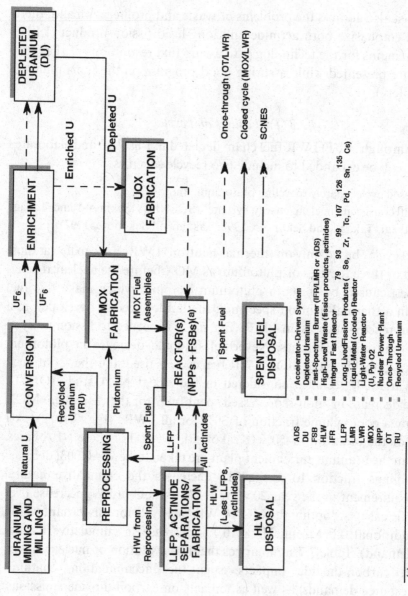

Figure 7.21 Once-through (OT/LWR) fuel cycle, plutonium recycle in LWRs (MOX/LWR), and an advanced actinide LLFP-burning fast-spectrum reactor [self-consistent nuclear-energy system, SCNES (Sekimoto and Takagi, 1991; Takagi and Sekimoto, 1992; Takagi et al., 1998)] fuel cycle.

Table 7.7 Summary of NEA/OECD (OECD, 1998a) NE Variant-I ("continued NE growth") parameter[a]

Year	1995	2000	2010	2020	2030	2040	2050
Capacity, P_E(GWe)	351	367	453	569	720	905	1120
Annual generation, GEN(TWeh)	2312	2450	3175	3988	5046	6453	7850
Percent primary energy	7						12
Percent electrical energy	17						35
Displaced CO_2 emission rate, ΔR_{CO_2} (GtonneC/yr)[b]	0.54	0.57	0.73	0.93	1.17	1.50	1.82
Displaced CO_2, $\Sigma \Delta R_{CO_2}$ (GtonneC)	—	2.8	9.3	17.6	28.1	39.4	56.0
Spent-fuel generation, R_{SF}(ktonneHM/yr)[c]	9.3	10.4	10.7	12.1	14.3	17.0	19.7
Cumulative spent fuel, ΣR_{SF}(ktonneHM)[d]	—	50.5	156.0	270.0	402.0	558.5	742.0
High-level waste generation, R_{HLW}(m³/yr)[e]	3510	3650	4530	5690	7200	9050	11200
Cumulative high-level waste, ΣR_{HLW}(10^3 m³)	—	17.9	58.8	109.9	174.4	255.6	356.9
Linear extent of waste, ℓ (m)[f]	—	26.2	38.9	47.9	55.9	63.5	70.9
Rate of uranium usage, R_U(ktonneU/yr)[g]	52	54.0	70.0	88.0	112.0	141.0	175.0
Cumulative uranium, ΣR_U(MtonneU)	—	0.34	0.94	1.75	2.75	4.0	5.6[h]

Notes:

[a] Variant I (OECD, 1998a) based on ecologically driven (Scenario C) (Nakicenovic, 1995, 1998).

[b] Based on 0.23 Gtonne C/1000 TWeh.

[c] Based on $FPY/(CBU \times \eta_{TH}) = 365/(40 \times 0.35) = 26.1$ tonne HM/GWe yr.

[d] Contains $\sim 0.9\%$ plutonium isotopes.

[e] Based on ~ 10 m³ HLW/GWe yr.

[f] Side of equivalent cube of HLW.

[g] Based on OT fuel cycle with 0.3% enrichment tailings.

[h] If enrichment tailings decreased from 0.3% to 0.15%, natural uranium requirement decreased ± 4.2 MtonneU ($\Delta M = 1.4$ MtonneU); plutonium recycle to 30% of all reactor cores leads to 0.6 Mtonne U cumulative savings; low tailings = MOX recycle leads to 2.0 MtonneU uranium savings (36%).

neutron economy in a critical thermal-spectrum reactor (where parasitic neutron absorption rates are high) is too precarious to deal with significant amounts of LLFPs. The use of a fast-spectrum reactor of the advanced liquid-metal reactor (ALMR) type can certainly address the consumption of all actinides, and possibly the transmutation of key LLFPs to shorter-lived species. Both actinide fissioning and LLFP transmutation can reduce repository requirements significantly. An ALMR can be used to reduce LWR-related repository demands while contributing electrical power to the grid (Wade, 1994; Brolin, 1993). Furthermore, the use of an integral-fast-reactor (IFR) pyrochemical processing does not create proliferation-prone separated (pure) plutonium, while promising significant reductions in waste streams relative to standard (PUREX) aqueous processing methods. In the pyrochemical process, chopped LWR oxide (or ALMR metal-alloy) fuel is electrochemically transferred in an electro-refiner to a molten-salt solution, with uranium being collected in a solid cadmium cathode, and actinides (including plutonium) are recovered in a cadmium anode. Because of the wider potential application of "dry" pyrochemical reprocessing in other re-engineered proliferation-resistant fuel-cycle (PRFC) configurations, this technology is receiving increased attention (Williamson, 1997).

The SCNES approach (Sekimoto and Takagi, 1991; Takagi *et al.*, 1998; Takagi and Sekimoto, 1992) incorporated in Fig. 7.21 targets the fissioning of all actinides (ACT) along with the transmutation of the seven main LLFPs listed on Fig. 7.14 to shorter-lived species. The overarching goal of the SCNES project is to target an equilibrium repository-specific inventory of ~3 tonne (LLFPs + ACT)/GWe. Since only a single (fast) reactor has been considered to date, and neutron balance remains a concern for three of the LLFPs because of the concentrations of stable isotopes (Zr, Cs, and Pd), isotopic separations are being considered. Each SCNES plant is envisaged to be fully integral, in that only natural uranium would enter the plant, and the toxicity of the material discharged from the plant to the repository would not exceed that of the feed (natural uranium) material. The severe specifications on decontamination factors (for ACT and LLFPs), as well as criticality considerations (even for a fast-spectrum reactor) and the potential costs of isotopic separations, have broadened the scope of the SCNES idea to include fusion- and accelerator-based neutron sources (Takagi, Takagi, and Sekimoto, 1998). While the long-term goals of the SCNES system(s) are challenging on physics, engineering, and economic fronts, the attractiveness of a nuclear energy waste stream that differs little in hazard and long-term impact from that of the relatively benign input streams needed to sustain the nuclear energy option can contribute to the resolution of the four cardinal issues in a way that can enhance public acceptance of nuclear energy.

The advantage of reduced cross sections and parasitic neutron capture associated with fast-spectrum reactors is countered by the need for large in-reactor inventories, and, hence, longer times are required to achieve fractional reductions in those large inventories (Pigford and Choi, 1991). The use of accelerator-produced neutrons to transmute LLFPs was proposed and analyzed more than four decades ago (Steinberg *et al.*, 1958). With advances in the prospects of high-current accelerators in recent years (Lawrence *et al.*, 1991), interest in accelerator transmutation of waste (ATW) rekindled (Sailor *et al.*, 1994; Bowman *et al.*, 1992; Venneri *et al.*, 1998). Early ATW concepts focussed on strong thermalization of the energetic spallation neutrons generated by the interaction of a ~1.0 GeV proton beam with a heavy-metal (W, Pb, U) target in order to transmute in the range of neutron energies where cross sections are high and inventories could be held to low values. Earlier (Sailor *et al.*, 1994) ATW concepts that consumed the actinides and two LLFPs (^{99}Tc and ^{129}I) output from a 1–GWe LWR, which at 0.27 GWe net power (per LWR serviced) from the ATW (20% recirculating power fraction needed to drive the high-power proton accelerator) give a support ratio of LWR/ATW $\simeq$ 3.7. These low-inventory, thermal-spectrum ATWs considered liquid-fuel blankets based either on (initially) aqueous slurries or (eventually) molten salts. More recent (conceptual) design directions for these ATW systems (Venneri *et al.*, 1998) are pointed towards the improved neutron economies associated with harder neutron spectra and higher active inventories. Figure 7.22 depicts the most recent embodiment of the ATW concept, which is centered around pyrochemical processes of the kind originally developed for the IFR (Laidler *et al.*, 1993) and more recently elaborated (Venneri *et al.*, 1998; Williamson, 1997) in the context of an ATW system designed to deal with the cumulating inventories of spent commercial nuclear fuels.

The appealing attributes of pyro-processing, as applied to IFR (Wade, 1994; Brolin, 1993; Laidler *et al.*, 1993) SCNES (Sekimoto and Takagi, 1991; Takagi *et al.*, 1998; Takagi and Sekimoto, 1992) or ATW, include (Venneri *et al.*, 1998):

- Proliferation resistance related to group rather than single-species separations;
- Processing media (molten salts and liquid metals) that can be exposed to multiple recycles/reuse;
- Relatively small process volumes and high throughputs, leading to short holdup times;
- No degradation of process media resulting either from decay heat or radiolysis; and
- Final products (e.g., actinides, LLFPs, disposable fission products, structural materials, recycled uranium) that are easily fabricated into either disposable or transmutable forms.

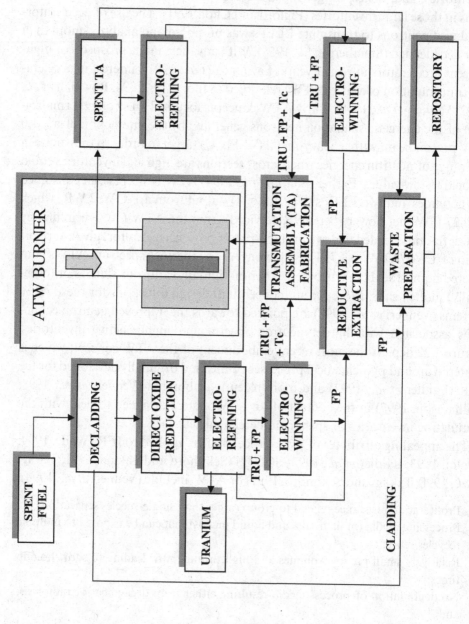

Figure 7.22 Recent (Venneri *et al.*, 1998) concepts in actinide and LLFP burning in accelerator-driven systems.

These attributes generally apply to IFR, SCNES, and/or ATW approaches in dealing with the proliferation and waste issues associated with the back-end of the nuclear fuel cycle. After mechanically removing the Zircalloy cladding, the uranium/plutonium/fission-product oxide fuel is reduced by calcium in a $CaCl_2$ flux, wherein the (chemically) active FPs (including iodine) are separated from the heavy metals (HM: U, Np, Pu, Am, and Cm). The heavy metals produced in the oxide reduction process are transferred to the electro-refining system. Upon application of a voltage to the HM-containing anode, uranium is transported through the eutectic NaCl–KCl molten salt and collected at the cathode, the more noble FPs remain in the "heel" at the anode, and transuranic elements (TRU) along with the rare-earth (RE) fission products partition to the molten salt. The anodic heel from the second stage of the electro-refining process (used to improve TRU recovery) is subsequently oxidized. Differences in the relative volatilities of the FP oxides carried in the anodic heel are used to separate technetium from the more noble FPs (Zr, Mo, and Ru). The TRU/RE-containing molten-salt is directed through an electro-winning process that separates TRUs from the REs. The TRUs are alloyed with Zr to produce ATW fuel. The REs are removed from the molten salt by a reductive extraction process, and the molten salt is recycled. While many of the chemical and process details remain to be resolved and/or optimized, particularly from the viewpoints of inventory and waste-stream reductions, key separations and the adherence to the philosophy of no separated plutonium and concomitant decrease in proliferation risk, while minimizing actinide and LLFP content in the waste streams, remain as guiding philosophies for this advanced, versatile fuel-cycle.

The Radkowsky thorium reactor (RTR) (Galperin, Reichert, and Radkowsky, 1997) portends the use of conventional LWR technology in a system that requires only thorium and LEU (20% ^{235}U) to achieve a power system with significantly reduced proliferation risk and increased access to the full and substantial thorium resource. The RTR systems diagram is depicted in Fig. 7.23. The RTP uses fuel assemblies in a conventional LWR (PWR or VVER) based on a (supercritical) "seed" (enriched uranium) region surrounded by and driving a (subcritical) "blanket" (breeding/fissioning thorium). The "seed" efficiently uses the uranium-based driver with reduced plutonium generation (minimize natural uranium requirement), and the blanket efficiently generates *in situ* fissioning of ^{233}U to maximize power generation. The claimed advantages of the RTR "seed-in-blanket" fuel assemblies include:

- Significant reduction (or possible elimination) of fuel-cycle proliferation risk;
- Reduced spent-fuel storage/disposal requirements;
- Fuel-cycle cost savings;

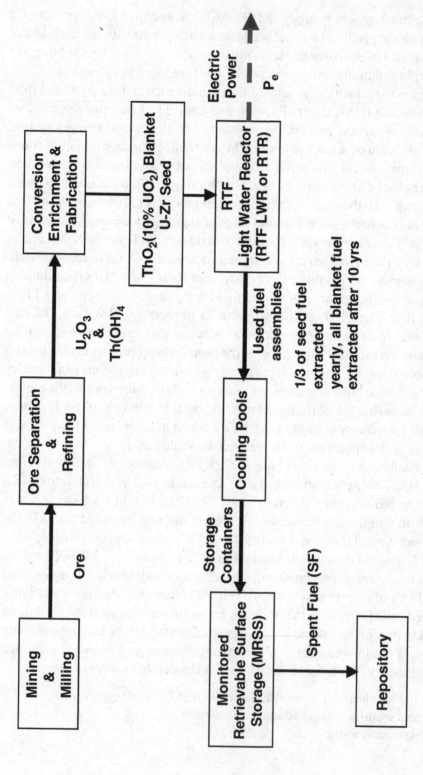

Figure 7.23　The Radkowsky thorium reactor (RTR) based on conventional LWRs　(Galperin et al., 1997).

- Use of new fuel in existing reactors [PWRs, VVERs, with minor hardware changes (fuel assemblies, control rods, burnable poisons)];
- Preservation of existing and proven safety and operational parameters; and
- Use of existing fuel (but not necessarily commercial) technologies.

Many of the above-listed attributes of the Th-U fuel cycle as applied to existing LWRs are also claimed by recent Russian studies (Marshalkin *et al.*, 1997; Murogov *et al.*, 1995) in a configuration that does not require a "seed-in-blanket" topology. Such a system would be initiated by burning excess plutonium stocks and eventually transition into a self-sustained Th-U fuel cycle without the generation of actinides.

7.4.3 *General prospects and directions*

7.4.3.1 *Five potential approaches*

Five potential (and incomplete) approaches to a nuclear future have been described above and map into the three phases described on Fig. 7.20 (Todreas, 1993):

- Once-through LWRs (OT/LWR), Phase I; and
- MOX recycle in LWRs (MOX/LWRs), Phase I and II.

The following Phase-III options have been suggested as routes to sustainable nuclear energy:

- Actinide (and possibly LLFP) destruction in ALMR/IFRs;
- IFR-based self-consistent nuclear energy systems (SCNES) which permit no actinide or LLFPs to leave the reactor site (Takagi, Takagi, and Sekimoto, 1998);
- Accelerator-driven systems for actinide fissioning and LLFP transmutation (ADSs or ATWs) (Venneri *et al.*, 1998); and
- Enriched-uranium (20%) driven thorium blankets assembled into existing LWRs, (Radkowsky) thorium reactors, RTRs (Galperin *et al.*, 1997), or homogeneous LWR Th-U [convertor/breeder (Marshalkin *et al.*, 1997; Murogov *et al.*, 1995)].

These generally partial/incomplete/disparate concepts, however (with the exception of the stand-alone OT/LWRs and MOX/LWRs), have the following common goals:

- Eliminate present stocks of separated plutonium through the generation of energy therefrom;
- Prevent the accumulation of future stocks of separated plutonium;
- Keep all but operationally necessary MOX inventories in strong intrinsic (fission products) and protecting radiation fields during all fuel cycle operations; and

- Reduce or eliminate the flow of:
    All actinides from the fuel cycle to the repository; and
    All LLFP from the fuel cycle to the repository.

These goals represent strategic elements of an architecture that offers a means to bridge to the sustainable future for nuclear power depicted in Fig. 7.20. This integrated actinide conversion system (IACS) (Arthur *et al.*, 1998) is described in Fig. 7.24 in terms of five essential components required to bridge an ELMR-based economy into a sustainable, plutonium- and/or uranium-breeding where no idle and/or inherently unprotected plutonium inventories in any form exist. The IACS is a philosophy and a means to this end. Possible roadmaps and sign-posts along the way needed to implement this philosophy are illustrated in Fig. 7.25 in terms of progress in each of the five component areas identified on Fig. 7.24:

- Eliminate separated plutonium in all forms;
- Improve ELWR safety and economics;
- Discharge all nuclear fuels into an IMRSS;
- Enlist all actinides in the generation of energy; and
- Create a repository that is free of actinides and LLFP which could be any one of the concepts described above.

7.4.3.2 Possible world futures

The rapid rate of growth of nuclear energy in the 1960s and 1970s has diminished considerably. The main growth and prospects for growth in nuclear energy is occurring in East and South Asia, with Western industrialized countries experiencing a period of stagnation. The prospects for nuclear energy can be assessed best by understanding the driving elements behind this stagnation. This slow down results from a complex mix of economic, environmental, and socio-political forces. Reduced to the essential elements, the main forces shaping public and market attitudes towards nuclear power plants are embodied in the four cardinal issues: waste, cost, (nuclear-weapons) proliferation, and safety. The priority of these concerns varies with personal and institutional tack [the relative (decreasing) order of import given above is that of the authors; albeit, all are important]. Additionally, each of these four cardinal issues shift in relative importance whether a short-term *versus* a long-term view is taken.

A number or recent studies (Fig. 7.19) address a range of nuclear energy futures (Beck, 1994; IAEA, 1997; NEI, 1998; Krakowski, 1999), although few if any of the above-described cardinal issues are incorporated directly. A quantitative picture of nuclear energy having to compete in a changing electric-supply-industry (ESI) market is given in Beck's study, wherein increased cost transparency, increased (short-term) market discipline, and the reduction of

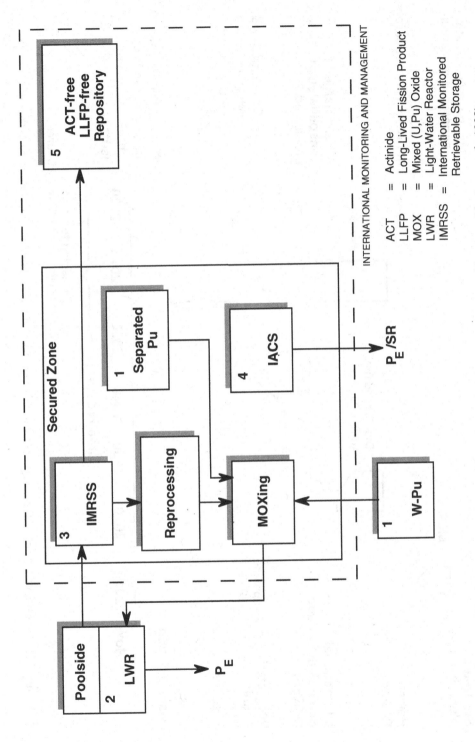

Figure 7.24 Integrated actinide conversion system (IACS) architecture (Arthur *et al.*, 1998).

INTERNATIONAL MONITORING AND MANAGEMENT

ACT = Actinide
LLFP = Long-Lived Fission Product
MOX = Mixed (U,Pu) Oxide
LWR = Light-Water Reactor
IMRSS = International Monitored
        Retrievable Storage

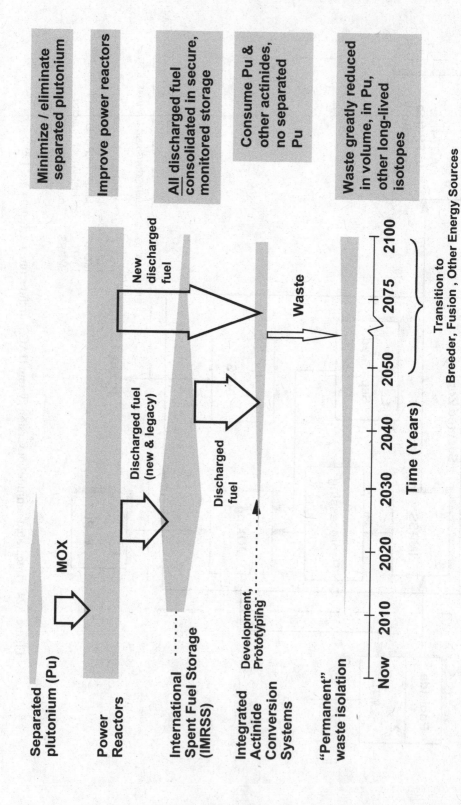

Figure 7.25 "Top-level" strategy for securing all nuclear materials and protecting the environment (Arthur *et al.*, 1998).

policy *per se* to a vestigial role creates an environment that is very different from that into which nuclear energy entered over two decades ago. The electric supply industry in general is facing significant and simultaneous changes that are vectored in a number of directions:

- organizational [structures, regulations, reduced intervention *via* policy routes, public *versus* private ownership, trends in the direction local–national–regional–global (multinational)];
- technological (impacts of emphasis on short-term competitiveness, e.g., CCGTs, small-to-medium sized units *versus* long-term sustainability);
- environmental (global climate change and possible restrictions on fossil-fuel use, pressures to close the nuclear fuel cycle, reductions in trans-boundary pollution, greenhouse-gas containment/sequestration); and
- economic (increasing imperatives on holding the line on energy costs, public–private ownership, increased cost transparency, internalization of energy costs and a drive towards sustainability).

The central issue is whether nuclear energy can find a niche in this changing market and the different view embodied in:

- public *versus* private financing and operations;
- public *versus* private perspectives on risk, risk *versus* benefit, and risk aversion; and
- the balance between short-term competitiveness *versus* long-term sustainability.

Furthermore, this situation is being defined under conditions of shifting (increases in) energy demand (to developing countries) and constrained fossil-energy resources, and whether the world can/should/needs to retain nuclear energy options.

While other studies (IAEA, 1997; OECD, 1998b; Krakowski, 1999) are more quantitatively consistent, Beck's more heuristic and intuitive study performed from a position on nuclear non-advocacy provides valuable insights that apply to all futuristic projections of these kinds, quantitative or not. Recognizing that nuclear technology cannot be "de-invented" and that the $\sim 100\,000$ tonnes of used nuclear fuel (and the ~ 1000 tonnes of plutonium and ~ 3000 tonnes of fission products contained therein, with the used fuel and plutonium inventories increasing to $300\,000$ tonnes and 2000 tonnes, respectively, by the year 2010) (Albright *et al.*, 1997) will not disappear, Beck (1994) suggests three (quantitatively) possible and encompassing world NE futures.

Beck's phase-out scenario (Case-1), for the reasons cited above as well as the local economic trauma that would ensue in countries that derive substantial fractions of their electrical needs from NPPs, is not considered realistic in terms of resolving on any reasonable time scale safety, waste, and proliferation issues. Maintenance of the health of the nuclear energy industry under Case-1 conditions also presents a large concern.

Beck's "muddle-along" (Case-2) is considered the most likely scenario, which assumes some countries will replace NPPs with NPPs, some countries will replace NPPs with non-NPPs, and some countries (particularly in East and South Asia) will add nuclear-energy capacity. During this no-growth period, the nuclear industry has an opportunity to address the four cardinal issues listed above and to make nuclear energy more acceptable to all stakeholders (e.g., governments, utilities and/or IPPs, and the general populace) in preparation for a period of strong growth sometime in the latter part of the 21st century. A similar late 21st-century re-invigoration of nuclear energy could be envisaged even for the Case-1 phase-out scenario, albeit the prospects for such a turnaround will depend critically on the condition/integrity/resiliency of a dwindling nuclear-energy infrastructure.

Beck's expansion scenario (Case-3, $\sim$25 GWe/yr) is predicated on a general (large) economic expansion that includes substantial increases in renewable energy sources to limit carbon-dioxide emissions by the year 2050 to 30% of present levels. In addition to waste-management and the possible need for reprocessing of used fuel for plutonium recycle, uranium resources may present an issue for the Case-3 growth scenario (a uranium resource of $\sim$18 Mtonne without breeder reactors).

The NEA/OECD (OECD, 1998b) futuristic study also considered three scenarios that are variations on those suggested by Beck, and are also depicted on Fig. 7.19. In addition to a continued growth scenario which is comparable to that suggested by Beck (1100 GWe in 2050 compared to 1400 Gwe), the NEA study (OECD, 1998b) considers a phase-out scenario along with a scenario that suggests a stagnation/revival. The NEA study departs from a more detailed model-based assessment performed by the IAEA (Wagner, 1997), which in turn utilizes the computational details of an earlier IIASA/WEC study (Nakicenovic, 1995). Figure 7.19 compares the IAEA high, medium, and low variants with an independent study conducted by Los Alamos (Krakowski, 1999); the high variant of the IAEA in the year 2050 is comparable with the projections of Beck (1994) and NEA (OECD, 1998b).

The goal of these kinds of long-term, multi-scenario studies is to project economic, energy (mixes, resources, etc.), and environmental (E^3) interactions/connectivities; Section 7.5 uses the Los Alamos study to elaborate on the role, goals, and results of these global E^3 studies, with an emphasis being placed on the role of nuclear energy in mitigating greenhouse gas (GHG) emissions. Section 7.5 elaborates on one of these global, long-term E^3 studies (Krakowski, 1999).

7.4.4 Maintaining equilibrium and reversing the trend

The suggestions/directions summarized in Fig. 7.20 (Todreas, 1993) and elaborated in Fig. 7.25 (Arthur, and Wagner, 1996; Arthur, Cunningham, and Wagner, 1998; Takagi, Takagi, and Sekimoto, 1998), coupled with the detailed recipe for regaining/restoring public confidence in nuclear energy outlined in Section 7.3.3 (Kasperson, 1993) are all essential elements in any strategy for maintaining (in some regions, restoring) developmental/implementational equilibrium for a period of public trial that might lead to circumstances that reverse the present global tendency towards a phase out of nuclear energy. A sustained record of worldwide NPP (and supporting facilities) safety and economic performance is essential to achieve the attributes claimed decades ago for this technology. These attributes increasingly must be demonstrated to publics that are becoming more and more adverse to the kinds of de-personalized risks associated with large technologies (Hiskes, 1998) that nuclear energy has come to represent. Many of the (governmental, industrial, and regulatory) management responses suggested in Section 7.4.1.3 will have to be successfully implemented for nuclear energy in a market environment that presents a range of economic and philosophical alternatives to meeting environmentally constrained energy needs.

7.5 Incorporating Nuclear Energy into Energy–Economic–Environmental (E³) Models

7.5.1 General setup and limitations of a model

Models of energy–economic–environmental (E^3) interactions provide a disciplined way in which long-term, global implications of a range of possible stories of the future might be evolved and examined. Rather than attempting to generate predictions of the future, these stories or scenarios are generated using relatively simple, transparent, and surprise-free deterministic models and assumptions to provide a future view that can be understood in terms of a relatively small set of input assumptions. In the present context of nuclear power and its potential for mitigating greenhouse-gas (GHG) emissions, the connectivities between the following elements must be understood:

- Energy (nuclear, fossil, and renewable) costs;
- Risks [e.g., global climate change (GCC) from use of fossil fuel or proliferation risk associated with nuclear energy, land use associated with some forms of renewable energies];

- Specific energy and support technologies adopted or evolved for each of these energy options;
- The desire and/or cost of diversity of energy supply;
- General global economic impacts (e.g., *per-capita* GNP as impacted by the cost of energy); and
- The long-term impacts of non-price-driven improvements in the energy costs of generating a given amount of GDP [anticipated decreases in primary-energy demand, PE(GJ/yr), per unit of GDP, or energy intensity, PE/GDP (MJ/$)].

These elements are of prime interest to those making energy decisions today for periods that extend far into the millennium. Global, long-term E^3 models, no matter how simplifying, are useful in developing an appreciation for these kinds of connectivities and the potential long-term impact of choices made or not made today.

Figure 7.26 depicts a simple input-output global E^3 model used in a recent IAEA-sponsored study of implications of a range of possible futures for nuclear energy (Krakowski, 1999; Bennett and Zaleski, 2000). This model was adopted from a established "top-down" (macroeconomic) global E^3 model (Edmonds and Reilly, 1985) to examine specifically the impacts of energy-generation costs (capital and fuel-cycle costs for nuclear energy, fuel cost for fossil energy *vis á vis* carbon taxes), GHG-mitigation potential and proliferation risk for nuclear energy, and broader economic issues (e.g., *per-capita* GDP) related to energy costs. The scenario attributes listed under 'Input' in Fig. 7.26 generally determine the long-term outcomes (energy mixes, GHG emissions and atmospheric accumulations, *per-capita* GNP growths, accumulations of radioactive and other nuclear materials, *etc.*). Under a set of assumptions for (regional) population growths (Bos *et al.*, 1995; DESA/UN, 1988; Lutz, 1996), economic productivity growths (Nakicenovic, 1995), energy resources (amounts *versus* cost and grade, and environmental impacts related thereto) (OECD, 1996; Rogner, 1997) and energy-generation costs (OECD, 1998a), and (particularly in the case of nuclear energy) the nature of the overall fuel cycle [e.g., once-through LWRs, MOX-recycle in LWRs, breeder reactors of given kinds, use of fast spectrum burners (FSBs) to reduce actinide and LLFP burdens on waste disposal] generally define a given view of the future attempted to be captured by a given scenario.

Typically, most-probable/credible projections for these "upper-level" scenario attributes listed on Fig. 7.26 are used to define a "point-of-departure" or "business-as-usual" (BAU) case against which other scenarios are compared. For example, in the recent IAEA E^3 Consultancy Study (IAEA, 1999; Krakowski, 1999) this BAU basis scenario invokes two nuclear-energy (sub)scenarios:

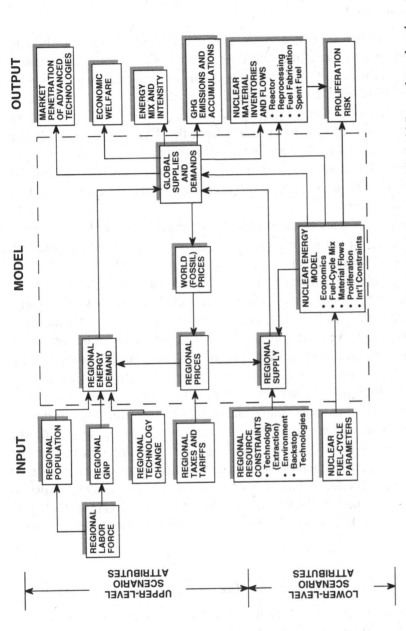

Figure 7.26 Structural layout of ERB global E³ model (Krakowski, 1999; Edmonds and Reilly, 1985) as adapted and modified for analysis support of Los Alamos Nuclear Vision Project (Arthur and Wagner, 1996); four main components comprise the ERB recursive economic-equilibrium model: energy demand; energy supply; energy balance; and greenhouse gas (GHG) emissions. Relationships between inputs and iterated (to common world fossil-fuel prices) outputs, as well as the addition of a higher-fidelity nuclear-energy model (e.g., resources, costs, nuclear-material flows, inventories, and proliferation risk) are also shown.

- A *basic options* (BO) scenario characteristic by nominal unit total cost (UTC ~2–2.4 $/We); and
- A case where nuclear energy was *phased out* (PO) through the imposition of very high capital costs, UTC.

An *ecologically driven* (ED) scenario was also considered, wherein a carbon tax was imposed on fossil fuels at a rate $CTAX$($/tonneC/15yr); the ED scenario also considered the two BO and PO (sub)scenarios for nuclear energy. The scenario approach adopted for the IAEA E^3 Consultancy Study, both in scope and assumption, follows that reported in Nakicenovic (1995). The following section summarizes selected ERB results from this IAEA study dealing with the potential impact of nuclear energy on GHG emissions, R_{CO_2} (GtonneC/yr), accumulations, W(GtonneC), and the resultant potential for atmospheric heating (Hasselmann *et al.*, 1995), ΔT(K), as UTC (BO $\rightarrow$ PO) and $CTAX$ (BAU $\rightarrow$ ED) are varied.

While tracking the parametric trade-offs occurring between these variables, the accumulation of plutonium in four forms (reactor, once-exposed spent fuel, multiply recycled spent fuel, and separated plutonium in both reprocessing and fuel-fabrication plants) is followed as a function of time for a thirteen-region model of the globe. Using a multi-attribute utility analysis, these inventories are related through a relative (0,1) proliferation risk index (PRI) (Krakowski, 1996), against which the ΔT(K) measure of GCC potential can be compared. Both PRI and ΔT are metrics that remain to be formulated in terms of a common and absolute damage function. This formulation is particularly difficult for proliferation in that PRI attempts to capture proliferation risks associated only with the civil nuclear fuel cycle, which is generally considered to present an unattractive source of nuclear-explosive material to a would-be proliferator compared to other sources [e.g., use of undeclared (not under international safeguards) facilities, particularly uranium-enrichment plants, or direct acquisition of smuggled nuclear-weapons components and materials]. In all instances, only the MOX/LWR ($f_{MOX} = 0.3$ MOX core fraction) fuel cycle was considered, although results from the OT/LWR fuel cycle are also reported.

7.5.2 Sample results: basis scenario (BAU/BO) demands and consequences

We report here the degree to which changes in the capital cost of nuclear energy, UTC($/We), and the fuel cost of fossil energy, through a carbon tax, UC_{TX}($/tonneC), imposed at a rate CTAX($/tonneC/15yr), impacts nuclear-energy demand, carbon-dioxide emissions, nuclear material accumulations, and GDP growth rates; these impacts are inter-connected through the modified

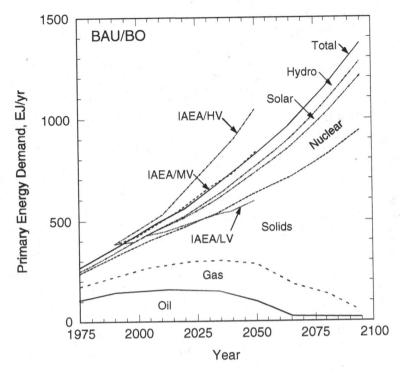

Figure 7.27 Time evolution of the six primary energies for the BAU/BO basis scenario, *showing each cumulatively*; a comparison with three IAEA scenarios (high variant, medium variant, and low variant) (Wagner, 1997) is shown.

ERB model described in Fig. 7.26. The primary-energy (PE) demand for each of six PE categories is given as a function of time on Fig. 7.27 for the BAU/BO basis scenario; a comparison with a recent IAEA study (Wagner, 1997), which largely followed the IIASA study (Nakicenovic, 1995), is also included. The demand for nuclear energy corresponding to this BAU/BO basis scenario is given in Fig. 7.19, which, in addition to comparing with the IAEA "new realities study" (Wagner, 1997), also gives results from a recent NEA (OECD, 1998a) study. After the year 2060, nuclear-energy demand is dominated by the developing (DEV) countries, with China being a major contributor to this demand. Generally, the world nuclear-energy demand for the BAU/BO basis scenario tracks closely that reported by the medium variant (MV) case considered by the IAEA "new realities study" (Wagner, 1997).

The global nuclear-materials (NM) accumulations and carbon-dioxide emissions consequences of this BAU/BO scenario are depicted in Figs. 7.28 and 7.29, respectively. The buildup of plutonium in the four forms described above is shown. Two key metrics adopted for this study are the accumulated

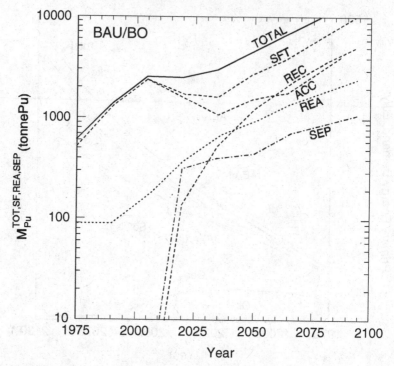

Figure 7.28   Evolution of world plutonium inventories for the BAU/BO basis scenario. Total inventory accumulations in four categories: once-exposed spent fuel, ACC; fully recycled ($N_{CYC}=3$) spent fuel, REC; in-reactor plutonium, REA, and separated plutonium, SEP = REP + FF; the time evolution of two (relative) proliferation metrics, $\langle u \rangle$ and PRI, are also shown for this $f_{MOX}=0.3$ basis scenario (Krakowski, 1999).

plutonium (according to time, form, and region; Fig. 7.28) and the accumulated atmospheric carbon dioxide, as estimated using integral response functions (Hasselmann *et al.*, 1995) in conjunction with the emission rates generated by the ERB model (Fig. 7.29). The correlation of world plutonium accumulation with atmospheric CO_2 accumulations ($W_0 = 594$ GtonneC in year ~1800) is shown in Fig. 7.30 and expressed in terms of the four plutonium categories. For the assumed (exogenous) growths in population (Bos *et al.*, 1995) and *per-capita* GDP (Nakicenovic, 1995), demands for both nuclear and fossil energy correspondingly increase for this BAU/BO basis scenario. Consequently, both plutonium and atmospheric carbon dioxide accumulate; Fig. 7.30 shows an "operating curve" that represents a "signature" for any given scenario (e.g., the transparently proclaimed set of scenario attributes in Fig. 7.26). At a lower level, the plutonium masses can be expressed in terms of PRI, and the impact of accumulated CO_2 can be estimated in terms of $\Delta T(K)$. Such a "second-level" correlation is given in Fig. 7.31, which also indicates how

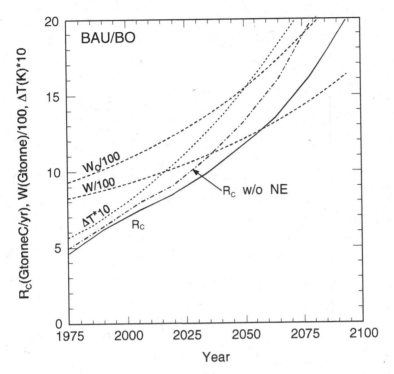

Figure 7.29 Evolution of world carbon-dioxide emission rate, integrated emissions, total atmospheric accumulations (Hasselmann *et al.*, 1995), and estimated global average temperature rise (Hasselmann *et al.*, 1995) for the BAU/BO basis scenario (Krakowski, 1999).

this scenario signature changes when the once-through fuel cycles ($f_{MOX} = 0.0$) for the otherwise BAU/BO basis scenario is considered. It is noted that the OT/LWR fuel cycle projects a somewhat lower cost of energy than for the $f_{MOX} = 0.3$ BAU/BO basis scenario, so that the OT/LWR scenario gains a slightly greater market share; ΔT in the year 2095 is thereby somewhat reduced, while the PRI is reduced by ~14% because of the reduction of (more heavily weighted) separated plutonium.

7.5.3 Sample results: capital cost variations

In searching for economic drivers needed to obtain a model-determined phase out of nuclear energy without the imposition of a carbon tax (e.g., the BAU/PO scenario), a programmed increase in the unit total cost for nuclear energy, UTC($/We), by a factor of f_{UTC} was exogenously imposed in the ERB model. The results of this parametric study are summarized in Fig. 7.32, which also includes comparisons made with the recent IAEA study (Wagner, 1997);

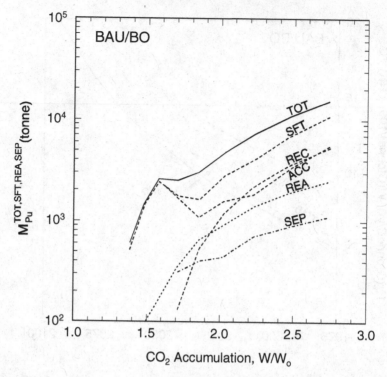

Figure 7.30 Correlation of world plutonium accumulation with atmospheric CO_2 accumulations ($W_o = 594$ GtonneC in year ~ 1800) expressed in four categories: once-exposed spent fuel, ACC; fully recycled ($N_{CYC} = 3$) spent fuel, REC; in-reactor plutonium, REA, and separated plutonium, SEP = REP + FF (Krakowski, 1999).

the case where $f^f_{UTC} = 3.0$ was adopted in the IAEA E[3] consultancy study (Krakowski, 1999; IAEA, 1999) as a description of the phase-out (PO) scenarios. The associated impacts on CO_2 emissions are shown in Fig. 7.33. An ultimate reduction of 75% in nuclear energy capital cost is sufficient to reach the IAEA/HV projection (Wagner, 1997). Cost increases by factors of 3–4 (relative to the BAU/BO conditions) are required to all but eliminate nuclear energy from the mix of primary energy suppliers. Even for these levels of cost increases, the nature of the logit-share functions used in the ERB model (Edmonds and Reilly, 1985) (contrary to "knife-edge" pricing decisions made under most linear-programming algorithms used in the technology or process-based "bottom-up" global E[3] models), and the increasing cost of fossil fuels driven by depletion of resources, together do not force a total foreclosure of the market share for nuclear energy. While CO_2 emissions rise somewhat when nuclear energy is all but removed from the energy mix (Fig. 7.33), this modest increase generally reflects the small share of nuclear energy in the primary-

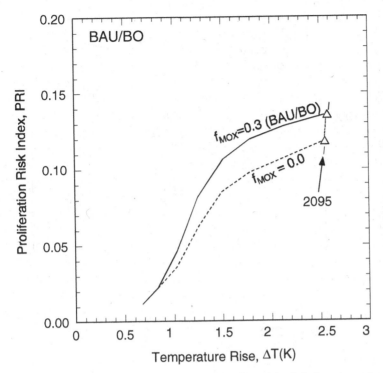

Figure 7.31   Correlation of proliferation-risk index with global-averaged surface temperature rise for BAU/BO scenarios ($f_{MOX} = 0.3$); the case of once-through fuel cycles ($f_{MOX} = 0.0$) for the BAU/BO basis scenario is also shown (Krakowski, 1999).

energy mix for this BAU scenario [i.e., relatively cheap fossil fuel (Edmonds and Reilly, 1985), but is not as cheap for the IIASA fossil-energy resource data base (Rogner, 1997)]. The impact of optimism in the fossil-fuel resource-base assumptions is reported elsewhere (Krakowski, 1999).

7.5.4 Sample results: carbon-tax variations

Carbon taxes are imposed at a linear rate CTAX($/tonneC/15yr) starting in the year 2005 (first "available" time after 1990 in the ERB model). The impact of increased fossil energy costs through carbon taxation on nuclear energy demand is shown parametrically in the CTAX parameter in Fig. 7.34. For the purposes of this study, the revenues collected from these taxes are assumed to "disappear" from the respective (13 regional) economies. The impact of more "revenue-neutral" schemes have been investigated (Krakowski, 1998a), but the ERB model [only one economic sector (energy) is modeled] is not sufficiently sophisticated to explore this problem with any degree of realism. Generally,

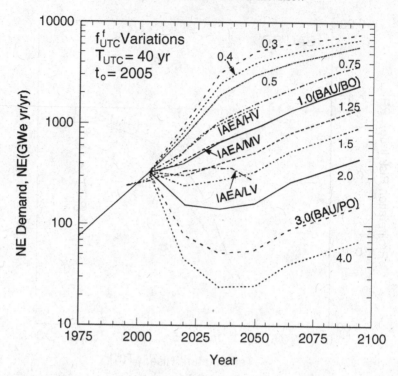

Figure 7.32   Impacts of increasing or decreasing the capital cost on nuclear-energy demand according to an exogenously determined schedule (Krakowski, 1999) have a time constant $T_{UTC} = 40$ yr; comparisons are made with recent IAEA high (HV), medium (MV), and low (LV) variant projections (Wagner, 1997), which are based on the IIASA/WEC study (Nakicenovic *et al.*, 1995, 1998); the case where $f^f_{UTC} = 3.0$ is adopted as a description of phase-out scenarios.

within the context of the ERB model, returning carbon-tax revenues to the GNP increases energy demand and makes such a tax less effective in stemming CO_2 emissions. Hence, the emission reductions depicted on Fig. 7.35 as CTAX is increased would not be as strong.

Figure 7.36 shows the correlation of decreased global temperature rise in the year 2095 with increased utilization of nuclear power induced through the imposition of a carbon tax starting in the year 2005 with the indicated constant rate CTAX($/tonneC/15yr). Also shown is the corresponding increase in absolute and relative (to the BAU/BO basis scenario) PRI. The impact of changing the rate at which SE → FE conversion efficiency improves with time, ε_k (yr^{-1}), is also shown; the BAU/BO basis scenario assumes ε_k is constant at 0.005 yr^{-1} up to the year 2005, and then linearly increases to 0.008 yr^{-1} by the year 2095.

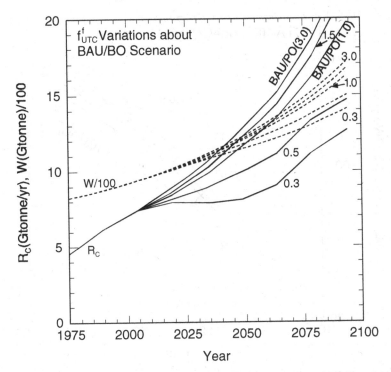

Figure 7.33 Impacts of increasing or decreasing the capital cost of nuclear energy according to an exponential schedule have a time constant $T_{UTC} = 40$ yr on CO_2 emission rates and atmospheric accumulations; both the BAU/BO ($f^f_{UTC} = 1.0$) and BAU/PO ($f^f_{UTC} = 3.0$) conditions are indicated (Krakowski, 1999).

7.5.5 Comparative summary of E³ modeling results

A key goal of a study of the kind reported by Krakowski (1999) is a quantified assessment of the long-term E³ impacts of nuclear energy for a range of scenarios, including a phase out. Identification of key developmental and operational characteristics for global nuclear power that will maximize global benefits while minimizing nuclear dangers is an equally important goal of this study. Fulfillment of each goal is crucial to understanding the future of nuclear power, particularly in identifying roles to be played in mitigating greenhouse warming in a way that avoids the creation of equally or more burdensome long-term problems.

Percentage changes in most of the key E³ parameters reported up to this point as nuclear energy is made cheaper or more expensive through the f^f_{UTC} cost algorithm have been collected in the form of a sensitivity diagram in

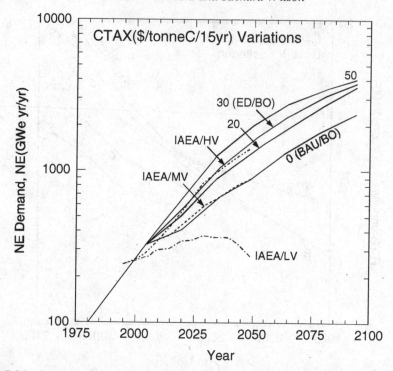

Figure 7.34   Impact of increasing the (linear) rate that carbon taxes are imposed after the year 2005 on nuclear-energy demand; comparisons are made with recent IAEA high (HV), medium (MV), and low (LV) variant projections (Krakowski, 1999; IAEA, 1999), which are based on an IIASA/WEC study (Nakicenovic *et al.*, 1995); the BAU/BO basis scenario corresponds to CTAX = 0.0 (Krakowski, 1999).

Fig. 7.37. All parameters give the percentage change for the year 2095 and are referenced to the $f^f_{UTC} = 1.0$ BAU/BO basis scenario. Table 7.8 defines key variables and lists the BAU/BO (2095) normalizing values for key parameters. The relative (percentage) change in CO_2 emission rates, ΔR (again in the year 2095), as well as the percentage change in atmospheric CO_2 inventories, ΔW_{CO_2} are given. The present value of world aggregated GNP, expressed relative to the basis scenario in the year 2095, is also given in Fig. 7. 37 as $\Delta GNP(PV)$. In magnitude, these GNP changes amount to fractions of a percent and are small compared to potential costs of global warming (Nordhaus, 1991a,b; Repetto and Austin, 1997), or the cost of direct carbon taxes to induce the use of reduced-carbon energy sources by making fossil fuels more expensive. Figure 7.37 also gives the relative (to the basis scenario) percentage variations of the fraction f_{NE} of PE comprised of nuclear energy, the fraction f_{EE} of PE that is converted to electricity, the global warming potential, ΔT, and proliferation-risk potential, ΔPRI. The dependence of the NE share, f_{NE}, on f_{UTC} indicates

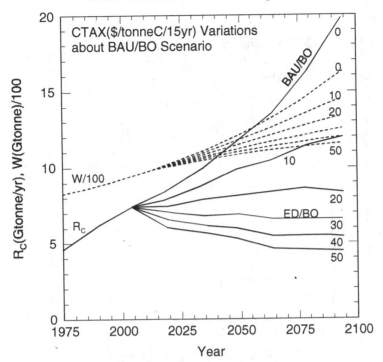

Figure 7.35   Impact of increasing the (linear) rate that carbon taxes are imposed after the year 2005 on CO_2 emission rates and atmospheric accumulations; the BAU/BO basis scenario corresponds to CTAX = 0.0, and CTAX = 30 \$/tonneC/15yr is adopted for the ED scenarios (Krakowski, 1999).

that a high effective "elasticity" of ~2.5 relates demand to capital cost of nuclear energy (Krakowski, 1999).

The direct correlation between accumulated CO_2 and total plutonium, and between the related ΔT and PRI metrics in the year 2095, as the carbon tax rate is varied, is shown in Fig. 7.36. To express parametric sensitivities of key system variables to the carbon-tax rate, CTAX(\$/tonneC/15yr), used to generate the ED/BO and ED/PO scenarios, the total present value (1990, $r = 0.04$ yr) of all collected carbon taxes is expressed relative to the total present value of the Gross World Product, with both being taken out to the year 2095. The resulting fraction, f_{TAX}, is used to express the sensitivities of the same variables displayed on Fig. 7.37 in terms of percentage changes in the year 2095 relative to the BAU/BO values (Table 7.8). The resulting sensitivity plot is given in Fig. 7.38, which also includes a curve that correlates f_{TAX} with the CTAX driving function. Figure 7.38 is the orthonormal component of Fig. 7.37 insofar as relating the key scenarios drivers used in this study.

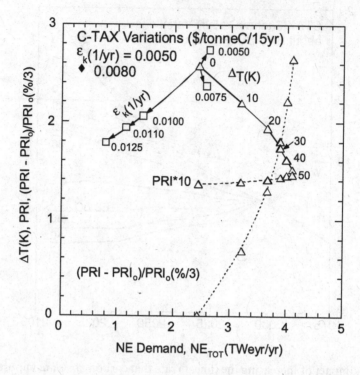

Figure 7.36   Correlation of decreased global temperature rise in the year 2095 with increased utilization of nuclear power induced through the imposition of a carbon tax starting in the year 2005 with the indicated constant rate CTAX($/tonneC/15yr). Also shown is the corresponding increase in absolute and relative (to the BAU/BO basis scenario) PRI. The impact of changing the rate at which SE → FE conversion efficiency improves with time, ε_k(1/yr), is also shown; the BAU/BO basis scenario assumes ε_k is constant at 0.005 1/yr up to the year 2005, and then linearly increases to 0.008 1/yr by 2095 (Krakowski, 1999).

7.6 Conclusion: A Possible Future for Nuclear Energy

Beck has noted that "The electric supply industry is the only market for civilian nuclear energy and in this regard the nuclear industry is on tap, but not on top" (Beck, 1994). If the world can satisfactorily address the public concerns about nuclear energy, and reverse the cost increases however caused, nuclear energy can begin to meet the promise of the optimists of 30 years ago and get back on top, instead of merely being on or under tap. It can go further and produce process heat, beginning perhaps in partnership with clean-coal technologies and eventually moving into hydrogen generation. Indeed, we do not have to go all the way back to the costs of 1973 (inflation corrected) for nuclear power to be the cost-effective method of generating electricity in many markets.

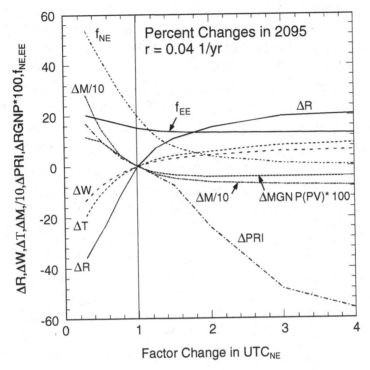

Figure 7.37 Summary of relative sensitivities of key metrics to NE capital-cost variations, as induced through the parameter f^f_{UTC}, measure in 2095 and referenced to the BAU/BO ($f^f_{UTC} = 1.0$, CTAX $= 0.0$ $/tonneC/yr) basis scenario in the year 2095; Table 7.8 gives values used to perform the normalizations, as well as key definitions (Krakowski, 1999).

Table 7.8 *Absolute values of key parameters used to generate normalized sensitivity diagrams (Figs. 7.37 and 7.38) for the year 2095*

Parameter	Value
Carbon-dioxide emission rate, R_{CO_2} (GtonneC/yr)	19.8
Atmospheric carbon dioxide inventory, W(GtonneC)	1631.6
Average global surface temperature increase, ΔT(K)	2.6
Total global plutonium, M_{Pu} (ktonnePu)	15.3
Proliferation risk index, PRI	0.14
Electrical energy as fraction of primary energy, EE/PE	0.15
Nuclear energy as fraction of primary energy, NE/PE	0.19

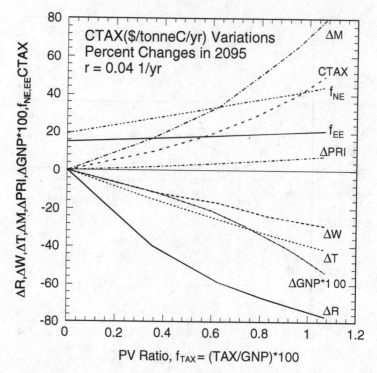

Figure 7.38   Summary of relative sensitivities of key metrics to (linear) carbon taxation, with the ratio of present-value of total carbon taxes to the present-value of gross world product, f_{TAX}, being used to express the impact of a range of (linear) carbon-tax rates, CTAX($/tonneC/15yr), and with metrics measure in the year 2095 and referenced to the BAU/BO ($f_{UTC}^f = 1.0$, CTAX = 0.0 $/tonneC/yr) basis scenario in the year 2095; Table 7.8 gives values used to perform the normalization, as well as key definitions (Krakowski, 1999).

This optimistic statement is contingent first upon continued operation of LWRs, with an open uranium fuel chain with ever-improving operating costs and decreased annualized capital-cost payments (Phases I and II of Sec. 7.4.2.1, Fig. 7.20). However, it is likely that one or more of the alternative reactors and fuel cycles discussed in Phase III (Fig. 7.20) will prove more economical of both electricity production, environmental impact, and, ultimately, of resource use, while further emphasizing the advantage of nuclear energy as a sustainable energy source.

It seems likely that a prerequisite for this optimism is a resolution of the public concerns discussed in Section 7.3.2. In particular, "re-engineering" of nuclear systems by itself will be ineffective in recovering public acceptance of this technology. The reality of, and the rate of approaching, any viable nuclear-energy future will depend on limits to growth as established by:

- The rate at which barriers to public acceptance of this technology are lowered;
- Energy demand shifts and growths at the global level; and
- Economic (financing) limitations.

Fuel resource limitations are often mentioned in this context, but Sections 7.1.4 and 7.2.6. show that real resource limitations do not exist for nuclear fission; for the kinds of growth scenarios depicted on Fig. 7.19, once-through use of low-enriched uranium should be adequate for the 21st century; any strong role for nuclear energy in significantly mitigating CO_2 emissions will require breeder reactors or economical seawater uranium coupled with FSBs (to control plutonium inventories), or the introduction of the Th–U fuel cycle (see Fig. 7.23).

The authors believe that a future for at least one hundred years is possible with present and evolutionary LWR technology (Phases I and II, Fig. 7.20), (Wilson, 2000). This future would involve a nuclear deployment rate of ~90 GWe/yr after the year 2030, leading to a final capacity of 5000 GWe by the year 2100. Nonetheless, the following approaches to a long-term nuclear energy future should be explored in Phase III, as nuclear energy strives for true sustainability:

- The long-term need for and economics of fissile-fuel breeders *versus* uranium-from-seawater/IACS (plutonium burning dedicated/LWR-supporting IFRs or ADSs);
- Use of the thorium resource in an LWR *versus* (plutonium) breeder reactors; and
- If a bridge to a viable nuclear energy cannot be constructed, the technological (nuclear-materials inventory), and overall infrastructural implications of a nuclear phase out (Beck, 1994; OECD, 1998b; Krakowski *et al.*, 1998b; Krakowski, 1999; IAEA, 1999) should be explored on both regional and global levels. This scenario presents a particularly significant challenge for those who wish to abandon nuclear energy.

If nuclear energy is to help in stabilizing CO_2 emissions to present rates (or below, if atmospheric concentrations are actually stabilized), the following will be required:

- Nuclear plant capacities of 5000 GWe by the year 2100, corresponding to deployment rates of 90 GWe/yr after ~2030;
- Depending on uranium resource assumptions, breeder reactors, based on either conventional plutonium breeders *or* a high-burn-up thorium cycle (most likely based on LWRs for electricity generation or HTGRs for process heat applications), will have to be deployed sometime after 2050 at a rate largely determined by the availability of startup plutonium; and
- Applications of nuclear energy for the production of transportable liquid fuels [clean(er) coal gasification first, leading ultimately to hydrogen generation, perhaps

using coal/CaO thermal-chemical cycles, along with CO_2 sequestration (Lackner and Ziock, 2001)] to satisfy a growing demand for non-electric energy represent future options and challenges.

Even in view of these projections, substantial increases in renewable energy sources, particularly solar and to a lesser extent biomass, will still be needed, depending largely on land-use competition.

An important issue is how to get there from here. One possible route might be to allow the nuclear industry effectively to die, and then to restart completely with new organizational infrastructure and new perspectives. This approach is similar to the "stagnate and revive" scenario suggested by the NEA/OECD study (Fig. 7.19 – NEA-III, OECD, 1998b). The build rates under "revive" conditions can be technically challenging (OECD, 1998b). Another approach might be to recognize that if the presently operating nuclear power plants can be kept running, the operating cost average of 1.9 cents per kWeh is much less than the "market entry" cost of 2.5 cents per kWeh of combined cycle natural gas. Brewer and Hanzlik (1999) have emphasized this possibility, and discussed who might pay for the return on capital already expended. A more likely scenario might be that nuclear power dies in some regions of the world, and flourishes in others, with past nuclear inventor-vendors becoming future customers of an irresistibly attractive new nuclear technology.

Finally we note that we can get close to meeting the Kyoto agreements if we wish to use nuclear energy and be in a position to make further reductions in CO_2 emissions thereafter. It need not be expensive. But we must be active and not merely talk about it. For nuclear energy to make significant contributions in a future diverse energy mix to mitigating global warming, strong progress in Phase III (Fig. 7.20) will be an absolute requirement (resolution of the front-end resource and back-end waste issues). Most importantly, ideas that recover and retain the competitiveness of safest and non-proliferating nuclear energy must be developed and connected to a global, long-term E^3 context; resolution of waste, economics, proliferation, and safety issues must be folded into an effective, self-consistent, and transparent package that the public can accept and trust.

List of Abbreviations

A	atomic mass
ACC	accelerator
ACRS	Advisory Committee on Reactor Safety
ACT	actinide
ADS	Accelerator-Driven Systems

AEC	Atomic Energy Commission
AEEI	Autonomous Energy Efficiency Improvement
ALMR	Advanced Liquid-Metal Reactor
ASCI	Advanced Super Computer Initiative
ATW	Accelerator Transmutation of (nuclear) Waste
BAU	Business-As-Usual
BU(GWtd/tonneHM)	Fission fuel burn-up
BWR	Boiling Water Reactor
c(m/s)	Speed of light in vacuum
CCGT	Combined-Cycle Gas Turbine
CHINA$^+$	China plus neighboring CPEs
CPE	Centrally Planned Economy
CTAX($/tonneC/15yr)	carbon tax rate
CY	Calendar Year
DEV	Developing regions (ME + CHINA$^+$ + NAFR + SAFR + LA + IND + SEA)
DF	Decontamination Factor (input/output)
DOE	Department of Energy
DOR	Direct Oxide Reduction
DP	DOE Defense Programs Office
DPY	Days Per Year
DU	Depleted Uranium
E (J)	energy
E^3	Energy–Economics-Environment
EE(EJ/yr)	Electrical Energy demand
EEU	Eastern Europe
EI(MJ/$)	Energy Intensity PE/GNP
ELWR	Evolutionary LWR
EPA	Environmental Protection Agency
ERB	Edmonds, Reilly, Barns model (Edmonds and Reilly, 1985)
FE	Fossil Energy
FP	Fission Product
FPY	Full-Power Year
FSB	Fast-Spectrum Burner
FSU	Former Soviet Union
f_{NE}	NE fraction of PE
f_{MOX}	(final) volume core fraction of LWR operated on MOX
f_{UTC}	(final) factor increase in UTC

GCR	Gas Cooled Reactor
GDP(B\$/yr)	Gross Domestic Product
GE	General Electric
GEN(TWeh)	annual eletricity generation
GHG	Green House Gas
GNP(B\$/yr)	Gross National Product
HLW	High Level Waste
HM	Heavy Metal
HV	High Variant
HVAC	Heating, Ventilation, Air Conditioning
IACS	Integrated Actinide Conversion System (Arthur *et al.*, 1996, 1998)
IAEA	International Atomic Energy Agency
IFR	Integral Fast Reactor
IIASA	International Institute for Applied Systems Analysis
IMRSS	International Monitored Retrievable Surface Storage
IND	India
IPCC	Intergovernmental Panel on Climate Control
IPP	Independent Power Producer
IRV	industrial revolution
JCAE	Joint Committee of Atomic Energy
L	Low
l(m)	scale length
LA	Latin America
LEU	Low-Enrichment Uranium (<20% ^{235}U)
LLFP	Long-Lived Fission Product
LV	Low Variant
LWR	Light-Water (fission) Reactor
M(kg)	atomic mass
ΔM(kg)	mass deficit
M_p(kg)	proton mass
M_n(kg)	neutron mass
M_{pu}(tonnePu)	total plutonium inventory ($M_{REA} + M_{ACC} + M_{REC} + M_{SEP}$)
M_{TAX}	carbon tax multiplier upon return to the economy
M_U(MtonneU)	integrated natural-uranium resource usage
ME	Middle East
MHTGR	Modular High-Temperature Gas (cooled) Reactor
MOX	Mixed (plutonium, uranium) Oxide fission fuel

MPC&A	Material Protection, Control and Accounting
MS	Molten Salt
MV	Medium Variant
N_{CYC}	number of plutonium recycles in LWRs
NAFR	North Africa
NDT	Non-Destructive Testing
NE(GWeyr/yr)	Nuclear Energy Demand
NEA	Nuclear Energy Agency
NPL	Nuclear Pumped Laser
NPP	Nuclear Power Plant
NPT	Nuclear Non-Proliferation Treaty
NRC	Nuclear Regulation Commission
NW	Nuclear Weapon
O&M	operating and maintenance
OECD	Organization for Economic Cooperation and Development (OECD-E + OECD-P)
OECD-E	OECD-Europe
OECD-P	OECD-Pacific
OT	Once-Through LWR
P_E(GWe)	NE capacity
PE(EJ/yr)	Primary Energy demand
PHWR	Pressurized Heavy-Water Reactor
PRA	Probability Risk Analysis
Prem	person rem
PRFC	Proliferation-Resistant Fuel Cycle
PRI	Proliferation Risk Index
PUREX	Plutonium-Uranium Recovery Extraction
PV	Present Value, photovoltaic (solar)
PWR	Pressurized-Water Reactor (LWR)
ppmv	parts per million by volume R_{CO_2} (GtonneC/yr) carbon dioxide emissions, same as R_C
R_{HLW}(m³/yr)	rate of HLW generation
R_{SF}(ktonneHM/yr)	rate of spent-fuel generation
R_U(ktonneU/yr)	rate of natural uranium use
RBMK	Russian water-cooled graphite-moderated reactor
RE	Renewable Energy, Rare Earth FP
REA	reactor plutonium
REC	plutonium in spent fuel not available for use in LWR
REF	reforming economies (FSU + EEU)
rem	Roetgen equivalent man

RG	Reactor Grade plutonium
RTP	Radkowsky thorium fuel reactor (Galperin *et al.*, 1997)
RU	recycled uranium
SAFR	Southern Africa
SCNES	self-consistent nuclear energy systems (Takagi *et al.*, 1998)
SE(EJ/yr)	Secondary Energy demand
SEA	South and East Asia
SEP	Separated Plutonium in Processing or fresh fuel
SF	Spent Fuel
SIPRI	Stockholm International Peace Research Institute
SLFP	Short-Lived Fission Product
T_{UTC}(yr)	time constant for UTC increase/decrease
TMI	Three Mile Island
TOT	total, world
TRU	transuranic elements
TX	carbon tax designator
$t_{o,f}$ (CY)	beginning, ending year for NE cost modifications
UC_{TX}($/tonneC)	unit carbon tax
UTC($/We)	Unit Total Cost for LWR
$\langle u \rangle$	global average utility function for nuclear proliferation
US	United States
UK	United Kingdom
VVER	Russian PWR
v(m/s)	speed
W(GtonneC)	integrated carbon dioxide emissions
W(GtonneC)	atmospheric carbon dioxide inventory, ppmv = $W/2.13$
W_{IRV}(GtonneC)	atmospheric carbon dioxide inventory in the year ~1800 (IRV = industrial revolution)
W_o(GtonneC)	integrated carbon dioxide emissions, same as W
$x_{p,f,t}$	^{235}U concentration in product (p), feed (f), and tailing (t) streams associated with uranium enrichment plant
	^{235}U, plutonium concentration in spent fuel from LWR.
Z	atomic number (change)
$\Delta R_C/R_C$	fractional reduction on carbon dioxide emissions

ΔR_{CO_2} (GtonneC/yr)	displaced CO_2 emission rate
$\Delta T(K)$	global average surface temperature rise
$\varepsilon_k(yr^{-1})$	annual improvement in secondary energy, energy service conversion, an AEEI-like parameter (Nordhaus, 1991a,b)
η_{TH}	thermal-to-electric conversion efficiency

References

Albright, D., Berkhout, F., and Walker, W. (1997), *Plutonium and Highly Enriched Uranium World Inventories, Capabilities and Policies*, 1996, Oxford University Press.

Arthur, E. D. and Wagner, R. L., Jr. (1996), *The Los Alamos Nuclear Vision Project*, Proc. Uranium Institute 21st Annual Symposium, London, UK (September 4–6).

Arthur, E. D., Cunningham, P. T., and Wagner, R. L., Jr. (1998), *Architecture for Nuclear Energy in the 21st Century*, Los Alamos National Laboratory report LA-UR-98–1931 (June 29).

Aston, F. W. (1919), A positive ray spectrograph, *Phil. Mag.* **38**: 707, and *Mass Spectra and Isotopes*, Edward Arnold, London, 1992.

Bashmakov, I. A. (1992), What are the current characteristics of global energy systems, in G. I. Pearman (Ed.), *Limiting the Greenhouse Effect: Options for Controlling Atmospheric CO_2 Accumulation*, Chap. 3, John Wiley and Sons, Ltd., New York, NY.

Beck, P. (1994), *Prospects and Strategy for Nuclear Power: Global Boon or Dangerous Diversion?*, Earthscan Publications, Inc.

Beller D. E. and Krakowski, R.A. (1999), *Burnup Dependence of Proliferation Attributes of Plutonium from Spent LWR Fuel*, Los Alamos National Laboratory document LA-UR-99–751 (February 12).

Benedict, M. (1971), Electric power from nuclear fission, *Technology Review* October/November.

Benedict, M., Pigford, T. H., and Levi, H. W. (1981), *Nuclear Chemical Engineering*, McGraw Hill, New York, NY.

Behnke, W. (1997), Communication from W. Benkhe, former C.E.O. of Commonwealth Edison Co., owner and operator of Dresden II and Dresden III.

Bennett, L. and Zaleski, C. P. (2000), Nuclear energy scenarios in the 21st century: potential for alleviating greenhouse gas emissions and saving fossil fuels, *Proc. Intern. Conf. on Global Warming and Energy Policy*, Global Foundation, Inc., Fort Lauderdale, Florida (26–28 October).

Bodansky, D. (1997), *Nuclear Energy*, American Institute of Physics, Woodbury, NY.

Bohr, N. and Wheeler, J.A. (1939), The mechanism of nuclear fission, *Phys. Rev.*, **56**:426.

Bowman, C. D. and Venneri, F. (1996), Underground supercriticality for plutonium and other fissile materials, *Science and Global Security*, **5(3)**, 279.

Bos, E.,Vu, My T., Massiah, E., and Bulatao, R. A. (1995), *1993: World Population Projection – Estimates and Projections with Related Demographic Statistics*, The World Bank, Johns Hopkins University Press.

Bowman, C. D., Arthur, E. D., Lisowski, R. W., Lawrence, G. P., Jensen, J. R. J.,

Anderson, L. *et al.* (1992), Nuclear energy generation and waste transmutation using an accelerator-driven intense thermal neutron source, *Nucl. Instr. and Meth.* **A320**, 336.

Brewer S. T. and Hanzlik, R. (1999), *Nuclear Power and the US Transition to a Restructured, Competitive Power Generation Sector*, B. Kursunoglu *et al.* (eds.) Global Foundation Conference Paris, October 22–23, 1998, Plenum Press, Inter. Conf. on Preparing the Ground for Renewal of Nuclear Power, New York, NY (1999).

Brolin, E. C. (1993), Factors affecting the next generation of nuclear power, *Proc. 2nd Intern. Conf. on the Next Generation of Nuclear Power Technology*, p. D-2, Massachusetts Institute of Technology report MIT-ANP-CP-002 (October 25–26).

Bunn, G. (1997), Strengthening international norms for physical protection on nuclear material, *IAEA Conf. on Physical Protection of Nuclear Materials: Experience in Regulation, Implementation, and Operations*, 17, (November 10–14, 1997).

CMP (1972), Report from Central Maine Power, Majority Owner of Maine Yankee.[10]

Cochran, R. G., and Tsoufanidis, N. T. (1990), *The Nuclear Fuel Cycle: Analysis and Management*, American Nuclear Society, La Grange Park, IL.

Cohen, B. L. (1977), High level waste for light-water reactors, *Revs. Mod. Phys.*, **49**:1.

Cottrell, Sir Alan (1981), *How Safe is Nuclear Energy?*, London; Exeter, N.H.: Heinemann.

Daniel, H. and Petrov Yu. V. (1993), *Feasibility Study of an Inherently Safe Subcritical Fission Reactor*, Petersburg Nuclear Physics Institute Preprint, PNPI-1992.

Daniel, H. and Petrov Yu. V. (1994), *Feasibility Study of a Subcritical Fission Reactor Driven by Low Power Accelerator*, Petersburg Nuclear Physics Institute Preprint, PNPI-1989.

Davis, Z. S. and B. Frankel (1993), *The Proliferation Puzzle*, Frank Cass and Company, Ltd., London, UK.

DESA (1988), *World Population Projections for 2150*, Population Division, Department of Economics and Social Affairs, United Nations Secretariat, New York, NY (February).

Diaz, Nils J. (1998), Nuclear technology: global accomplishments and opportunities, *Nuclear News*, **36** (May).

Douglas, M. T. (1970), *Natural Symbols*, Barrie and Rockliff, London, UK (1970).

Edmonds J. and Reilly, J. M. (1985), *Global Energy: Assessing the Future*, Oxford University Press, New York, NY.

EPA (1998), *Guidelines for Preparing Economic Analysis* (draft), US Environmental Protection Agency (July 17, 1998).

EPA (2000) US Environmental Protection Agency, 40 CFR Parts 141 and 142, *National Primary Drinking Water Regulations; Arsenic and Clarifications to Compliance and New Source Contaminants Monitoring*, Federal Register: 65 (121): 38887–38983 (2000).

Fang, D., Lew, D., Li, P., Kammen, D. M., and Wilson, R. (1998), *Options for*

---

[10] This cost does not include a (later) cost of $20 million to remove a causeway and improve tidal flow in the coolant estuary (which many experts thought was unnecessary and certainly would *not* have been demanded of a fossil fuel plant).

Reducing CO2: (i) Improving Energy Efficiency, (ii) Alternative Fuels in Reconciling Economic Growth and Environmental Protection in China, M. McElroy, Harvard University Press, Cambridge, MA 02138.

Federal Register (1986), 51 FR 28044.

Fisher, J. C. and Pry, R. H. (1971), A simple substitution model of technology change, *Technical Forecast and Social Change*, **3**, 75 (1471).

Galperin, A., Reichert, P., and Radkowsky, R. (1997), Thorium fuel for light water reactors – reducing proliferation potential for nuclear power fuel cycle, *Science and Global Security*, **6**, 225.

Gardner, G. T. (1994), *Nuclear Non Proliferation: a Primer*, Lynne Reiner, Boulder, CO.

Goldschmidt, B. (1982), *The Atomic Complex: A Worldwide Political History of Nuclear Energy*, American Nuclear Society, La Grange Park, IL.

Hafele, W. (1981), *Energy in a Finite World: Paths to a Sustainable Future*, International Institute for Applied Systems Analysis (IIASA), Ballinger Press, Cambridge.

Hahn, O., and Strassman, F. (1939), Uber den Nachweis und das Verhalten der bei der Bestahlumg des Urans mittels Neutronen entstehenden Erdalkalimetalle, *Naturwiss.*, **27**:11

Hasselmann, K., Hasselmann, S., Giering, R., and Ocana, V. (1995), Optimization of CO_2 emissions using coupled integral climate response and simplified cost models: a sensitivity study, in *Climate Change: Integrating Science Economics, and Policy*, Nakicenovic, N., Nordhaus W. D., Richels R., and Toth, F. L. (Eds.) International Institute of Applied Systems Analysis Report CP-96–1.

Hill, J. (1997), communication to the author by Sir John Hill, Chairman of the UK Atomic Energy Authority at the time of the Thorpe construction.

Hiskes, R. P. (1998), *Democracy, Risk, and Community*, Oxford University Press, Oxford, UK.

Holloway, D. (1994), *Stalin and The Bomb: The Soviet Union and Atomic Energy, 1939–1956*, Yale University Press.

IAEA (1991), *Electricity and the Environment*, Proceedings of the senior expert symposium, Helsinki, Finland, 13–17 May 1991, IAEA, Vienna.

IAEA (1997), *Nuclear Fuel Cycle and Reactor Strategies: Adjusting to New Realities*, International Atomic Energy Agency.

Jackson, S. A. (1998), Transitioning to risk-informed regulation: the role of research, *Nuclear News*, p. 29, January.

Jasper, J. M. (1990), *Nuclear Politics. Energy and the State in the United States, Sweden, and France*, Princeton University Press, Princeton, NJ.

Johansson, T. B., Kelly, H., Reddy, A. K. N., and Williams, R. H. (Eds.) (1993), *Renewable Energy: Sources for Fuels and Electricity*, Island Press, Washington, DC.

Joliot, F., von Halban, H., and Kowarski L. (1939), Number of neutrons liberated in a nuclear fission of uranium, *Nature*, **143**:680.

Kasperson, R. E. (1992), The social amplification of risk: progress in developing an integrative framework, *Social Theories of Risk*, S. Krimsky and D. Golding (Eds.), pp. 153–158, Praeger, Westport, CT.

Kasperson, R. E. (1993), Can nuclear power gain public acceptance?, *Proc. 2nd Intern. Conf. on the Next Generation of Nuclear Power Technology*, p. 3–2, Massachusetts Institute of Technology report MIT-ANP-CP-002 (October 25–26).

Kastenberg, W. E., Peterson, P. F., Ahn, J., Burch, J., Casher, G., Chambre, P., Greenspan, E., Olander, D. R., Vujic, J., Bessinger, B., Cook, N. G. W., Doyle, F. M., and Hilbert, L. B., (1996), Considerations of autocatalytic criticality of fissile materials in geologic repositories, *Nucl. Sci. and Technol.*, **115**, 298.

Krakowski, R. A. (1996), *A Multi-Attribute Utility Approach to Generating Proliferation-Risk Metrics*, Los Alamos National Laboratory document LA-UR-96–3620 (October 11).

Krakowski, R. A. (1998a), *Energy-Economic-Environment (E³) Modeling Activities/ Capabilities at Los Alamos National Laboratory*, Los Alamos National Laboratory document LA-UR-98–945 (March 8).

Krakowski, R. A. (1998b), *Preliminary Parametric Studies Using the ERB Model in Search of Demand Scenarios for Use by IAEA Consultancy E³ Study Team*, Los Alamos National Laboratory document LA-UR-98–2252 (May 18).

Krakowski, R.A. (1998c), *The Role of Nuclear Energy in Mitigating Greenhouse Warming*, International Conf. on Environment and Nuclear Energy, Washington DC (October 27–28, 1997), Plenum Press.

Krakowski, R. A. (1999), *Los Alamos Contributions to the IAEA Overall Comparative Assessment of Different Energy Systems and Their Potential Role in Long-Term Sustainable Energy Mixes*, Los Alamos National Laboratory document LA-UR-99–627 (February 3).

Krakowski, R. A. and Bathke, C. G. (1997a), *Long-Term Nuclear Energy and Fuel Cycle Strategies*, Los Alamos National Laboratory document LA-UR-97–3826 (September 24).

Krakowski, R. A., Davidson, J. W., Bathke, C. G., Arthur, E. D., and Wagner, R. L, Jr. (1997b), Global economic/energy/environmental (E³) modeling of long-term nuclear futures, *Proc. Global '97 International Conf. on Future Nuclear Systems*, p. 885, Yokohama, Japan (October 5–10).

Krakowski, R. A., Davidson, J. W., Bathke, C. G., Arthur, E. D., and Wagner, R. L, Jr. (1998a), Nuclear energy and materials in the 21st century, *International Symp. on Nuclear Fuel Cycle and Reactor Strategy: Adjusting to New Realities*, Vienna, Austria (June 3–6, 1997).

Krakowski, R. A., Bennett, L., and Bertel, E. (1998b), Nuclear fission for safe, globally sustainable, proliferation resistant and cost effective energy, *Inter. Conf. on Preparing the Ground for Renewal of Nuclear Power: Global Foundation*, Paris (October 22–23, 1998), Plenum Press, New York, NY.

Krakowski, R. A., and Bathke, C. G (1998c), *Long-Term Global Nuclear Energy and Fuel-Cycle Strategies*, Los Alamos National Laboratory document LA-UR-97–3836 (September 24).

Kress, T. (1994), Report to Nuclear Regulatory Commission from the Advisory Committee on Reactor Safeguards.

Kurihara, H. (1977), Strengthening physical protection of nuclear material, *IAEA Conf. on Physical Protection of Nuclear Materials: Experience in Regulation, Implementation, and Operations*, 9 (November 10–14).

Lackner, K. S., Wendy, C. H., Butt, D. P., Joyce, E. I., and Sharp, D. M. (1995), Carbon dioxide disposal in carbonate minerals, *Energy*, **20**(11), 1153.

Lackner, K. S. and Ziock, H. (2001), How coal could fuel the 21st century. Talk at Columbia University.

Laidler, J. J., Battles, J. E., Miller, W. E., and Gay, E. C. (1993), Development of IFR pyroprocessing technology, *Proc. Int. Conf. on Future Nuclear Systems: Emerging Fuel Cycles and Waste Disposal Options Global '93*, pp. 1061–1065, Seattle, WA.

LaPorte, T. R. and Metla, D. S. (1996), Hazards and institutional trustworthiness – facing a deficit of trust, *Public Admin. Rev.*, **56**(4), 341.

Lawrence, G. P., Jameson, R. A., and Schriber, S. O. (1991), Accelerator technology for Los Alamos nuclear-waste transmutation and energy-production concepts, *Proc. Intern. Conf. on Emerging Nuclear Energy Systems*, Monterey, CA (June 16–21).

Lindstrom, S. (1992), The brave music of a distant drum: Sweden's nuclear phase out, *Energy Policy*, **20**, 623–631.

Lochard, J. (Chm.) (1997), Key issues paper no. 4: safety, health, and environment implications of the different fuel cycles, *International Symposium on Nuclear Fuel Cycle and Reactor Strategies: Adjusting to New Realities*, Vienna, Austria (June 3–6, 1997), International Atomic Energy Agency, Vienna (1998), 191.

Longworth, R. C. (1998), *Global Squeeze: The Coming Crisis for First-World Nations*, Contemporary Books, Chicago, IL (1998).

Lutz, W. (Ed.) (1996), *The Future Population of the World: What Can We Assume Today?*, Earthscan, London.

Manne, A. S. and Richels, R. G. (1992), *Buying Greenhouse Insurance: The Economic Costs of Carbon Dioxide Emission Limits*, The MIT Press, Cambridge, MA.

Marshalkin, V. E., Pavyshev, V. M., Trutnev, Ju. A. (1997), On solving the fissionable materials non-proliferation problem in the closed uranium–thorium cycle, in *Advanced Nuclear Systems Consuming Excess Plutonium*, E. R. Merz and C. E. Walter (Eds.) Kluwer Academic Publ., 237–257.

Marshall, W. (Ed.) (1983), *Nuclear Power Technology*, Clarendon Press, Oxford.

Masters, C. D., Attanasi, E. D., Dietzman, W. D., Meyer, R. F., and Mitchell, R. W. (1987), World resources of crude oil, natural gas, natural bitumen, and oil shale, *Proc. 12th World Petroleum Congress*, Houston, TX.

Masters, C. D., Attanasi, E. D., and Root, D. H. (1994), World petroleum assessment, *Proc. 14th World Petroleum Congress*, Stavanger, Norway (1994).

McCoy, B. (1998), talk by vice president of Commonwealth Edison, Kennedy School of Government.

Meneley, D. (Chm.) (1997), Key issues paper no. 3: future fuel cycle and reactor strategies, *International Symposium on Nuclear Fuel Cycle and Reactor Strategies: Adjusting to New Realities*, Vienna, Austria (June 3–6, 1997), International Atomic Energy Agency, Vienna (1998).

Messner, S. (1997), *Endogenized Technological Learning in an Energy Systems Model*, International Institute for Applied Systems Analysis document RR-97–15 (November).

Messner, S. and Strubegger, M. (1995), *User's Guide for MESSAGE III*, International Institute for Applied Systems Analysis report WP-95–69.

Meyer, S. M. (1984), *The Dynamics of Nuclear Proliferation*, Univ. Chicago Press, Chicago, IL.

Murogov, V. M., Dubinin, A. A., Zyablitsev, D. N. *et al.* (1995), *Uranium–thorium Fuel Cycle – Its Advantages and a Perspective of the Nuclear Power Development on Its Base*, Institute for Physics and Power Engineering (IPPE) preprint 2448, Obninsk, Russia.

Nakicenovic, N. (Study Director) (1995), *Global Energy Perspectives to 2050 and Beyond*, International Institute for Applied Systems Analysis (IIASA) and World Energy Council (WEC) report (1995); recently re-issued by N. Nakicenovic, A. Grubler, and A. McDonald (Eds.), *Global Energy: Perspectives*, Cambridge University Press, Cambridge, UK (1998).

Nakicenovic, W. D., Nordhaus, R., Richels, and Toth, F. L. (Eds.) (1995), International Institute for Applied Systems Studies report CP-96–1.

NAS (1991), *Nuclear Power: Technical and Institutional Options for the Future*, National Research Council, J. F. Ahearn (Chm.); National Academy Press.

NCRP (1990), *Implementation of the Principle of As Low As Reasonably Achievable (ALARA) for Medical and Dental Personnel*, National Council on Radiation Protection and Measurements, Report No. 107.

NEI (1998), *Nuclear Energy: Perception Gap*, Nuclear Energy Insight, Nuclear Energy Institute, 8, (March).

Nelkin, D. (1971), Scientists in an environmental controversy, *Science Studies*, 1:245.

Nelkin, D. (1974), The role of experts in a nuclear siting controversy, *Bulletin of the Atomic Scientists*, **30(9)**:29–36.

Nelkin, D. and Pollak, M. (1981), *The atom besieged: extra-parliamentary dissent in France and Germany*, MIT Press, Cambridge, Mass.

NRC (1975), *Rule making RM-30-2*, Nuclear Regulatory Commission.

NRC (1990), *Severe Accident Risks: An Assessment for Five US Nuclear Power Plants*, Nuclear Regulatory Commission report NUREG 1150.

Nordhaus, W. D. (1991a), To slow or not to slow: the economics of the greenhouse effect, *The Economic Journal*, **101**:929.

Nordhaus, W. D. (1991b), The cost of slowing climate change: a survey, *The Energy Journal*, **12**, 37.

Nuclear News (1992), The new reactors, **35**(12), 65 (September).

NYT (1996), Figures reported in the New York Times.

OECD (1996), *Uranium: 1995 Resources, Production, and Demand*, Nuclear Energy Agency of the Organization for Economic Co-operation and Development (NEA/OECD) and the International Atomic Energy Agency (IAEA) joint report, OECD, Paris.

OECD (1998a), *Projected Costs of Generating Electricity: Update 1997*, OECD/NEA-IEA report.

OECD (1998b), *Nuclear Power and Climate Change*, Nuclear Energy Agency (NEA)/Organization for Economic Co-operation Development (OECD) report (April).

OTA (1977), *Nuclear Proliferation and Safeguards*, Office of Technology Assessment, Praeger Publishers, New York, NY (1977).

Parkinson, C. N., 1957, *Parkinson's Law*, 12, Houghton Mifflin, Boston.

Pigford, T. H. and Choi, J. S. (1991), Inventory reduction factors for actinide-burning liquid-metal reactors, *Trans. Amer. Nucl. Soc.*, **64**, 123.

Price, T. (1990), *Political Electricity: What Future for Nuclear Energy*, Oxford University Press.

Reiss, M. (1995), *Bridled Ambitions: Why Countries Constrain Their Nuclear Capabilities*, Woodrow Wilson/Johns Hopkins University Press, Washington, DC.

Repetto, R. and Austin, D. (1997), *The Costs of Climate Protection: A Guide for the Perplexed*, World Resources Institute report.

Rochlin, G. I. (1993), Nuclear technology and social culture, *Proc. 2nd Intern. Conf. on the Next Generation of Nuclear Power Technology*, p. 7–11, Massachusetts Institute of Technology Report MIT-ANP-CP-002 (October 25–26).

Rochlin, G. I. and von Meier, A. (1994), Nuclear power operations: a cross-cultural perspective, *Ann. Rev. Energy and Environment*, **19**, 153.

Rochlin, G. I. and Suchard, A. (1994), Nuclear power operations: a cross-cultural perspective, *Ann. Rev. Energy and Environment*.

Rogner, H. H. (1996), Hydrogen technologies and the technology learning curve, *Proc. 11th World Hydrogen Energy Conf.*, **2**, 1839, Stuttgart, Germany (June 23–28).

Rogner, H. H. (1997), An assessment of world hydrocarbon resources, Annual Review, *Energy and Environment*, **22**, 217.

Sailor, W. C., Beard, C. A., Venneri, F., and Davidson, J. W. (1994), Comparison of accelerator-based with reactor-based nuclear waste transmutation schemes, *Prog. in Nuclear Energy*, **28**(4), 359.

Sakharov, A. M. (1978), Nuclear power and freedom of the West, *Bulletin of the Atomic Scientists*, pp. 12–14.

Seaborg, G. (1968) Report to Joint Committee on Atomic Energy, *Congressional Record*.

Sekimoto, H. and Takagi, N. (1991), Preliminary study on future society in nuclear quasi-equilibrium, *J. Nucl. Sci. and Technol.*, **28**(10), 941.

Shlyakhter, A. I. and Wilson, R. (1992), Chernobyl: the inevitable result of secrecy, *Public Understanding of Science*, **1**:251–259.

Shlyakhter, A., Stadie, K., and Wilson, R. (1995), *Constraints limiting the expansion of Nuclear Energy*, Energy, Environment, and Economics Program, US Global Strategy Council, Washington, DC.

Silvennoinen, P. (1982), *Nuclear Fuel Cycle Optimization: Method and Modeling Techniques*, Pergamon Press, Oxford, UK.

Slovic, P. (1993), Perceived risk, trust and democracy: a system perspective, *Risk Analysis* **13**:675–682.

Slovic, P., Fischhoff, B., and Lichtenstein, S. (1982), Risk-aversion, social values, and nuclear safety goals, *Trans. Amer. Nucl. Soc.*, **41**, 448.

Steinberg, M., Wotzak, G., and Manowitz, B. (1958), *Neutron Burning of Long-Lived Fission Products for Waste Disposal*, Brookhaven National Laboratory Report BNL-8558.

Takagi, N. and H. Sekimoto (1992), Feasibility of fast fission system confining long-lived nuclides, *J. Nucl. Sci. and Technol.*, **29**(3), 276.

Takagi, N., Takagi, R., and Sekimoto, H. (1998), Effect of decontamination factor of recycled actinide and fission products on the characteristics of SCNES, *Prog. in Nuclear Energy*, **32**(3/4), 441.

Taylor, T. B. and Willrich, M. (1974), *Nuclear Theft: Risks and Safeguards*, Ballinger, Cambridge, Massachusetts.

Tengs, T. O., Adams, M. E., Pliskin, J. S., Safran, D. G., Siegel, J. E., Weinstein, M. C., and Graham, J. D. (1995), Five hundred life saving interventions and their cost effectiveness, *Risk Analysis*, **15**:369.

Thompson, M., Ellis, R., and Wildavsky, A. (1990), *Cultural Theory*, Westview Press, Boulder, CO.

Todreas, N. E. (1993), What should our future nuclear energy strategy be?, *Proc. 2nd Intern. Conf. on the Next Generation of Nuclear Power Technology*, p. 7–2, Massachusetts Institute of Technology Report MIT-ANP-CP-002 (October 25–26).

Towers Perrin (1994), *Nuclear Regulatory Review Study*, Final Report (October), Report to Nuclear Energy Institute, Washington DC.

Venneri, F., Ning, Li, Williamson, M., Houts, M., and Lawrence, G. (1998), Disposition of nuclear waste using sub-critical accelerator-driven systems: technology choices and implementation scenario, *Proc. 6th Intern. Conf. on Nuclear Engineering (ICONE-6)*, ASME Publication (May 10–15).

Wade, W. D. (1994), Management of transuranics using the integral fast reactor

(IFR) fuel cycle, *Proc. Intern. Conf. on Reactor Physics and Reactor Computations*, 104, Tel-Aviv, Israel (January 23–26).

Wagner, H. F. (Chm.) (1997), Key issues paper no. 1: global energy outlook, *International Symposium on Nuclear Fuel Cycle and Reactor Strategies: Adjusting to New Realities*, Vienna, Austria (June 3–6, 1997), International Atomic Energy Agency, Vienna (1998).

Way, K. and Wigner, E. P. (1948), The rate of decay of fission products, *Phys. Rev.*, **73**:1318.

Webster (1972), Letter from William Webster, President of NE Electric System (operator of Connecticut Yankee) to Richard Wilson.

WEC (1995), *Survey of Energy Resources*, World Energy Council report.

Weinberg, A. M. and Wigner, E. P. (1958), *The Physical Theory of Neutron Chain Reactors*, University of Chicago Press, Chicago, IL.

West, J. (2000), Presentation to Special Committee on ANL (Argonne National Laboratory) Nuclear Technology Division, (October 17, 2001).

Williamson, M. A. (1997), Chemistry technology base and fuel cycle of the Los Alamos accelerator-driven transmutation system, *Proc. Global '97 Conf.*, Yokohama, Japan.

Willrich, M. and Taylor, T. B. (1974), *Nuclear Theft: Risks and Safeguards*, Ballinger, Cambridge, MA.

Wilson, R. (1989), Global energy use: a quantitative analysis, in *Global Climate Change Linkages*, J. C. White (Ed.), Elsevier, NY.

Wilson, R. (1992), The future of nuclear power, *Env. Sci. Technol.*, **26**, 1116–1120.

Wilson, R. (1994a), The potential for nuclear power, in *Global Energy Strategies: Living with Restricted Greenhouse Gas Emissions*, White, J. C. (Ed.), Plenum Press, NY. pp. 27–45.

Wilson, R. (1994b), The potential environmental effects of nuclear power, in *Environmental Contaminants and Health*, S. K. Majumdar, F. J. Brenner, E. W. Miller, and L. M. Rosenfeld, (eds.) *Pennsylvania Academy of Sciences* (in press).

Wilson R (1998a), *Over-regulation and Other Problems of Nuclear Power*, presented at the Global Foundation Conference, Washington DC (October 27–29, 1997). Plenum Press, New York, NY.

Wilson, R. (1998b), *Remembering How to Make Cheap Nuclear Electricity*, testimony to Subcommittee on Energy and Water Development, United States Senate Committee on Appropriations (May 19), Washington DC.

Wilson, R. (1998c), *Public Acceptance of Nuclear Energy: Regulatory Issues and a Critique of Accelerator Driven Reactors*, Energy, Environment and Economy Program Report.

Wilson, R. (1999), Restoring the balance in safety regulation, *Inter. Conf. on Preparing the Ground for Renewal of Nuclear Power*: Global Foundation, Paris, (October 22–23, 1998), Plenum Press, New York.

Wilson, R. and Spengler, J. D. (Eds.) (1996), *Particles in Our Air: Concentrations and Health Effects*, Harvard University Press, Cambridge, MA 02138.

Wilson R. (2000), The changing need for a breeder reactor, *Nuclear Energy*, **39**:99–106.

Appendix 7.A *Abbreviated chronology of nuclear energy and nuclear weapons development*[a]

Date(s)	Event(s)
1934	Discovery of artificial radioactivity in Paris by Frédéric and Irene Joliot-Curie.
1938	Discovery of nuclear fission in Berlin by Otto Hahn, Lisa Meitner, and Fritz Strassmann.
1938	Measurement of average number of neutrons released per fission by J. Huban and Kasanski (1939).
1939	Neils Bohr and John Archibald Wheeler predict that only very heavy nuclei containing an odd number of neutrons would be fissile to neutrons of all energies, even down to nearly zero energy.
1939	Secret memorandum by refugee German physicists Rudolf Peierls and Otto Frisch asserted the extraordinary power of a fission based on ^{235}U, set out the mechanism for such a weapon, and described its effects.
1939	Leo Szilard suggests colleagues among the Allies cease open-literature publication of findings from uranium research. Alexander Sachs passes on letter from Albert Einstein and an affixed report by Leo Szilard to Franklin D. Roosevelt; research secrecy under the Manhattan Project is born.
December 1942	Demonstration of controlled nuclear fission reactor by Enrico Fermi at Stagg Field, University of Chicago.
August 6 and 9, 1945	Japan struck by first atomic bombs used in warfare.
November 1945	Tripartite Declaration by United States, United Kingdom, and Canada that "industrial applications" of nuclear energy should not be shared among nations until adequate safeguards and international controls were in place.
1946	Baruch Plan presented to United Nations and called for the creation of an International Atomic Development Authority that would be entrusted with all phases of the development and use of atomic energy. All nuclear weapons would be destroyed under this revolutionary form of world government, and the development of nuclear energy would only then proceed. Soviet Union wanted first the destruction of all nuclear weapons (e.g., the US nuclear arsenal must be eliminated first).
1946	Atomic Energy Act of 1946 (McMahon Act) established the civilian US Atomic Energy Commission, made secret all information on the use of fissionable material for the production of electrical energy. The government–industry–university partnership created under the Manhattan Project remained intact to develop both a nuclear arsenal and nuclear energy.
1948	Formation of Westinghouse Atomic Power Division (WAPD) at Bettis, Pittsburgh, to work jointly with the AEC and Argonne National Laboratory to design/construct/operate the Submarine Thermal Reactor (STR) based on the pressurized-water reactor (PWR) concept.

Appendix 7.A (*cont.*)

Date(s)	Event(s)
1949	Soviet Union tests nuclear explosive, after a 1942 decision to pursue the development of a nuclear weapon, which did not become a high priority until 1945.
1952	United Kingdom tests nuclear explosive, followed by a period (1952–57) of nuclear-weapons buildup at a rate constrained by nuclear-weapons *versus* nuclear-energy economic trade-off.
ca. 1953	Major task was to demonstrate nuclear power could be practical and economic; the prize was cheap power at home and substantial sales abroad. United States, United Kingdom, France, and Canada (e.g., the 1945 Tripartite Declarers plus France) gave nuclear power development a high priority.
1954	US submarine *Nautilus* launched powered by the STR2 (PWR, highly enriched uranium core) reactor.
December 8, 1954	Dwight D. Eisenhower makes "Atoms for Peace" speech to the United Nations, which suggested the establishment of an International Atomic Energy Agency to be responsible for contributions of fissionable material made by the United States, Soviet Union, and other nations, and to devise ways and means to allocate fissionable materials for peaceful purposes and "to provide abundant electrical energy to the power-starved areas of the world". The Soviet Union responded with curt negativity (e.g., the proposal would do little in reducing the danger of nuclear war); the proposal, however, enjoyed strong worldwide support, and presented a conjunction of ideas and interest that reordered priorities away from international inspections and towards peaceful applications of nuclear energy: • US Non-Proliferation Policy: Allies could move under the US Nuclear Umbrella, while benefiting from the development of a peaceful use of atomic energy. • Nuclear Options: The common path to nuclear weapons and nuclear energy could be traversed by nations close to the point of divergence without fear of reprimand or reprisal. • Nuclear Science: Global scientific exchange or nuclear information not related to weapons was encouraged, expanded, and supported. Cooperation and competition flourished. • Industrial Interests: Also flourished under private ownership of nuclear enterprises under substantial government support of a promising, but risky, business. • Government Interests: Assured a strong US position in the world nuclear-energy market, particularly in areas of expensive energy costs (Europe, developing countries).
1954	Atomic Energy Act of 1954 changed domestic law to the extent needed to implement the Atoms for Peace proposals of Dwight D. Eisenhower, including allowing the US to become a member of the (future) IAEA. Impacted (US) government–industry relationships with regard to things nuclear; openness *versus* secrecy swung to the former; strong support of industry by government to promote nuclear energy.

Appendix 7.A (*cont.*)

Date(s)	Event(s)
1955	Soviet Union first used nuclear reactor for electricity production from a 5–MWe experimental reactor.
1956	United Kingdom uses electricity from four 50–MWe carbon-dioxide-cooled, natural-uranium-fueled graphite-moderated reactors (at Calder Hall). These "dual-use" reactors were built for weapons-plutonium production and were uneconomic for electricity production alone; the French followed the same route at Marcoule.[b]
1956	US bilateral agreements on nuclear energy flourished in the form of both outright gifts for nuclear research (small research reactors and required fuel) and for research and power applications.
1957	The United States contributes the light-water-cooled Shippingport (Pennsylvania) as a 60–MWe small-scale demonstration of nuclear electric power generation. Joint project between AEC and Duquesne Light Company; power later increased to 90 MWe; used later as a test bed for the U–Th-based light-water breeder reactor (LWBR).
1957	Secret Chinese–Soviet agreement on sharing in nuclear science and engineering.
1957	IAEA becomes new international organization with potential for global membership. Idea of the IAEA as nuclear materials bank never took hold; became a framework for technical assistance that is aimed primarily at providing guidelines for safeguarding special nuclear material and providing safeguard-related inspections.
1958	Dresden-1(180 MWe) was the first commercial boiling-water reactor (BWR) commissioned after a turnkey contract was signed (1954) between General Electric Company and Commonwealth Edison of Chicago.
1958	Formation of the European Atomic Energy Community (Euratom) within the European Economic Community (began with formation of the European Coal and Steel Community in 1952) to foster rapid growth of nuclear industries in Europe; US bilateral nuclear agreements ultimately transferred to/through Euratom.
1960	France tests nuclear weapon, after a development period (1956–60) that largely tracked that of the United Kingdom.
December 1963	Jersey Central Power and Light Company announces intention to build the 560–MWe LWR at Oyster Creek without government assistance with the promise of producing cheaper electricity than a fossil-fuel plant of similar capacity; orders for nuclear power plants dramatically increased.
1963	Limited Test Ban Treaty (LTBT) concluded; limited nuclear weapons testing to under ground.
1964	China tests nuclear explosive.
1964	Private ownership of fissionable materials in the United States was authorized by law.

Appendix 7.A (*cont.*)

Date(s)	Event(s)
1967	Treaty of Tlatelolco completed; prohibited nuclear weapons in Latin America; impelled by the Cuban missile crisis.
ca. 1967	Decision to scale US light-water reactors up in size by roughly a factor of two to $> \sim 1000$ MWe to gain competitive edge with fossil-fuel electric generation stations.
ca. 1968	Canada advances technologies based on heavy-water-moderated, natural-uranium-fueled, light-water-cooled CANDU reactors, which find domestic markets as well as markets in less-developed countries (India, Pakistan).
1968	Non-Proliferation Treaty (NPT) completed.
1969[b]	UK switches from Calder Hall type reactors to AGRs, and France adopts LWRs, both operating on low-enriched uranium fuel.
1971	Zangger Committee (dozen nuclear technology exporting nations) formed to control exports of nuclear technologies by creating "trigger list" of nuclear technologies requiring safeguards if exported to non-nuclear-weapon nations.
June 1972	First published domestic regulation specifically addressing physical protection of nuclear material in response to increasing terrorists activities in the early 1970s; US Regulation 10 CFR Part 73, Physical Protection of Plant and Materials.
May 18, 1974	India explodes an experimental (physics) plutonium nuclear device in Rajasthan desert near Pakistan (Peaceful Nuclear Explosive, PNE) using plutonium generated in its CIRUS research reactor.
1974	US Nuclear Regulatory Commission publishes a comprehensive study on the safety of light-water reactors (US DOE report WASH-1400, the Rasmussen Report).
1975	IAEA responds to terrorist-related concerns in 1972 by establishing Advisory Group that issued first set of international guidelines (June 1972) that set the basis for more elaborate international guidelines issued in 1975 as INFCIRC/225, The Physical Protection of Nuclear Materials, again in response to increased terrorist activities; sequentially broadened/strengthened/updated: INFCIRC/225/Rev.2(1989),INFCIRC/225/Rev.3(1993), and INFCIRC/225/Rev.4(1998)[c,d]
1975	Nuclear Suppliers Group (NSG, or "London Club") formed to consider/implement further restrictions on trade in nuclear technologies; similar membership as Zangger Committee (1971), but included France.
1977	Issue US Congress Office of Technology Assessment report, "Nuclear Proliferation and Safeguards".
1978	US Nuclear Non-Proliferation Act passed (NNPA); US would no longer reprocess spent fuel or export uranium enrichment or reprocessing technologies.

Appendix 7.A (*cont.*)

Date(s)	Event(s)
	• End nuclear trade with non-nuclear-weapon states with nuclear facilities not subject to full-scale safeguards.
	• Requirement of US permission to reprocess, enrich, or re-export nuclear materials received from the US.
	• Prohibited export of nuclear technologies to "non-nuclear-weapons" nations that detonated a nuclear explosive.
	• Re-negotiation of all contracts to assure compliance with the last three requirements.
1979	Core meltdown at Unit 2 of the Three Mile Island PWR generating station.
1980	IAEA report of the International Nuclear Fuel Cycle Evaluation (INFCE).
June 7, 1981	Israel destroys Iraqi research reactor at the Tammuz nuclear center at El-Tuwaitha (outside Baghdad).
1983	Fast-neutron, liquid-metal-cooled, 1200–MWe, breeder reactor, Super Phénix is completed.
1985	Treaty of Rarotonga completed; non-nuclear-weapons zone established for the South Pacific, while requiring that some exports to nuclear-weapons states be safeguarded.
1986	Former technician at Dimona nuclear facility reveals strong evidence of Israeli nuclear weapons program, which began with the importation of French research reactor and reprocessing equipment in 1956.
March 28, 1986	RBMK reactor at Chernobyl experiences core meltdown/burn/dispersal.
1988	South Africa claims capability to manufacture nuclear weapons (began in early 1970s; six uranium nuclear weapons in 1990; program terminated in 1990 and weapons destroyed).
1992	Nuclear Suppliers Group (NSG) adopts full-scale safeguards and expands trigger to include "dual-use" items.
1995	Non-Proliferation Treaty renewal/extension.

Notes:
[a] Willrich and Taylor (1974), Goldschmidt (1982), Gardner (1994).
[b] By ~1969, the British, who at that time had the largest nuclear power capacity in the world, had shifted to low-enriched uranium fuel used in advanced gas reactor (AGR) design. The French also shifted to light-water reactors (LWR) using low-enriched uranium that were manufactured at that time under license from the US and/or German firms.
[c] Kurihara (1977).
[d] Bunn (1997).

Appendix 7.B *Design characteristi*

Design characteristics	Evolved LWR				Evolutio
	System 80+	ABWR	SBWR	APWR 1000	APWR 13
Reactor type	Pressurized water reactor	Boiling water reactor	Boiling water reactor	Pressurized water reactor	Pressurized water reacto
Lead designer	ABB-CE	GE	GE	Westinghouse	Westinghous
MWe (net)	1300	1300	640	1050	1300
MWt	3817	3926	2000	3150	3900
Coolant type	Light water	Light water	Light water	Light water	Light water
Moderator type	Light water	Light water	Light water	Light water	Light water
Fuel material	UO_2 and/or PuO_2	UO_2	UO_2	UO_2	UO_2
Cladding material	Zircaloy-4	Zircaloy-2	Zircaloy-2	Zircaloy-4	Zircaloy-4
Fuel geometry	16×16	8×8 or 9×9	8×8 or 9×9	17×17	19×19
Number of fuel assemblies	241	872	732	193	193
Fuel pin diameter, mm	9.7	12.3	12.3	9.5	10.3
Number of control rods	93 control element assemblies (CEAs)	205	177	53 black rods; 16 gray rods	69 black rods 28 gray rods 88 displacer rods
Control rod material	48 B_4C CEAs; 20 Ag-In-Cd CEAs; 25 Inconel 625 CEAs	B_4C	B_4C	Black–Ag-In-Cd; Gray–stainless steel	Black–B_4C/Ag-In-Cd hybrid; gray–stainless steel; displacer–zirconium
Control rod form (i.e., drive power source)	Magnetic jack	Electric/ hydraulic	Electric/ hydraulic	Magnetic jack	Magnetic jack for black and gray rods; hydraulic for displacer rods
Active fuel length, mm	3810	3708	2743	3658	3900
Equivalent core diameter, mm	3650	5164	4880	3370	4000
Thermal-neutron or fast-neutron reactor	Thermal	Thermal	Thermal	Thermal	Thermal
Vessel height/diameter, m	15.3/4.6 (I.D.)	21.0/7.1	24.5/6.0	12.06 (outside)/ 4.47 (inside)	16.4 (outside)/ 5.1 (inside)

new nuclear reactors (Nuclear News, 1992)

R	AP600	EPR	PIUS	Other		New LMR	
				MHTGR	CANDU-3	EFR	ALMR
surized r reactor	Pressurized water reactor	Pressurized water reactor	Gas-cooled reactor	Pressurized heavy-water reactor	Liquid-metal fast breeder reactor	Liquid-metal fast breeder reactor	
inghouse	NPI	ABB Atom	General Atomics	AECL CANDU	EFR Associates	GE/Argonne	
	1450	640	538 (4 modules)	450 (nominal)	1450	1440 (3 power blocks)	
	~1450	2000	1400 (4 modules)	1440.7	3600	4245 (3 power blocks)	
t water	Light water	Light water	Helium	Heavy water	Liquid sodium	Liquid sodium	
t water	Light water	Light water	Graphite	Heavy water	None	None	
	UO_2 or UO_2/PuO_2	UO_2	UCO fissile, THO_2 fertile	Natural UO_2	Mixed UO_2 & PuO_2	U/25%-Pu/ 10%-Zr metal alloy, by wt.	
aloy	Zircaloy	Zircaloy-4	Refractory coated particles	Zircaloy-4	Austenitic stainless steel AIM1 & Nimonic PE16	HT-9 ferritic alloy	
17	17×17	18×18	Hexagonal graphite blocks	37-element fuel bundle	331 pins in triangle, in hex steel envelope	217 pins in hexagonal bundle	
	205	213	660	232 fuel channel assemblies	387	66 driver fuel assemblies	
	9.5	9.5	13	13.1	8.2	7.2	
control s; 16 y rods	69	None	30	24	33	6	
In-Cd/ SS	Hybrid: B_4C & Ag-In-Cd	NA	B_4C compacts	Stainless steel sheathed cadmium	Boron carbide	Natural boron carbide	
gnetic jack	Gravity	NA	Electric	Mechanical (gravity & spring)	Raised & lowered by electric motors; free fall by gravity	Gravity and fast drive-in motor	
58	4200	2500	7925	5944	1000	1350	
22	~3470	3760	1650 I.D., 3500 O.D. annular	4912	4000	1570	
ermal	Thermal	Thermal	Thermal	Thermal	Fast	Fast	
59/4.39	12.8/5.25 (O.D. of core shell)	Overall height– 43; width– 27×27; cavity diameter–12; depth–38	22/6.8	NA/NA	17/17.2	18.7/5.7	

Appendix 7

Design characteristics	Evolved LWR				Evolution
	System 80+	ABWR	SBWR	APWR 1000	APWR 13
MW/m³	95.5	50.6	41.0	96.2	80.0
Linear heating rate, kW/m	7.1	5.8	4.7	6.8	6.7
Coolant inlet temperature, °F (°C)	558 (292)	420 (216)	420 (216)	548 (287)	558 (292)
Coolant outlet temperature, °F (°C)	615 (324)	550 (288)	550 (288)	616 (325)	621 (327)
Main coolant system pressure, psig (MPa)	2235 (15.41)	1025 (7.07)	1025 (7.07)	2250 (15.51)	2250 (15.51)
Containment	Dual: spherical steel w.concrete shield building	Pressure suppression/ reinforced concrete	Pressure suppression/ reinforced concrete	Cylindrical steel	Cylindrical st
Operating cycle between refuelings, startup to shutdown, months	18 to 24	18	24	17	16.5
Refueling outage duration	0.56 months	45 days	45 days	30 days	45 days
Estimated yearly total occupational radiation exposure person-rem/yr/ reactor	<70	<100	<100	<100	<100

Note:
[a] Based on 4254/3 MWt per power block, 217(16 driver fuel pins, and 70% to total thermal power generated in the driver fuel pins.

nt.)

R	AP600	EPR	PIUS	Other		New LMR	
				MHTGR	CANDU-3	EFR	ALMR
		~107	72.3	5.9	12.78	290	1258[a]
		7.6	5.1	0.8	1.7	15.3	51.2[a]
	(276)	~555 (~291)	500 (260)	497 (258)	515 (268)	743 (395)	640 (338)
	(312)	~617 (~325)	554 (290)	1268 (687)	590 (310)	1013 (545)	905 (485)
	5 (15.41)	2250 (15.51)	1305 (8.99)	925 (6.38)	1436 (9.90)	Unpressurized	Unpressurized
	l	Yes	Yes	None	Yes	Yes	Yes
	r 24	12 to 18	11–12; can go to 24	19.2	NA (on-power refueling)	12	24
	days ueling & ntenance)	~1 month	<1 month (2–3 weeks expected)	0.5 months	NA	0.6 months	0.6 months/ reactor
		<100	100	40	40 in initial years, 75 in final years	20	20

8

Nuclear Fusion Energy

8.1 Introduction

Fusion could become the energy source of choice early in the third millennium – it is a long-term, inexhaustible energy option that offers the possibility of generating power in an economically and environmentally attractive system, which is compact relative to renewable energy power plants, and does not emit carbon dioxide or other greenhouse gases. The small loss of mass when light nuclei fuse into heavier nuclei provides the source of energy. In one sense it is the only long-term option other than fission, since the sun, which is the source of all forms of renewable energy, is powered by fusion reactions. In this chapter, we will present our personal vision of fusion's potential, illustrated with examples of innovative concepts at an early stage of development, and grounded in the dramatic progress towards fusion energy demonstrated by the more traditional concepts.

Fusion occurs when two positively charged nuclei approach closely enough for the attractive short-range nuclear forces to overcome the repulsive Coulomb force. Getting two nuclei this close requires that they have a high energy or temperature of about ten thousand electron volts (10 keV ≈ 100 million Kelvin). (At temperatures above 0.01 keV, the atoms have such a high velocity that the electrons are knocked off by collisions. This results in the mixture of equal amounts of positively charged ions and negatively charged electrons that is known as a plasma.) In addition to needing a high temperature for fusion, the hot ions must be held at a high enough density for a long enough time so that they release more energy through fusion reactions than went into heating them and than they lose by radiation and plasma-particle loss. This density–time–temperature product has a required threshold magnitude of 10^{21} m^{-3}s keV for a fusion power plant fueled by deuterium–tritium. These requirements of density, confinement time, and temperature are why fusion has been difficult to demonstrate in the laboratory.

Table 8.1 *Candidate fusion reactions*

Fuel	Reaction
d-t	$^2H + {}^3H \rightarrow {}^1n + {}^4He + 17.6$ MeV
d-d	$^2H + {}^2H \rightarrow {}^1H + {}^3H + 4.0$ MeV
	$^2H + {}^2H \rightarrow {}^1n + {}^3He + 3.3$ MeV
d-^{3}He	$^2H + {}^3He \rightarrow {}^1H + {}^4He + 18.7$ MeV
p-^{6}Li	$^1H + {}^6Li \rightarrow {}^3He + {}^4He + 3.9$ MeV
d-^{6}Li	$^2H + {}^6Li \rightarrow {}^1H + {}^7Li + 4.9$ MeV
p-^{11}B	$^1H + {}^{11}B \rightarrow 3{}^4He + 8.7$ MeV

Note: Courtesy of the University of California, Lawrence Livermore National Laboratory, and the Department of Energy under whose auspices the work was performed.

The difficulty of fusion has a positive side: inherent-passive safety – reactions will not, and cannot, runaway. (Passive safety means that an off-normal condition will correct itself, rather than requiring positive action, as in "active safety", to prevent an accident.) The fuel in the core of a fusion power plant is sufficient, at most, for only a few seconds of operation, and needs to be continually replenished. Fusion energy is distinguished from fission in which a heavy nuclei splits or fissions into two or more lighter nuclei, that are frequently radioactive. Fusion reaction products are non-radioactive, a significant environmental advantage. Fission releases energy when a critical mass is gathered into a volume that is small enough so that a neutron released by a fission reaction will cause another fission reaction before escaping from the fissionable material. Passive safety is much more difficult to achieve with fission reactors, because the stored energy in the fuel of a fission core is sufficient for about two years of operation.

Several possible fusion fuels exist among low atomic number nuclei. These are listed in Table 8.1, along with the reactions. The deuterium–tritium reaction has a larger cross-section (i.e., larger reactivity) at lower energy than the others do, so it is the nearest term fusion fuel. However, 80% of the output energy from the fusion of deuterium and tritium is carried by 14 MeV neutrons, which can activate the power plant structure. The remaining 3.5 MeV is carried by the ^{4}He or alpha particle. The four reactions at the bottom of Table 8.1 produce few or no neutrons, and nearly all their energy is carried by energetic ions. These reactions have the double advantage of reducing or eliminating activation of the power plant structure, and of offering the possibility of directly converting the reaction product energy to electricity by the interaction of the energetic ion reaction products with electric or magnetic fields. Direct conversion can, in principle, achieve efficiency levels greater than obtained with a thermal cycle. These ions, however, have a higher charge, and hence require not only a higher plasma temperature to overcome the Coulomb repulsive force but also a greater density-confinement time product to overcome smaller cross-sections. These considerations strongly encourage the use of deuterium–tritium in first generation fusion power plants. And, as we will discuss, concepts exist to achieve low-activation, even with high neutron production.

Significant progress towards fusion energy has been achieved: the density–time–temperature product has increased by more than eight orders of magnitude in 30 years to $>10^{21}$ m^{-3}s^{-1}keV (Ishida, 1999). Twelve megawatts of fusion power has been achieved in the laboratory, almost equal to the 18 MW of heating power (Watkins, 1999). The progress by the 1980s was sufficient to initiate the design of the ITER (International Thermonuclear Experimental Reactor) as a joint project between the US, Russia, Europe, and Japan. ITER, a large tokamak, was intended to study plasmas producing fusion power greater than the external power used to heat the plasma, and to begin studying power plant engineering. The engineering design has been completed (Shimomura, 1999), but construction at this time is doubtful: the costs grew to exceed $10 billion, the US has withdrawn as a major player, Russia's faltering economy has forced it into a more minor role, and Europe and Japan have not committed to assuming the full load by themselves (Glanz and Lawler, 1998).

8.2 Potential of Fusion

The attractiveness of fusion is determined by its potential in the areas of fuel resources, cost, safety, environment, plant size, and products. In the competitive energy marketplace envisioned for the twenty-first century, fusion must be attractive in every area to succeed (Perkins, 1998). These areas are discussed below.

Fuel resources of deuterium are effectively unlimited. Water, by weight, contains 75 times as much energy as gasoline for the deuterium–deuterium reaction. (This corresponds to 10 billion years of primary energy at 1990 "burn rate" of 11 TW (10^{12} W), which is virtually unlimited.) For deuterium–tritium, assuming that half of the 17.6 MeV of the reaction is due to deuterium, water, by weight, contains 330 times as much fusion energy as the chemical energy in gasoline. Lithium, the source of tritium for first-generation fusion power plants, is significantly more abundant in the earth's crust than are the fission fuels, uranium or thorium. It is also about 50 times more abundant than uranium in seawater. The energy resources for several energy candidates are listed in Table 8.2. More detailed data are available in Perkins (1998).

Table 8.2 *Lifetime of candidate energy resources*

Candidate	Lifetime (years)
Breeder fission ($^{238}U + ^{232}Th$)	1000s → million's*
Solar electric	Unlimited
Natural gas	50 → 1000s*
(Coal)	(200 – 500)
Fusion: DT (Li)	10000s → 100 million*
DD	Unlimited

Notes:

* Additional occurrences which are believed to exist but for which extraction technology does not yet exist and/or may not be economically viable, e.g. U, Li in seawater, gas-hydrates in clathrates.

Courtesy of the University of California, Lawrence Livermore National Laboratory, and the Department of Energy under whose auspices the work was performed.

Costs of fusion power are estimated to be in the \$0.04–0.10/kWh for 1000 MWe size plants (Delene, 1994; Delene *et al.*, 2000; Krakowski, 1995; Najmabadi, 1998). Doubling the plant size to 2000 MWe can reduce the COE (cost-of-electricity) further by about 25% due to economy-of-scale (Logan *et al.*, 1995). The lower range is competitive with other advanced forms of energy, with the exception of today's natural gas without carbon sequestration. However, at present, the lower cost range is populated only by more innovative concepts that involve more unknowns than does the conventionally-engineered tokamak concept which is predicted to occupy the higher cost range. An increasing number of fusion researchers are identifying cost as a go no-go issue (Delene, 1994; Delene *et al.*, 2000; Krakowski, 1995). They are achieving some success in reducing the costs with innovative concepts, by simplifying old concepts or devising cheaper manufacturing techniques to reduce capital costs, and by introducing concepts that reduce the down time for periodic mainten-ance – for example, by using a flowing liquid that is continuously replaced for the tritium-generating lithium-containing blankets rather than needing to shut down a power plant to replace solid blanket modules (Moir, 1994, 1996, 1997). As development of fusion energy continues, the uncertainties that limit our confidence in present cost estimates will be reduced.

Most fusion power plant concepts are passively safe. If we show that they meet the "no-evacuation in case of an accident" requirements, as expected, then we could site them in highly populated areas and tap the market for low-grade heat for space-heating applications. Since the amount of fuel within a fusion power plant will not sustain operation for more than a few seconds at most, there is minimal stored nuclear energy available to trigger severe accidents.

The strong safety characteristics can be maintained by care in design. For example, all other sources of stored energy (chemical, magnetic, thermal) must be contained so that they cannot drive an accident; and materials need to be chosen that minimize activation of the power plant. It is particularly important to minimize the volatile hazardous materials that could be released in an accident. The materials forming the containment vessels must themselves be shielded from radiation to prevent degradation of their properties. With careful and thorough design and construction, radioactive releases are not expected to be a problem. While first generation power plants using deuterium–tritium must contain the tritium, the reaction products are not radioactive. Low activation also minimizes after-heat issues and the need for emergency cooling systems. These issues address local safety in the vicinity of a power plant.

Fusion is also attractive on the global safety issue: nuclear weapons proliferation – which can be minimized by ensuring that fissionable materials are absent. If, however, the world is using a significant amount of fission energy, then a fission–fusion hybrid power plant – consisting of a fissionable blanket surrounding a fusion core – could be a highly effective fission breeder (Maniscalco *et al.*, 1981). Each hybrid power plant could supply fuel for several non-breeder fission power plants. Preventing nuclear weapons proliferation would then be more difficult than in a pure fusion scenario, but similar to the safeguards needed for fission. Fusion, as well as fission, power plants should be operated under international supervision to prevent proliferation.

Fusion is environmentally attractive due to no carbon dioxide or other greenhouse gas emission, small land use compared with renewables, and non-radioactive reaction products. The major issue to be addressed is the activation level, volume, and lifetime of the power plant chamber components under bombardment by 14 MeV primary neutrons from deuterium–tritium fusion, plus the scattered neutrons that are degraded in energy. The use of thick (0.5 to 1 m) liquid walls or jets, composed of low-atomic-number materials, is one method that reduces activation to low levels and eliminates the damage suffered by solid materials from atomic displacements under neutron bombardment. Liquid walls also promise significant reductions in the volume of activated waste due to operation at higher power density, which allows smaller wall area (Moir, 1997). A second method, recycling of waste, promises further reduction in waste volumes. A third method, development work on long-lived and low-activation solid materials, is also important for two reasons. First, some solid materials are exposed to neutron bombardment either because the materials penetrate the liquid walls or because they are visible through gaps in the liquid walls. Second, thick liquid walls may not be compatible with all fusion concepts. For this reason it is important to develop the facilities to irradiate materials with a fusion spectrum of neutrons (a continuous distribution of neutron energies up to, but not exceeding, 14 MeV). Such facilities are needed to measure the damage and activation of candidate materials, in order to develop long-lifetime low-activation materials and concepts (see, e.g. Perkins *et al.*, 2000).

The industrial ecology concept of recycling materials and components, rather than disposing of them (Frosch, 1992), is beginning to be applied to fusion power plants. Extensive recycling requires the use of relatively low-activation materials; "low-activation" in this context meaning that remote-handling and accident containment and cleanup are feasible during reprocessing; or better yet, that activation is so low that hands-on manipulations are feasible during reprocessing. It is not necessary to separate all activated materials during reprocessing; for example, replacement components for use in radioactive environments could use activated materials. Wherever possible, highly activated materials with inconveniently long half-lives would be separated for specialized uses or disposal. Recycling has the potential of diverting and re-using much of the waste stream, greatly reducing the amount of waste that requires disposal.

The high power density of fusion requires little land for the plant, compared with the requirements for the various forms of renewable energy. The space occupied can be made even smaller if we bury the larger elements deep enough for roads and agriculture to exist above it. This has been done at major high-energy physics accelerators, such as the Stanford Linear Accelerator (SLAC) near Palo Alto, California, and the European Laboratory for Particle Physics (CERN) accelerator complexes near Geneva, Switzerland. Such dual use of land will become increasingly important as population growth results in an increased need to use all the available land for agriculture and minimize that devoted to energy generation.

Electricity and heat are the primary products of fusion power plants. The heat can have direct value in industrial processes, or space heating for cities – if safety constraints can be met as we think is possible. Locating fusion power plants near the largest electricity and heat consumers will reduce costs by minimizing electrical transmission lines in addition to increasing the income through the sale of heat. Other products are also possible. The power can be applied to generating hydrogen or other synthetic fuels for efficient long distance power transmission through pipelines and for portable needs such as transportation, but competing in this area requires very low COE (Logan *et al.*, 1995). Like fission, fusion is appropriate for recharging hydropower or other energy storage during off-peak demand periods. Radioisotope production and transmutation of fission waste would take advantage of the intense flux of neutrons available, but would also raise issues that need to be addressed: nuclear proliferation and possible additional activation of the fusion plant.

8.3 Approaches to Fusion Energy

Historically, two approaches to harnessing fusion power for energy production have been followed. The first is magnetic fusion energy (MFE) which creates "magnetic bottles" that need to hold the plasma for a time of order one second (Chen, 1974; Sheffield, 1994). The best known example is the tokamak, a toroidal geometry. (A toroid is a donut shaped configuration, which is preferred for most magnetic fusion applications, because magnetic field lines can encircle the toroid nearly endlessly without intercepting a wall. Plasma ions and electrons flow along magnetic field lines much more easily than they flow across them, so plasma loss rates to walls are reduced in toroids.) The second approach is inertial fusion energy (IFE) which today uses powerful lasers, which compress a millimeter-sized capsule of fusion fuel to the high densities and temperatures needed for fusion to occur (Hogan, 1995). For IFE, the confinement time is the time for the capsule to blow itself apart, of order nanoseconds (10^{-9} s): hence the name "inertial" fusion. These approaches are at roughly similar stages of scientific development: magnetic fusion has demonstrated up to 60% as much fusion power produced as heating power injected into the plasma in the JET (Joint European Torus) tokamak (Watkins, 1999). Even better confinement has been demonstrated in the JT-60 tokamak in Japan, but without the use of deuterium–tritium fuel, resulting in much less generation of fusion energy (Ishida, 1999). The inertial fusion program is constructing the NIF, National Ignition Facility, with a goal of demonstrating ignition and energy gain of ten near 2100 (Kilkenny *et al.*, 1999). Ignition means that the 3.5 MeV alpha particles from fusion reactions deposit enough energy in nearby fuel to heat it to fusion energies, analogous to heating a log so that it ignites and burns. Energy gain is the ratio of fusion energy out to energy in.

A generic fusion power plant, Figure 8.1(a), consists of a central reaction chamber (the core), surrounded by an approximately 1-m thick blanket that contains lithium. The blanket performs two functions: it generates the tritium fuel from neutron interactions with the lithium, and it shields the surrounding structure from neutron bombardment. Penetrations through the blanket must

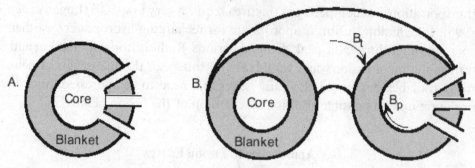

Figure 8.1 (a)   Generic fusion core with blanket/shield. (b) Toroidal fusion device. Courtesy of the University of California, Lawrence Livermore National Laboratory, and the Department of Energy under whose auspices the work was performed.

be provided for injecting fuel and energy for either MFE or IFE. These may be from one side as shown, but more generally are from opposite sides or even uniformly distributed around the periphery. Outside the blanket are superconducting magnets for magnetic fusion, and beamlines for inertial fusion. The power plant in Figure 8.1(a) could be spherical or cylindrical. The other major geometry studied is the torus shown in Figure 8.1(b).

8.3.1 Magnetic fusion energy

Many concepts in magnetic fusion have been studied during the last four decades, with a strong emphasis on the tokamak during the last two decades. The tokamak consists of an externally applied toroidal magnetic field B_t, and a toroidal plasma current to create a poloidal magnetic field, B_p, Figure 8.1(b). A transformer drives the plasma current in nearly all tokamaks, with auxiliary current drive by radio-frequency waves or neutral-atom beams on many tokamaks. Additional externally applied magnetic fields provide vertical fields for plasma-position control and shaping, and can provide divertor geometries, where the outermost field lines strike a divertor plate to allow the removal of impurities from the edge plasma. Within the last decade, innovations in tokamaks have dramatically increased their ability to confine plasma. The ARIES-RS, Figure 8.2, applies these innovations to improving the power plant core (Najmabadi, 1998).

Tokamaks keep refusing to be written off as too large and complicated, and have reached performance levels that would allow the construction of a burning plasma experiment. (A burning plasma is one in which the fusion power in alpha particles heats the plasma at a rate exceeding the external heating power.) Potential experiments under consideration encompass short-

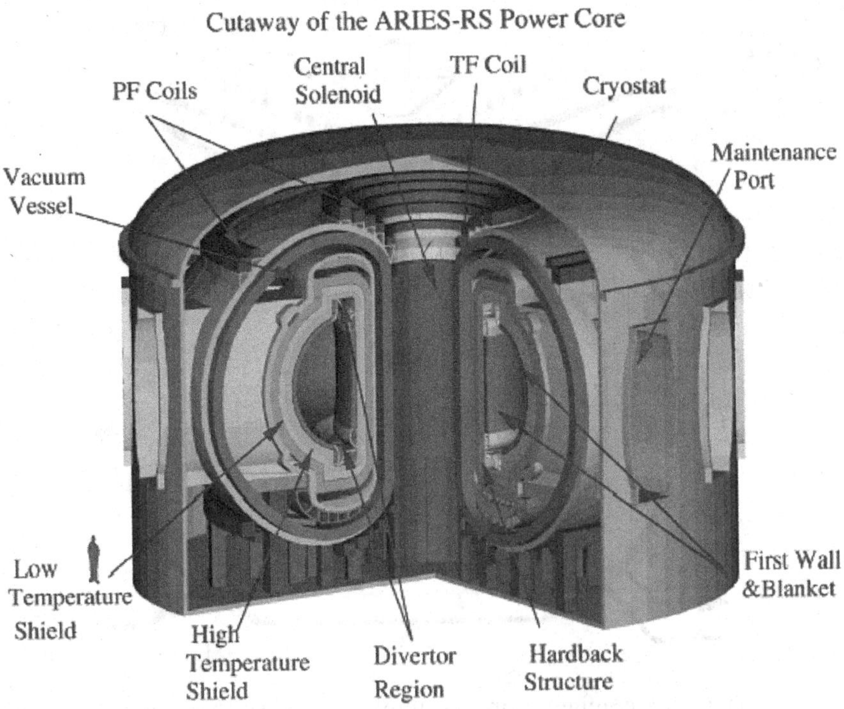

Cutaway of the ARIES-RS Power Core

Figure 8.2   ARIES-RS Tokamak power plant. Courtesy of Farrokh Najmabadi, University of California, San Diego.

pulse (few seconds) machines with normal-conducting magnets and long-pulse (thousands of seconds) machines with superconducting magnets. An example of the former is FIRE (Fusion Ignition Research Experiment) (PPPL, 2000) with a price tag of around $1–1.5B. The latter type of machine is exemplified by ITER (the International Thermonuclear Experimental Reactor) (Shimomura, 1999) under study by the international fusion community. ITER would be a full engineering test reactor capable of testing engineering components and, accordingly, carries a price-tag of ~$5–10B depending on performance requirements. Such burning plasma experiments will enable many of the remaining physics issues for a power plant to be studied. However, the present lack of low-activation materials prevents the conventional tokamak from being on an environmentally attractive power plant path. At present, therefore, it should be considered as a physics and technology experiment, not a power plant prototype.

The stellarator is a toroidal magnetic configuration in which all confining magnetic fields are imposed by external coils. It has the advantages over a tokamak of not needing current drive, and not disrupting (in a disruption, the

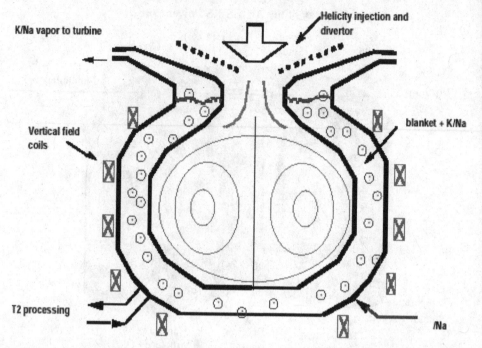

Figure 8.3   Spheromak configuration has simpler geometry, that is more compatible with liquid walls, than devices with toroidal blankets. Courtesy of the University of California, Lawrence Livermore National Laboratory, and the Department of Energy under whose auspices the work was performed.

plasma current decays within a few tens of milliseconds, much faster than usual, and exerts large forces on the walls). Stellarator coils are more complicated; but, in the modular type, the multiple versions of potato-chip-shaped coils are not trapped or linked by other coils and so can be more easily disassembled for maintenance. As a power plant, it shares most of the other characteristic of a tokamak, and is expected to yield a similar cost-of-electricity.

Other magnetic configurations are described as self-organizing, because more of the currents generating the magnetic fields are carried by the plasma itself (Riordan, 1999). For example, the spheromak has both toroidal and poloidal currents to create both the poloidal and toroidal magnetic fields respectively. A spheromak plasma will persist for some time when placed inside a can with conducting walls to carry the image currents. The pot blanket spheromak, Figure 8.3, is very simple, but uses a metal wall to separate the liquid from the plasma; this wall will limit the power density and will be damaged and activated by neutrons. In other concepts, a swirling liquid is against the plasma. Another magnetic configuration is the field-reversed-configuration (FRC) which generates only a poloidal field with plasma currents, and which is immersed in an axial magnetic field.

We have just classified magnetic bottles by the degree to which the magnetic field is created by currents in external coils versus currents flowing in the plasma. Another useful classification criterion is the topology of the chamber: a cylinder, sphere, or toroid; and if the latter, the amount of structure required in the central hole of the toroidal geometry. The minimum size of the fusion core of a power plant is greatly increased by such structure, because it generally includes superconducting magnet windings that must be shielded against 14 MeV neutrons. Neutron shields are about 1 m thick, Figure 8.1, to reduce the refrigeration requirements at 4 K, as well as to reduce neutron damage, thereby extending the life of the windings. Therefore, the central hole must have a minimum diameter of about 2 m, and small, high-power-density fusion cores are not possible with such geometries.

Some toroidal concepts, such as the spherical tokamak, avoid this restriction by using unshielded room-temperature magnet coils in the center leg. This allows the center leg to be much smaller than one meter in diameter. In determining the attractiveness of the spherical tokamak, one needs to evaluate the impact of the circulating power to drive the center leg current, and the economic and environmental consequences of regularly replacing the activated center conductor.

We now use these classifications to further discuss two potentially attractive examples of confinement configurations. A more compact structure with no structure in the center, such as the spheromak or the FRC, would provide a lower-cost development path, lower cost-of-electricity and smaller volume of activated material, if adequate confinement could be achieved. Our present knowledge of confinement properties is poorer for this class. The fusion program today is building small experiments that will test these concepts with better power plant potential – to determine whether they can be made to confine plasma sufficiently well. This broadens the policy of the last 15 years of developing only the tokamak because it has the best confinement and is best understood, so that its confinement and performance at power plant scale can be confidently predicted.

The reversed-field pinch (RFP) is also largely self-organized, but requires magnet windings within the donut hole. It therefore requires a larger fusion core and looks more like a tokamak, except for a lower applied toroidal magnetic field. Although RFPs may have less potential as power plants, they have proven to be very productive in increasing our physics understanding of these self-organized configurations (Prager, 1999).

The spheromak and FRC are the MFE configurations most likely to be compatible with thick liquid walls; because their simpler boundary shape makes formation of liquid walls less difficult, and their high plasma density is more likely to withstand the high vapor pressure of the liquids. As with many configurations, spheromaks gain stability from the plasma pressure against the walls; whether liquid walls are too easily pushed aside by the plasma remains to be determined. Even if steady-state solutions do not exist, pulsed configurations may be attractive, if their repetition rate is high enough to minimize thermal cycling in the liquid that operates the steam cycle (Fowler *et al.*, 1999).

The extreme pulsed limit of magnetic fusion is magnetized target fusion, MTF (Drake *et al.*, 1996), which fits a largely unexplored niche between MFE and IFE. MTF employs magnetic fields to reduce electron thermal conduction, thereby allowing plasma densities orders of magnitude lower than with IFE, but with plasma density and magnetic fields exerting pressures too large to be restrained in a steady-state device. As a result, some of the structure must be replaced each shot. The structure to hold the plasma is small and relatively inexpensive, allowing multiple confinement geometries to be evaluated in one facility. The development path to ignition may be the least expensive of any fusion concept. The path to a power plant is less clear, but might involve replaceable structures of frozen liquid and/or closely coupled liquid walls, fusion yields exceeding a gigajoule, and repetition rates of slower than one per second, to allow time for reloading.

8.3.2 *Inertial fusion energy*

Inertial fusion (Hogan, 1995) requires depositing energy in a short time (less than 10 nanoseconds – the time for light to travel 3 m or 10 ft.) on the outside of a millimeter-sized capsule. The outside of the capsule blows off at high velocity, causing the rest of the capsule to rocket inwards. This compresses the fuel to high density and temperature, with the goal of reaching sufficiently high values to ignite a fusion burn.

Inertial fusion is characterized as direct- or indirect-drive, Figure 8.4. For direct drive, the energy to compress the capsule of fusion fuel is incident directly on the outside of the capsule. Experiments with lasers have demonstrated that the energy must be extremely uniform over the surface of the capsule in order to have the entire surface rocket inwards at the same velocity, a requirement if the capsule is to remain spherical during compression. Variations in velocity from nonuniform illumination drive large amplitude waves, the so-called Rayleigh–Taylor instabilities on the surface. Such waves prevent the capsule from compressing as a sphere by the 10–30 fold in radius that is required to obtain fusion energy.

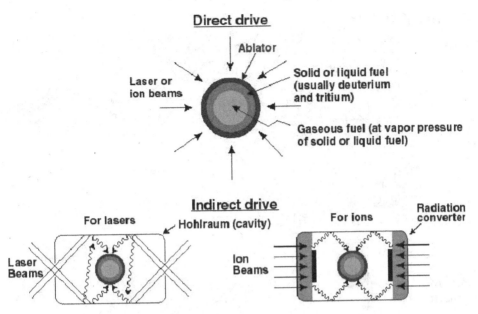

Figure 8.4   Direct drive energy is directly incident on the outside of the capsule. Indirect drive energy is deposited on the walls of a hohlraum, which heats up and emits x-rays that illuminate the capsule. Either can be driven by lasers or ion beams. Courtesy of the University of California, Lawrence Livermore National Laboratory, and the Department of Energy under whose auspices the work was performed.

For indirect drive, Figure 8.4, the energy is deposited on the inside of a can or hohlraum (German for "radiation-room"), which contains the capsule. The energy creates soft x-rays of a few hundred electron volts (few million Kelvin) energy, that reflect off the walls of the can and uniformly illuminate the capsule. This is analogous to indirect lighting, that creates a relatively uniform, diffuse illumination, as contrasted with direct illumination which is more efficient but creates a higher contrast lighting. Direct drive is similarly more efficient, since energy is not expended in heating the inside of the hohlraum. Indirect drive is more compatible with liquid-wall fusion chamber design, because the beams can be clustered to come in from one or two sides, rather than distributed uniformly over the surface of a spherical chamber.

Inertial fusion is also categorized by the type of driver it uses. Experiments to date have used lasers; most of these are flashlamp-pumped solid state lasers that have low electrical efficiencies (of order 10^{-3} to 10^{-2}) and low repetition rates of ~ 10 shots per day. A few experiments have also used z-pinch radiation sources or light ions. Other drivers are being developed. DPSSLs (diode-pumped solid state lasers) should deliver efficiencies near 10%, and repetition rates of 10 Hz. Electron-beam pumped gas lasers, such as KrF (krypton

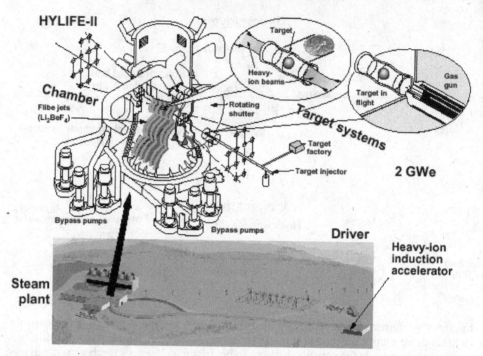

Figure 8.5 HYLIFE-II heavy-ion driven, inertial fusion power plant. Plant layout shows the driver, target and chamber. The use of liquid jets to protect the fusion chamber results in long lifetime, low cost, and low environmental impact. Courtesy of the University of California, Lawrence Livermore National Laboratory, and the Department of Energy under whose auspices the work was performed.

fluoride), have nearly as high efficiency. Heavy-ion accelerators are attractive driver candidates for inertial-fusion energy, building on extensive experience with high-energy and nuclear physics facilities. They promise high efficiency, high repetition rate, long life, and magnetic final optics to focus the beam onto the target (Bangerter, 1999). Compared with the mirrors or lenses that lasers need, magnetic optics are relatively immune to the effects of target explosions, because neutrons, x-rays, and debris can pass through the aperture of the magnet, while the magnet windings are shielded. The high beam currents required with heavy-ion fusion are a new and challenging element for both accelerators and focusing systems. Target designs have been developed for both laser and ion drivers, as indicated in Figure 8.4, which are predicted with 2–D codes to have reasonable gains (near 100) for a power plant.

A heavy-ion driven, inertial fusion power plant is shown in Figure 8.5. It is characterized by modularity, where the driver, chamber, target fabrication plant, and target injector are all separated. Thick liquid walls within the chamber, composed of jets of Flibe, enclose the reaction region, and protect the solid chamber walls from neutrons and shock waves.

IFE provides a paradigm shift from MFE tokamaks in two areas. First, it places most of its complexity in the driver, which is decoupled from the fusion chamber, whereas MFE incorporates most of its complexity immediately around the fusion chamber. This provides IFE with potential advantages of (a) reducing development costs by allowing the chamber and driver modules to be upgraded independently, and (b) reducing operational costs by delivering a higher availability to utilities. Higher availability is possible because the complexity within the fusion chamber is minimized, and repairs to systems outside the neutron shield can be made more rapidly, in some cases even while the driver continues to operate (Moir *et al.*, 1996). Second, indirect-drive IFE (heavy-ions, or possibly lasers) offers the potential for lifetime fusion chambers with renewable liquid coolants facing the targets, instead of solid, vacuum-tight walls that could be damaged by heat and radiation. Protected in this way, all of the chamber structural materials would be lifetime components. Their minimal residual radioactivity would mean that at the end of the fusion plant's life, most of the materials could be recycled, rather than requiring deep underground disposal.

However, much of the technology and engineering capability for conventional tokamaks is near the levels required for a power plant. This is not the case for IFE, nor MFE alternative concepts, where today's experiments are much further from power-plant technology. A consequence of the philosophy presented here, is that in seeking more attractive concepts for fusion, we push fusion energy further into the future. We believe that further into the future is preferable to never. "Never" is the likely result of trying to build power plants on which the public is not sold.

A concept that could dramatically reduce the driver energy is the fast ignitor (Tabak *et al.*, 1994). With conventional IFE, the capsule is heated by a high degree of compression to ignite at the center, Figure 8.6a. With the fast ignitor, the capsule is compressed to moderately high density but low temperature. A portion of it is then heated to ignition by a very short energy pulse with duration measured in picoseconds (millionth of a millionth of a second). The hot spot heats the surrounding region, and the fusion burn propagates through the fuel, Figure 8.6b. This concept can give gains (the ratio of fusion energy out to driver energy in) of a factor of 5 to 10 higher than conventional IFE. Relaxing the compression ratio allows a lower energy and therefore less expensive driver, resulting in a lower cost-of-electricity (COE). However, several difficulties must be overcome first: the driver must bore a hole through the plasma surrounding the capsule, then deliver the fast-ignitor pulse through the hole and focused to

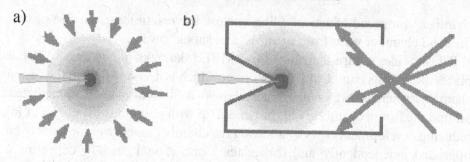

Figure 8.6 Inertial confinement fusion targets using the "fast ignition" concept, showing laser or heavy-ion driver beams (arrows) and fast ignitor laser focused on the target from the left. The illumination geometries are (a) direct drive symmetrically illuminated and (b) indirect drive, illuminated from one end. In (a) the ignition laser has to penetrate an overlying plasma corona before heating the target. In (b) using cone focusing within a hohlraum, the plasma blowoff is excluded from the ignitor laser path and the driver lasers can be concentrated at one end of the chamber. Courtesy of the University of California, Lawrence Livermore National Laboratory, and the Department of Energy under whose auspices the work was performed.

a very small size, at precisely the right time, with tolerances of less than 100 picoseconds (100×10^{-12} s).

The fast ignitor in IFE and the several new MFE concepts mentioned are examples of innovations that may significantly increase the attractiveness and lower the cost of electricity from a fusion power plant. Even nearly 50 years into the fusion program, new ideas continue to emerge; and old, previously rejected, ideas become attractive with new technology.

Scientific understanding of inertial-confinement fusion will be significantly advanced early in the next century by the completion and operation of the NIF (National Ignition Facility) at Lawrence Livermore National Laboratory in the US, and the Laser MegaJoule (LMJ) facility in France (Kilkenny). The NIF is likely to be the first laboratory device to realize fusion ignition – when its remaining issues, that currently cloud the cost and schedule for achieving ignition, are solved (Malakoff and Lawler, 1999). Although the primary missions of both the NIF and the LMJ are defense related, an important spin-off benefit is to show that inertial-fusion energy is feasible. To capitalize on this demonstration requires a parallel effort in developing suitable and cost-effective drivers and chambers for IFE.

8.4 Where to from Here?

For both magnetic and inertial fusion, new capabilities have been developed. Sophisticated theory, diagnostics to measure fine details of plasma parameters, and experimental techniques to control and alter plasma parameters and the geometry in which they are imbedded have provided increased understanding and the ability to dramatically improve plasma performance over what was possible 10 or 20 years ago. Some examples are: control of the current profile changes the magnetic field produced by the current, and can produce reversed central shear in tokamaks – this has improved confinement to the theoretical limits, although still on a transient basis (Ishida, 1999; Watkins, 1999). The beta (plasma pressure normalized to the magnetic field pressure) of magnetically confined plasma has been increased to theoretical limits, and in some cases has broken through into a new second stability regime producing higher limits in tokamaks. Rayleigh–Taylor instability growth rates of planar targets have been measured as a function of small imperfections in the surface of the target or the laser uniformity, and have been found to agree with theory. Rayleigh–Taylor instabilities also limit the compression of thin spherical shells, but to a lesser degree than predicted by early theories: additional stabilizing effects that relax the Rayleigh–Taylor constraints have been predicted and experimentally verified.

Today, fusion energy is at a crossroads – the $10 billion price tag of the ITER has dismayed many supporters (Glanz and Lawler, 1998). A vigorous re-evaluation of magnetic and inertial concepts is currently underway (Glanz, 1999). A consensus is growing for developing multiple concepts with low-level funding at the exploratory level. Those that appear attractive will be funded at the higher level proof-of-principle stage, and the best of those will reach the performance extension stage. Beyond that, one or two will enter the engineering test facility phase, and probably only one will enter the demonstration power plant phase, but only if the costs are much less than that of ITER. Most of the world's magnetic fusion programs are still studying tokamaks, but the emphasis is now on advanced modes of operation that improve the power plant potential of the tokamak. A few alternative concept experiments are just beginning operation, with more following.

The emerging new phase bears a cursory resemblance to the early days of fusion when a large zoo of magnetic configurations was studied experimentally, with only tenuous ties to theory. This apparent resemblance is misleading – with the theoretical and experimental tools that have been developed and honed, researchers are able to ask more subtle questions and obtain answers. Many of the techniques learned in studying tokamaks are applicable to other configurations.

Our view is that the ITER Project has demonstrated one route to a fusion power plant that would work – we now have an existence theorem for fusion energy in the form of engineering drawings for a device that would have a high probability of working as intended. This enables us to turn with renewed confidence to exploring other routes to fusion that will be more cost effective, and that will be better able to realize the potential of fusion energy. An innovative fusion energy research program, focused on the critical issues, will maximize the probability of achieving the full potential of fusion energy. The development of a virtually limitless energy source will provide a profound benefit to future humanity.

List of Abbreviations

COE   cost-of-electricity
DPSSL   diode-pumped solid-state laser
FIRE   Fusion Ignition Research Experiment
FRC   field reversed configuration
IFE   inertial fusion energy
ITER   International Thermonuclear Experimental Reactor
JET   Joint European Torus
JT-60(U)   Japan, largest Tokamak, slightly larger than JET
KrF   Krypton–fluoride laser
LMJ   laser megajoule
MFE   magnetic fusion energy
MTF   magnetized target fusion (or magnetically-insulated IFE)
NIF   National Ignition Facility
RFP   reversed field pinch

References

Bangerter, R. (1999), Ion Beam Fusion, *Phil. Trans. Royal Soc. London Series A – Mathematical Physical and Engineering Sciences*, 357:575.

Chen, F.F. (1974), *Introduction to Plasma Physics*, Ed. 1., Plenum Press, Ch. 9.

Delene, J. G. (1994), Advanced fission and fossil plant economics – implications for fusion, *Fusion Technology*, 26:1105.

Delene, J. G., Sheffield, J., Williams, K. A., Reid, A. L., and Hadley, S. (2000), An assessment of the economics of future electric power generation options and the implications for fusion, *Fusion Technology*, 39, 228 (2001).

Drake, R. P., Hammer, J. H., Hartman, C. W., Perkins, L. J., and others, (1996), Submegajoule linear implosion of a closed field line configuration, *Fusion Technology*, 30:310.

Fowler, T. K., Hua, D. D., Hooper, E. B., Moir, R. W., and Pearlstein, L. D. (1999), Pulsed spheromak fusion reactors, *Comments on Plasma Physics and Controlled Fusion, Comments on Modern Physics*, 1:83.

Frosch, R. A. (1992), Industrial ecology: a philosophical introduction, *Proc. Nat. Acad. Sci. USA*, 89:800–803.

Glanz, J. (1999), Common ground for fusion, *Science*, **285**:820.

Glanz, J. and Lawler, A. (1998), Planning a future without ITER, *Science*, **279**:20.

Hogan, W. J. (1995), *Energy from Inertial Fusion*, IAEA International Atomic Energy Agency, Vienna.

Ishida, S., JT-60 Team, (1999), JT-60U high performance regimes, *Nuclear Fusion*, **39**:1211.

Kilkenny, J. D., Campbell, E. M., Lindl, J. D., Logan, B. G. *et al.* (1999), The role of the National Ignition Facility in energy production from inertial fusion, *Phil. Trans. Royal Soc. London Series A – Mathematical Physical and Engineering Sciences*, **357**:533.

Krakowski, R. A. (1995), Simplified fusion power plant costing – a general prognosis and call for new think, *Fusion Technology*, **27**:135.

Logan, B. G., Moir, R. W., and Hoffman, M. A. (1995), Requirements for low-cost electricity and hydrogen fuel production from multiunit inertial fusion energy plants with a shared driver and target factory, *Fusion Technology*, **28**:1674.

Malakoff, D. and Lawler, A. (1999), Laser fusion – a less powerful NIF will still cost more, *Science*, **285**:1831.

Maniscalco, J. A., Berwald, D. H., Campbell, R. B., Moir, R. W., and Lee, J. D. (1981), Recent progress in fusion–fission hybrid reactor design studies, *Nuclear Technology/Fusion*, **1**:419.

Moir, R. W. (1996), Liquid wall inertial fusion energy power plants, *Fusion Engineering And Design*, **32–33**:93.

Moir, R. W. (1997), Liquid first walls for magnetic fusion energy configurations, *Nuclear Fusion*, **37**:557.

Moir, R. W., Bieri, R. L., Chen, X. M., Dolan, T. J. *et al.* (1994), HYLIFE-II – a molten-salt inertial fusion energy power plant design – final report, *Fusion Technology*, **25**:5.

Najmabadi, F. (1998), Overview of ARIES-RS tokamak fusion power plant, *Fusion Engineering and Design*, **41**:365.

Perkins, L. J. (1998), The role of inertial fusion energy in the energy marketplace of the 21st century and beyond, *Nuclear Instruments & Methods in Physics Research Section A – Accelerators Spectrometers Detectors and Associated Equipment*, **415**:44.

Perkins, L. J., Logan, B. G., Rosen, M. D., Perry, M. D., de la Rubia, T. D., Ghoneim, N. M., Ditmire, T., Springer, P. T., and Wilks, S. C. (2000), High-intensity-laser-driven micro neutron sources for fusion materials applications at high fluence, *Nuclear Fusion*, **40**:1.

PPPL (2000), fire.pppl.gov

Prager, S. C. (1999), Dynamo and anomalous transport in the reversed field pinch, *Plasma Physics and Controlled Fusion*, **41**:A129.

Riordon, J., (1999) Many shapes for a fusion machine, *Science*, **285**:822.

Sheffield, J., (1994) The physics of magnetic fusion reactors, *Rev. Mod. Phys.*, **66**:1015.

Shimomura, Y., Aymar, R., Chuyanov, V., Huguet, M., Parker, R., ITER Joint Central Team (1999), ITER overview, *Nuclear Fusion*, **39**:1295.

Tabak, M., Hammer, J., Glinsky, M. E., Kruer, W. L., Wilks, S. C., Woodworth, J., Campbell, E. M., Perry, M. D., and Mason, R. J. (1994), Ignition and high gain with ultrapowerful lasers, *Phys. Plasmas*, **1**:1626.

Watkins, M. L. and JET Team, (1999), Physics of high performance JET plasmas in DT, *Nuclear Fusion*, **39**:1227.

9

Energy Prosperity within the Twenty-first Century and Beyond: Options and the Unique Roles of the Sun and the Moon

9.0 Summary

How much and what kind of commercial energy is needed to enable global energy prosperity, and possibly global economic prosperity, by the middle of the 21st century? Economic prosperity requires approximately 6 kWt of thermal commercial power per person or ~2 kWe of electric power per person. A prosperous world of 10 billion people in 2050 will require ~60 terawatts (TWt) of commercial thermal power or 20 TWe of electric power. What are the options for providing the necessary power and energy by the middle of the 21st century and for centuries thereafter? The twenty three options analyzed for commercial power fall under the five general categories of (1) mixed and carbon-based, (2) terrestrial renewable, (3) terrestrial solar, (4) nuclear fission and fusion, and (5) space and lunar solar power systems. It is argued that the only practical and acceptable option for providing such large flows of commercial power is to develop the Moon as the platform for gathering solar energy and supplying that energy, via low-intensity beams of microwaves, to receivers, termed "rectennas", on Earth. The rectennas will output clean and affordable electric power to local grids. No pollution (greenhouse, ash, acids, radioactive wastes, dust) will be produced. All energy inputs to the biosphere of the net new electric power can be completely balanced on a global basis without "greenhouse-like" heating of the biosphere.

9.1 Twenty-first Century Challenges: People, Power and Energy

At the end of the 20th century, the 0.9 billion people of the economically Developed Nations of the Organization of Economic Cooperation and Development (OECD) used ~6.8 kWt/person of thermal power. The 5.1

345

billion people of the Developing Nations use ~1.6 kWt/person (Nakicenovic *et al.*, 1998). If the large per capita use of power by former states of the Soviet Union is subtracted, the other non-OECD nations use less than 1 kWt/person of commercial power (Criswell, 1998). The majority of people in the Developing Nations have very limited, if any, access to commercial power and essentially no access to electric power. It is commonly stated that the world has adequate fossil-fuel resources for many centuries. This is because virtually all projections of global energy consumption assume restricted economic growth in the Developing Nations. Such studies usually project accumulated global consumption of carbon fuels to be less than 2000 TWt-y over the 21st century. That is only true if most of the people in the world stay energy and economically poor throughout the 21st century.

What scale of commercial power is required by the year 2050, and beyond, to provide ten billion people with sufficient clean commercial energy to enable global energy and human prosperity? Western Europe and Japan now use ~ 6 kWt/person. Analyses of the mid-1960s United States and world economies revealed that ~ 6 kWt/person, or in the 21st century ~2 kWe/person of electric power, can enable economic prosperity (Goeller and Weinberg, 1976; Criswell and Waldron, 1990; Criswell, 1994 and references therein). This level of commercial power enables the provision of goods and services adequate to the present standard of living in Western Europe or Japan. All industrially and agriculturally significant minerals and chemicals can be extracted from the common materials of the crust of the Earth. Fresh water can be obtained from desalting seawater and brackish water. Adequate power is provided to operate industries, support services, and provide fuels and electricity for transportation and residential functions. Global power prosperity by 2050, two generations into the 21st century, requires ~ 60 terawatt of thermal power (60 TWt = 60×10^{12} Wt = 6 kWt/person $\times 10 \times 10^9$ people). With reasonable technology advancement, ~2–3 kWe/person can provide these same goods and services.

From 1850 to 2000, humankind consumed ~ 500 TWt-y of non-renewable fuels. During the 20th century, commercial power increased from ~2 TWt to 14 TWt. Power prosperity by 2050 requires an increase from ~14 to 60 TWt. The total increase of 46 TWt is 3.3 times present global capacity and requires the installation of ~0.9 TWt of new capacity per year starting in 2010. This is 7.5 times greater than the rate of commercial power installation over the 20th century. Sixty terawatts by 2050 is two to three times higher than considered by the United Nations Framework Convention on Climate Change (Hoffert *et al.*, 1998). It is also higher than is projected by recent detailed studies.

The World Energy Council sponsored a series of studies projecting world energy usage and supply options over the 21st century. The International

Institute for Applied Systems Analysis (IIASA) conducted the studies and reported the results at the 17th World Energy Congress in Houston (Nakicenovic *et al.*, 1998). The models are constrained, in part, by the capital required to install the new power systems. The ability of Developing Nations to purchase fuels is a limitation. Power capacity is also limited by operating costs of the systems and *externality* costs such as for environmental remediation and degradation of human health. Providing adequate power by 2050 requires systems that are lower in cost to build, operate, and phase out than present fossil systems.

Nakicenovic *et al.* (1998) developed three general models for the growth of commercial power during the 21st century that are consistent with present use of power in the Developed and Developing Nations. Interactions between the rates of growth of commercial power, populations, and national economies were modeled. Their *Case A2*, adapted to Table 9.1, projects the greatest increase in commercial power over the 21st century. *Case A2* projects the most aggressive development of coal, oil, and natural gas and assumes the least environmental and economic impacts from burning these fossil fuels. By 2050, per capita power use rises to 8.8 kWt/person in the Developed Nations and to 2.5 kWt/person in the Developing Nations. By 2100 the per capita power usage of Developed and Developing Nations converge to 5.5 kWt/person and energy prosperity is achieved. Increasing economic productivity in the use of thermal power is assumed over the 21st century. This enables the decrease in per capita power use in the Developed Nations between 2050 and 2100.

The "All Carbon" dashed curve in Figure 9.1 depicts the total global energy consumed by the Developed and the Developing Nation under *Case A2* as if all the commercial energy were provided from fossil fuels. The curve is negative because the non-renewable fossil fuels are consumed. Fuels consumed prior to the year 2000 are not included. A total of 3600 TWt-y of fossil fuel is consumed between 2000 and 2100. This corresponds to ~2700 billion tons of equivalent oil (GToe) or ~3900 GTce of equivalent coal. The horizontal lines indicate the estimated quantities of economically *ultimately recoverable* (UR) conventional oil, conventional gas, unconventional oil, and coal and lignite. Coal and lignite are the dominant sources of commercial fossil fuels over the 21st century. Global energy prosperity quickly depletes the oils and natural gases. Given the uncertainties in estimates of *ultimately recoverable* coal and lignite, it is conceivable that they could be depleted within the 21st century. Major technological advances in coal mining technology are required once near-surface deposits are exhausted (Bockris, 1980). Coal and lignite would certainly be consumed by a 64 TWt economy a few decades into the 22nd century.

At the beginning of the 21st century, ~1.2 TWt of commercial power is

Table 9.1 *21st century power, energy, and GWP models*

Year	2000	2050	2100
Nakicenovic *et al.*, 1998: Case A2 (mixed system)			
Commercial power (TWt)	14.2	33	64
Total energy consumed after 2000 (TWt-y)	—	1200	3600
Per capita power: GLOBAL (kWt/person)	2.3	3.1	5.5
Developed Nations (OECD) (kWt/person)	6.9	8.8	5.5
Developing Nations (non-OECD) (kWt/person)	1.6	2.5	5.5
Population: Developed Nations ($\times 10^9$)	0.9	1.0	1.07
Developing Nations ($\times 10^9$)	5.1	9.6	10.10
Gross World (Domestic) Product (T$/y)	26	100	300
Summed GWP after 2000 (T$)	—	2800	12 000
Energy sector investment over 21st century (T$)	—	—	120
Fuels costs to users @ 4$\times$shadow cost	—	—	830
Externality costs @ 4$\times$shadow cost	—	—	800
Lunar solar power system			
Commercial power (TWe) "e" = electric	—	20	20
Total NEW LSP energy consumed after 2000 (TWe-y)	—	520	1520
Per capita power: GLOBAL (kWe/person)	Above	2	2
Gross World (Domestic) Product (T$/y)*	26	319	425
Summed GWP after 2000 (T$)	—	7400	25 800
Energy Sector Investment over 21st century (T$) for LSP(Ref) and *LSP(No EO)*	—	—	60 to *300*
Fuels costs to users	—	0	0
Externality costs	—	0	0

Note:
* Economic output of unit of electric energy increases @ 1 %/y.

produced from renewable sources. It comes primarily from burning wood and secondarily from hydroelectric installations. The upward directed curve in Figure 9.1 depicts the cumulative energy supplied by a new source of renewable energy. For renewable energy the curve is positive because net new energy is being provided to the biosphere. No significant terrestrial resources are depleted to provide the energy. This upward curve assumes that a new renewable power system is initiated in 2010 and that it rapidly grows, by 2050, to the functional equivalent in output of ~60 TWt. The renewable system operates at the equivalent of ~60 TWt level thereafter. By 2100, the renewable system has contributed the equivalent of 4500 TWt-y of net new commercial thermal

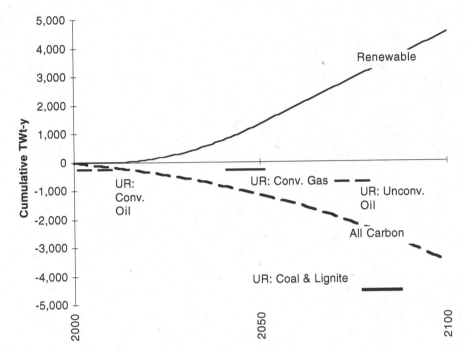

Figure 9.1   Cumulative energy utilized after the year 2000 and ultimately recoverable (UR) fuels.

energy to humankind's commercial activities on Earth. Thereafter, the renewable energy system contributes the equivalent of 6000 TWt-y per century. What are the options for providing 60 TWt, or the equivalent of ~20 TWe, of commercial power by 2050 and for centuries thereafter?

9.2 Sources to Supply 60 TWt or 20 TWe of Commercial Power by 2050

Columns 1 through 9 of Tables 9.2–9.6 summarize the characteristics of conventional and unconventional systems in 2050. Most, such as biomass, fossil, and nuclear, primarily yield thermal power (TWt) and thermal energy (TWt-y). Some, such as wind and hydroelectric turbines and solar photovoltaic cells, are sources of electrical energy and are normally rated in terms of electric power (TWe) and electric energy (TWe-y). The last column provides an estimate of the feasible level of electric power each power system can supply, as constrained by technical considerations or by the funds available for their development and operation. Potential technical capacity can be much larger than what is economically feasible. For example, over 167 000 TWt of sunlight intersects the disk of Earth. However, at the surface of Earth sunlight is

Table 9.2 *Mixed and carbon-based sources of thermal and electric power in 2050*

1. Power System	2. Maximum energy inventory on Earth (TWt-y)	3. Annual renewal rate (TWt)	4. Key non-technical issues @ ≤20 TWe	5. Limiting technol. factors @ 20 TWe	6. Deplete or exhaust (Y) @ 20 TWe or 60 TWt	7. Pollution products	8. Long-term trend of total costs @ 20 TWe	9. Feasible electric output by 2050 in TWe
1. Mixed System (Case A2)	Non-renew ≤3200 @ 2050	7.7 System output	• All issues (#2–19)	• All issues (#2–19)	<100 for coal @ 2050	• All issues (#2–19)	• Rising • All new systems by 2150	~11 Case A2 used in #1–#19
2. Bio-resources	<230 @ 2000	<50 (primarily wood)	• Cost • Less bio-diversity • Political objections	• Supply • Mass handling • Nutrients • Water • Land Use	≤3	• Smoke • Methane • Diseases • Erosion • Increased CO_2	• NA (Not applicable)	<0.2
3. Peat	<60 @ 2000	~0	• Destroy – wetlands – Ag. uses	• Supply • Transport (<100 km)	<1	• Dust • Fire • Ash	• NA	~0
4. Coal	<4500 @ 2000	0	• Coal lost to future • Land recovery • Environmental impacts	• Supply • Pollution control	≤100	• CO_2 • Ash, acids, heavy metals • Waste heat	• NA	≤4 Steady to decreasing
5. Oils/Gas	<1300 @ 2000	0	• HCs lost to future • CO_2	• Supply	≤30	• CO_2, acids • Waste heat	• NA	≤8 Sharply decreasing

| 5.1 Natural gas hydrates | ≥10000 | Not well mapped | ~0 | • Seabed disruption • Cost • #4 & #5 above | TBD | • Diffuse resource • Efficient separation | TBD | • Greenhouse gas • Natural releases • CO_2 | TBD | TBD |

Notes:
Column descriptions (same for Tables 9.2–9.6)

#1. Name of the large scale power system and primary energy source.

#2. Total quantity of energy that can be reasonably extracted over the life of the energy source. Terawatt-year of thermal energy (TWt-y) is used unless noted otherwise in the text.

#3. Annual power rate, in terawatts of thermal power (TWt), at which the source of energy is renewed. Other power units may be noted in the text.

#4. Lists the major non-technical factors limiting 20 terawatts (TWe) of electrical power.

#5. Lists the major technical factors that limit production of 20 TWe of electric power.

#6. Estimates the lifetime of each energy source, in years, at 20 TWe of electric output.

#7. Lists the major pollution products of each power system.

#8. Indicates the long-term trend in cost for producing 20 TWe of power.

#9. Estimates the maximum power production, in TWe, for each power system.

Table 9.3 Renewable terrestrial systems in 2050

1. Power System	2. Maximum energy inventory on Earth (TWt-y)	3. Annual renewal rate (TWt)	4. Key non-technical issues @ ≤20 TWe	5. Limiting technical factors @ 20 TWe	6. Deplete or exhaust (Y) @ 20 TWe or 60 TWt	7. Pollution products	8. Long-term trend of total costs @ 20 TWe	9. Feasible electric output by 2050 in TWe
6. Hydro-electric	<14	<5	• Costs • Multi-use – Site – Fresh water	• Sites • Rainfall *NSA (Not stand-alone)	<1	• Sediment • Flue water • Dam failure	• NA	<1.6
6.1 Salinity-gradient to – seawater – brine	1700 to seawater 24000 to brine	0.5 to 7	• Blocking hydrologic cycles • Brine pond area	• Conversion means • Brine production	0.02 (rivers) to 700 (polar caps)	• Restrict river flows • Membranes, brine	• NA	<0.3
7. Tides	0	<0.07 (tech. feasible)	• Costs • Shoreline effects	• Sites • Input • NSA	<0.01	• Change local tides Fish kills?	• NA	≤0.02
8. Waves	0	1 to 10 (global deep waters)	• Costs • Shore processes • Navigation	• NSA • Good sites	<0.1	• Transfer – Gases – Nutrients – Heat – Biota	• NA	<0.1 or much less
9. Ocean thermal	~2×10^5 but perhaps ≤100 affordable to access	~2100 but ≤0.04% is likely useful	• Costs • Ocean local and global circulation • Cooling surface waters	• Sites • Low effic. • Bio-fowling • Transmission to shore	<800 @ 7% conv. effic. But locally perhaps <1	• #8 above • OTEC – mass – rusts – fouling • La Nina effects	• NA	<0.1

10. Geothermal	≤9×10^6 (global in top 7 km)	<30 global Mostly low grade	• Costs • Geologic risks • Reinjection effects?	<1 @ 10% effic.	• Local depletion • Flow resistance • Efficiency	• Waste – heat – minerals	• NA	<0.5 on continents @ 10% effic.
11. Wind	0	<100 on land ~200 off-shore	• Costs • Intrusiveness • Biota hazards	≥10^9	• Diffuse & irregular • NSA • ≤10 MWe/km^2 • Storage	• Land use • Noise • Modify winds (?) – local – global	• Possibly down • Requires low cost storage & transmiss.	≤6

Table 9.4 *Terrestrial solar power systems*

1. Power System	2. Maximum energy inventory on Earth (TWt-y)	3. Annual renewal rate (TWt)	4. Key non-technical issues @ ≤20 TWe	5. Limiting technol. factors @ 20 TWe	6. Deplete or exhaust (Y) @ 20 TWe or 60 TWt	7. Pollution products	8. Long-term trend of total costs @ 20 TWe	9. Feasible electric output by 2050 in TWe
12. Terrestrial solar power (thermal)	0	≤1 to 20 MWe/km² output of regional system	• Very high systems cost • Local climate change • Weather	• Irregular flux • NSA	≥10⁹	• Waste heat • Induced climates • Production wastes • Land use	• Possibly down • Slow learning	≤3.3 • Sum of 12 and 13
13. Terrestrial solar power (photo-voltaic)	0	Above	Above	Above	Above	Above	Above	• Above (#12)

Table 9.5 *Nuclear power systems*

1. Power System	2. Maximum energy inventory on Earth (TWt-y)	3. Annual renewal rate (TWt)	4. Key non-technical issues @ ≤20 TWe	5. Limiting technol. factors @ 20 TWe	6. Deplete or exhaust (Y) @ 20 TWe or 60 TWt	7. Pollution products	8. Long-term trend of total costs @ 20 TWe	9. Feasible electric output by 2050 in TWe
14. Nuclear fission (No breeder)	<430 @ <130 $ per kg ^{235}U	0	• Full life cycle costs • Political acceptance • Health and safety	• Wastes control • Reactor life time	≤7	• Radioactive – fuels – parts – wastes	• NA	≤1.5
15. Nuclear breeder (^{238}U/Th)	≤33000	0	• Above • Proliferation	• Above	≤550	• Above • Weapons grade materials	• Perhaps constant or decreasing	• Contribution to #14
16. Nuclear breeder (U in sea water)	≤6×10^6 @ 3.3 ppb of ^{238}U	0	• Above • Higher uses	• Above	≤300000	• Above	• Above (#15)	• Contribution to #14
17. Nuclear fusion-fission or accelerator (D-T with ^{238}U-Th)	<6×10^9	0	• Above	• Above • Rate of fuel production per unit of power	TBD	• Above • Radioactivity (much lower)	• Possibly decreasing	• Contribution to #14
18. Nuclear Fusion (D-T)	≥1×10^9	0	• Above	• Practical fusion • Reactor life time	>1×10^9 • Lithium limited (tbd)	• Above (#17) • Tritium • Waste heat	• TBD	• 0 likely
19. Nuclear Fusion (D-^{3}He lunar)	≤100 to 1×10^5	~ 0 (9.5 kg/y)	• Lunar mining • Gas release	• Above (#18) • ^{3}He inventory	≤1 to 5000	• Above	• TBD	• 0 likely

Table 9.6 *Space and lunar power systems*

1. Power System	2. Maximum energy inventory on Earth (TWt-y)	3. Annual renewal rate (TWt)	4. Key non-technical issues @ ≤20 TWe	5. Limiting technol. factors @ 20 TWe	6. Deplete or exhaust (Y) @ 20 TWe or 60 TWt	7. Pollution products	8. Long-term trend of total costs @ 20 TWe	9. Feasible electric output by 2050 in TWe

20. Geo-space solar power sats (from Earth)	• 0 with power relay satellites
	• ~0.01 with storage
		20 to 250 We/m² times rectenna area	• Life cycle costs
			• Fleet
			  – visible
			  – variable
			  – life
			• System likely NSA
				• Geo-arc length
				• Managing
				  – satellites
				  – shadows
				• Load following
				• Spectrum availability
					$>10^9$	• Microwave noise
						• Transport
						  – noise
						  – exhaust
						• New sky objects
						• Orbital debris
						• Shadowing Earth
							• NA
							• Down from very high initial cost
								≤1
								Even with ~100 decrease in Earth-to-orbit transport costs

21. LEO/MEO – solar power sats	• 0 with sat to satellite re-beaming
	• 0.01–0.05 with storage
		≤250×D We/m² times rectenna area
		• D = Duty cycle
		0.01 ≤ D ≤ 0.3
			• Above (#20)
			• NSA
				• Managing
				  – satellites
				  – shadows
				  – debris
				• Load following
				• Spectrum availability
				• Duty cycle
					$>10^9$	• Above (#20)
						• Earth illumination
							Up
							Due to maintenance for debris
								≤0.1
								Even with ~100 decrease in Earth-to-orbit transport costs

22. Space solar sats in deep space	• 0 to 0.01 with excess capacity in space	20 to 250×D We/m² times rectenna area 0.3 < D ≤ 1	• Above (#20) • NSA	• Very large deep space industry • Power use on Earth	$>10^9$	• Microwave noise • New sky objects	Down	≤1
23. Lunar solar power system	• 0 with EO beam redirectors • 0.01 Moon eclipse • 0.04 with no EOs	20 to 250×D We/m² times rectenna area 0.3 < D ≤ 1	• Life cycle costs	• Power use on Earth • Area of Moon • EO beam redirector satellite	$>10^9$	• Microwave noise • Debris of redirectors	Potentially ~0.1 ¢/kWe-h	≥20 to ~1000 in 22nd century

diffuse, degraded in intensity, interrupted by the day–night cycle and the atmosphere, and ≤25% intersects the populated continents. Systems to gather solar power on Earth and transfer it to diverse final users are very expensive. Thus, commercial solar power is limited to niche markets. An earlier version of Tables 9.2–9.6 was first published by Criswell and Waldron (1991) and slightly revised by Criswell (1998b). Hoffert and Potter (1997) have also explored these topics.

9.2.1 Mixed and carbon systems extrapolated from current practice

Mixed systems

The 14.2 TWt global power system of the year 2000 is a mixed system (Nakicenovic *et al.*, 1998). It is fueled primarily by carbon (fossil coal and renewable wood) and hydrocarbons (fossil oils and gases). Nuclear and hydro-electric contribute ~10% of the primary energy. Mixed global power systems can consist of enormously different combinations of primary energy sources and options for conversion, storage, and distribution of the commercial energy. There are strong motivations to extend the use of existing power systems and practices. This extension minimizes needed investments for increased capacity, takes advantage of locally attractive "gifts of nature", such as hydropower or biomass in Developing Nations, can stretch the lifetime of non-renewable sources, utilizes current business practices and labor skills, and can be pursued by existing businesses.

Case A2 of Nakicenovic *et al.* (1998: p. 69–71, p. 118–124, p.134) projects the highest level of world economic growth. Table 9.1 summarizes the growing power use projected by *Case A2* from 2000 to 2100. Global power production is 33 TWt in 2050. In 2050, coal produces 10.6 TWt. At this rate coal reserves are projected to last ~170 years. Oils and gases produce 13.6 TWt and they are projected to last for ~100 years. Conventional nuclear provides 1.3 TWt and reserves are adequate for ~280 years.

Total power increases to 64 TWt by 2100. Between 2000 and 2100 this mixed system consumes 3600 TWt-y of primary energy, mostly fossil and nuclear. By 2100, the fossil fuels and biomass provide 65% of primary energy. Projected lifetime of coal reserves at 2100 is ~50 years. Gas and oil use is dropping rapidly as they approach exhaustion. Consumption of coal and biomass is rising along with that of nuclear. Carbon dioxide and other emission of fossil fuels rise at a rapid rate throughout the 21st century. Carbon dioxide production reaches 20 GtC/y in 2100 and cumulative emissions from 1990 are ~1500 GtC. Atmospheric CO_2 approaches 750 ppmv, or 2.8 times the pre-industrial value, and global warming the order of 3 to 4.5°C is projected. There is increasing confidence that greenhouse warming is occurring (Kerr, 2000a).

Investment in this mixed system is 0.2 T\$/y in 1990. It is projected to be 1.2 T\$/y by 2050 and assumed, for this illustration, to rise to 2.3 T\$/y by 2100 in order to install 64 TWt of capacity. Total investment from 2000 to 2100 is ~120 T\$. See Table 9.1.

Both the fuel users and producers must deal with externality cost created by the use of these non-renewable fuels. Externality cost arises from the greenhouse effects of carbon dioxide, neutralizing acids and ash, suppressing dust, and the effects of uncertainties in energy supplies. Other factors such as the costs to human health of mining and emissions and defense of primary energy sources must be included. For discussion, assume the externality cost equals the price of the primary fuels. Total externality cost is then ~800 T\$. Total cost of this *Case A2* global power system, from 2000 to 2100, is ~1800 T\$. Thus, total cost of energy to users is projected to be ~13% of integral GDP. The oil supply disruptions of the 1970s, which increased oil prices by a factor of two to three, slowed global per-capita growth for a decade. If externality cost was ~15 times the price of the fuels, all economic gains over the 21st century would be wiped out.

There is a long standing debate about whether or not the use of a depletable resource (fossil and nuclear fuels) in a core economic activity (production of commercial power) leads to the creation of "net new wealth" for the human economy inside the biosphere. Solar energy from facilities beyond Earth offers a clear alternative to depletable fossil and nuclear fuels. The solar energy for facilities in space definitely provide "net new energy" to Earth. It is arguable that space and lunar solar power systems offer expedient means to supply dependable and clean renewable power at attractive commercial rates.

Tables 9.2–9.6 characterize the options for a global power system (column #1) that might be utilized to achieve 60 TWt by 2050 and maintain that level thereafter. This scenario requires a more aggressive development of commercial energy than is projected in *Case A2* of the WEC study and delivers ~600 TWt-y more thermal-equivalent energy over the 21st century. Refer to Row 1 of Table 9.2 and column 2. For *Case A2*, ~3200 TWt-y of non-renewable fuels are available in 2050. *Case A2* analyses project that renewable commercial power systems produce 7.7 TWt in 2050 (Column #3). Column #4 summarizes the key non-technical issues that will limit the production of 20 TWe, or 60 TWt, by 2050. Specifics for the mixed systems of *Case A2* are deferred to the discussion of each major potential element of the mixed power system (rows #2–#19). The same is true for columns #5 and #7. Column #6 provides an estimate of the lifetime of the fuel resources at the year 2050 for *Case A2* at their burn rate in 2050 of 24 TWt. Non-renewable fuels will be exhausted by ~2180. *Case A2* projects 64 TWt by 2100. Thus by 2120 the fossil fuels will be depleted.

The cost of energy from the mixed system is likely to tend upward. Rather than focusing capital on the most cost-effective power systems, it will be necessary to provide R&D, construction, and maintenance funds to a wide range of systems. The costs of non-renewable fuels will increase as they are depleted, and, very likely, the cost of measures to protect the environment will also increase. Column #8 of Tables 9.2–9.6 indicates a rising cost, driven in part by the need to replace most capital equipment and systems before 2100. The total power production of *Case A2* is equivalent to 33 TWt in 2050. Total electric output would be only 11 TWe (Column #9). *Case A2* does not provide the 20 TWe required in 2050 for an energy-prosperous world.

Nakicenovic *et al.* (1998) also consider power systems that are more environmentally friendly than *Case A2*. *Case C* assumes extensive conservation of energy, greatly expanded use of renewable sources of power, and a reduced rate of growth of the world economy. In *Case C* global power may be as low as 19 TWt in 2100, or ~1.7 kWt/person. Integral GWP (2000 to 2100) is ~10 000 T$. Neither energy nor economic prosperity is achieved on a global scale. *Case C* is closer to the power and economic profiles considered by the Intergovernmental Panels on Climate Change.

Carbon-based power systems

The mixed-power system in row #1 uses contributions from each of the next 18 types of power sources. Each of these is examined in terms of its ability to provide 60 TWt or 20 TWe by 2050 and indefinitely thereafter.

Bioresources (#2)

Bioresources is used to provide more detail as to the analysis approach. In *Column #2* the energy inventory available on Earth, for this and all following power options, is described in terms of terawatt-years of total thermal power, whether the primary energy source provides thermal, nuclear, or electric energy. To a first approximation, 1 TWt of thermal power yields approximately 0.33 TWe of electric power. Most useful biomass is available on land in the form of trees, with a total thermal energy inventory of ~230 TWt-y. Primary estimates for biofuels are from Trinnaman and Clarke (1998: 213, 124), Criswell (1994, 1998b), Criswell and Waldron (1991), and references therein. Ten billion people will ingest ~0.003 TWt, or 3 GWt, of power in their food.

Column #3 provides an estimate of the rate at which the primary energy resource is renewed within the biosphere of Earth (TWt-y/y or TWt). Annual production of dry biomass is approximately equally divided between the oceans and land. However, the primary ocean biomass is immediately lost to the ocean depths or consumed in the food cycle. New tree growth provides most of the new useful biomass each year. The renewal rate is approximately 50 TWt-y/y or 50 TWt of power.

Columns #4 and #5 identify the major non-technical and technical issues relevant to the energy source providing 20 terawatts of electric power (20 TWe) by 2050. For bioresources, costs will be high because of gathering, transportation, and drying of biomass that has a relatively low fuel density per unit of mass. The continents will be stripped of trees, grasses, and fuel crops, biodiversity will be sharply reduced, and great political conflicts will ensue. New nutrients will be required as most biomass is removed from fuel farms. Massive irrigation will be required and land use will be dominated by growth of fuel-wood. Agriculture will compete with fuel production for land, water, nutrients, labor, and energy for the production processes.

Column #6 estimates the time in years that a particular energy source will be depleted if it provides 20 TWe or $\sim$ 60 TWt. In the case of biomass, the global inventory of biofuels will be depleted in less than three years. Because the net-energy content of dried biomass is low, it is assumed that 90 TWt of biomass fuel enables only 20 TWe. The renewal rate and energy content of biofuels are so low that they cannot provide 60 TWt or 20 TWe on a sustainable basis.

The primary pollution products of biomass are summarized in *Column #7*. For biofuels these include obvious products such as smoke. However, methane and CO_2 will be released from decaying biomass and disturbed soils. Recycling of CO_2 to oxygen will be reduced by at least a factor of two until forests recover. Erosion will be increased. Diseases will be liberated as animals are driven from protected areas.

Column #8 indicates the long-term trend in cost *if the final electric power is provided exclusively* by the given source of energy. Bioresources are unable to provide the 20 TWe, or 60 TWt, on a sustainable basis. Thus, Bioresources are NOT APPLICABLE (NA). *Column 9* estimates the feasible electric output of each energy source by 2050. Sustainable power output can be limited by the size of the resources base (biofuels, oils and natural gas, coal), pollution products (coal), the cost that society can afford, availability of technology (controlled nuclear fusion), or other factors. *Case A2* projects $\leq$0.2 TWt as the limit on power production from biomass.

Peat (#3), coal (#4), oils, gas (#5), and natural gas hydrates (5.1)

The 60 TWt system requires 4500 TWt-y of input energy through the year 2100, and 6000 TWt-y through the 22nd century. If only peat, oil, and natural gas are used, they will be exhausted well before the end of the 21st century (columns #2, #3, #6). Coal would be exhausted early in the 22nd century. See Trinnaman and Clarke (1998, p. 205, peat). For coal (#4) and oil and gas (#5) see Nakicenovic *et al.* (1998, p. 69 – Cases *A1, A2*, and *C*). The thermal-to-electric conversion efficiencies are assumed to be 33.3% for coal and 45% for oil and gas. Column #9 uses the output of power systems of *Case A2* to estimate the feasible electric output by 2050 of coal and oil/gas (Nakicenovic *et al.*, 1998).

Natural gas hydrates were discovered in marine sediments in the 1970s and are considered to represent an immense but largely unmapped source of fuels (Haq, 1999). Global marine deposits of the frozen methane hydrates may exceed 10 000 gigatons in carbon content. Assuming 45 GJ of net thermal energy per ton of natural gas liquids, this corresponds to 14 000 TWt-y of energy or more than twice the estimated stores of coal, oils, and natural gas. However, the marine deposits are present in relatively thin and discontinuous layers at greater than 500 meters depth. There is little commercial interest at this time because cost-effective recovery may not be possible.

There is growing evidence that enormous quantities of methane can be released to the atmosphere as the hydrates in the deep ocean unfreeze due to undersea avalanches and increasing deep sea temperature (Blunier, 2000; Kennett *et al.*, 2000; Stevens, 1999; Dickens, 1999; Norris and Röhl, 1999). Such releases are associated with sudden shifts from glacial to interglacial climate. Large-scale hydrate mining and warming of the deep waters by OTEC systems (next section) could release large quantities of methane.

9.2.2 Renewable terrestrial systems

Renewable terrestrial power systems, except for Tidal (#7) and Geothermal (#10), are driven indirectly as the Sun heats the oceans and land. Conventional hydroelectric dams can provide only 1.6 TWe by 2050 because of a lack of suitable sites. See Trinnaman and Clarke (1998, p. 167) and Criswell (1994, 1998b). Tides (#7) and Waves (#8) are very small power sources (Trinnaman and Clarke, 1998).

Hydroelectric (#6) and not stand-alone (NSA)

Hydroelectric facilities are generally considered to be dependable sources of power for local or regional users. They are considered "stand-alone". However, even major facilities can decrease in output. In the case of the Grand Coulee Dam this is occasionally caused by lack of regional rainfall and insufficient stream flow through the Columbia River Basin of Washington State. Under these conditions even major hydroelectric dams can become not stand-alone (NSA). Their power output must be augmented by fossil fuel or nuclear power plants attached to the same power grid. As regional and global power needs increase, hydroelectric systems are less able to provide dependable power on demand. Backup systems, such as fossil fuel power plants, must be provided. At this time, the electric grids of conventional power systems can support ~20% of their capacity in the form of NSA power sources such as hydro and the more quickly varying wind and solar.

Alternatively, NSA systems could be distributed across large regions, even on different continents, to average out variations in power supplied to the system. Massive systems must be established to transmit power over long distances, possibly worldwide. Power storage must be provided close to major users. Unfortunately, it is impossible to predict the longest time required for adequate power storage. Such ancillary systems, especially when employed at a low duty cycle, greatly increase the cost of a unit of electric energy. Unit cost of power will undoubtedly be higher than for a more cost-effective stand-alone system. For these reasons, the "feasible" power capacity of renewable systems tends to be substantially less than the potentially available power.

Salinity-gradients (#6.1). Isaacs and Schmitt (1980) provided one of the first comprehensive reviews of potentially useful power sources. They noted that energy can be recovered, in principle, from the salinity-gradient between fresh or brackish water and seawater. They note that fresh river water flowing into the sea has an energy density equivalent to the flow of water through a 240 meter high dam or ~2.3 MW/(ton/second). They mention five conversion techniques and note also that any reversible desalination technique can be considered. They caution that none are likely economically feasible. Laboratory experiments in the 1970s demonstrated the generation of 7 We/m^2 across copper heat exchange surfaces at 60% conversion efficiency. Using the above engineering numbers, the capture of all accessible fresh water run-off from the continents, ~6.8 × 10^{13} tons/y (Postel *et al.*, 1996), is projected to yield ~0.3 TWe. The fresh and salt waters must pass through ~4 × 10^5 km^2 of copper heat exchangers. The polar ice caps and glaciers, 2.4 × 10^{16} tons, are the major stores of fresh water. Polar ice melt worked against seawater can release ~1700 TW-y of total energy, or, using the above numbers ~1000 TWe-y.

Fresh and ocean water mixing into a coastal brine pond can potentially be the power equivalent to a dam 3500 meters high. Given solar-powered brine ponds of sufficient total area, the above power and energy inventories could be increased by a factor of 14. Maintaining 20 TWe output requires the production of $\sim 3.3 \times 10^{13}$ tons/y of brine. The evaporation and transpiration of water from all land is $\sim 7 \times 10^{13}$ tons/year. This implies that 50% of all land, or 100% of lower latitude land, would be given over to brine production. It is worth reading Isaacs and Schmitt (1980) to expand one's mind to possible energy sources such as volcanic detonations, brine in salt domes, tabular-iceberg thermal sinks, tornadoes and thunderstorms, and other smaller sources of averaged power.

Ocean thermal energy conversion (#9)

"The oceans are the world's largest solar collector" (Twidell and Weir, 1986). The top 100 meters of tropical waters are 20–24 °C warmer than waters ~ 1 km to >7 km below the surface (~ 5 to 4 °C). Approximately 25% of the mass of tropical ocean waters has a difference of ~ 24 °C between the surface and deep waters. Approximately 1% has a temperature difference of ~ 28 °C. Thermal energy of the surface waters is renewed daily by sunlight. Cold water renewal is primarily through the sinking of waters in the high latitude oceans, primarily in the southern hemisphere, and the release of ~ 1600 TWt through evaporation of water to the atmosphere (Hoffert and Potter, 1997). Secondary cooling of waters in the North Atlantic releases ~ 500 TWt of power that heats the air that streams eastward and heats northern and Western Europe (Broecker, 1997). Thousands of years are required to produce $\sim 1 \times 10^{18}$ tons of deep cold ocean waters.

Ocean Thermal Energy Conversion (OTEC) systems mine the energy of the temperature difference between the cold waters of the deep tropical oceans, and the warm surface waters. The cold waters are pumped upward 1 to 6 kilometers and are used to condense the working fluids of engines driven by the hot waters above. Engineering models indicate $\sim 7\%$ efficiency is possible in the conversion of thermal to electric power (Avery and Wu, 1994). Prototype OTEC plants demonstrate an efficiency of 3%. Small demonstration plants have demonstrated net electric power outputs of 15 kWe and 31.5 kWe. This net output is slightly more than 30% of the gross electric output of each plant, respectively 31.5 kWe and 52 kWe. Twidell and Weir (1986) note that pumping of cold seawater from and to depth will likely absorb $\sim 50\%$ of the gross electric output of a large OTEC plant.

An upward flow of 2×10^{15} tons/y of deep water is required to produce 20 TWe of net electric output over a temperature difference of 20 °C. This implies that a maximum of $\sim 2 \times 10^5$ TWt-y, or $\sim 2 \times 10^4$ TWe-y, of energy can be extracted over a "short" time from the ocean. However, when 20 TWe of commercial power is considered, several factors combine to significantly decrease the extractable energy.

OTEC systems are projected to have high capital costs. A current challenge is to reduce the cost of just the heat exchangers to less than 1500 $/kWe capacity. Offshore installations will be far more expensive than onshore installations. Offshore installations require platforms, means of transmitting energy or power to shore, and more expensive support operations. Producing intermediate products such as hydrogen decreases overall efficiency and increases costs. Trinnaman and Clarke (1998: p. 332–334) suggest an OTEC potential ≤0.02 TWe by 2010. To grow significantly by 2050 the costs of OTEC plants must be minimized. This requires onshore construction. Thus, only a fraction, possibly ≤1%, of the coldest deep waters and warmest surface waters can be economically accessed.

The warmest tropical waters extend ~100 meters in depth from the surface. A 20 TWe OTEC system processes this mass of water in ~1 year. La Niña-like events might be enhanced or created by the outflow of cold, deep waters from a fleet of OTEC installations located in the equatorial waters of the eastern Pacific. Levitus *et al.* (2000) have discovered that the heat content of the oceans has increased by $\sim 2 \times 10^{23}$ J, or 6300 TWt-y, from 1948 to 1998. This corresponds to a warming rate of 0.3 W/m^2 as averaged over the surface of the Earth and likely accounts for most of the "missing" energy expected to be associated with greenhouse heating since the 1940s (Kerr, 2000). Note that the change in ocean heat content since the 1940s is of the same magnitude as associated with extracting 20 TWe of electric energy from the oceans over a 50 year period. See Watts (1985) for a discussion of the potential effects of small natural variations in deep-water formation on global climate.

The major ocean currents convey enormous quantities of water and thermal power from low to high latitudes. Production of cold and higher density and salinity water in the North Atlantic plays a key enabling role in the present general circulation. However, the relative roles of salinity differences, wind, and tides as the driving forces of the circulation have not been clear. New data indicates that winds and lunar tides transfer ~2 TWm of mechanical power to the oceans to drive the large scale ocean currents and the associated transfer of ocean thermal energy between the cold waters of the high latitudes and the warm waters of the low latitudes. The estimate of lunar tidal power is based on recent analyses of seven years of data on the height of the ocean, obtained by means of an altimeter onboard the TOPEX/POSEIDON satellite. According to Egbert and Ray (2000) and Wunsch (2000), the winds across the ocean provide ~60% of the power that drives ocean circulation. The dissipation of lunar tidal forces in selected portions of the deep ocean provides ~40% of the driving power. Wunsch maintains that the ocean would fill to near the top with cold water and the conveyor would shut down without the ~1 TWm of lunar tidal power to drive the circulation of the ocean (Kerr, 2000b).

An OTEC system with a net output of 20 TWe, using the demonstration data and calculations mentioned above, will require a gross electrical output of approximately 60 TWe. Twenty terawatts of the additional 40 TWe is directed into the operation of the plant and approximately 20 TWe into the pumping of cold waters from great depth and returning the heated water to depth. The 20 TWe of pumping power is ~20 times the tidal power the Moon places into the general circulation of the deep waters. Given the complexity of real, versus averaged, ocean currents it seems inevitable that the OTEC pumping power will modify the circulation of the ocean. Highly accurate and trustworthy models of ocean circulation must be demonstrated and the effects of large scale OTEC systems included before major commitments are made to large OTEC systems. What is large for OTEC? A first estimate can be made by assuming that the pumping power of an OTEC system is restricted to ≤10% of the lunar tidal power. This implies a gross OTEC electric output of ≤0.1 TWe or 100 GWe. Using the above engineering estimates for OTEC implies a maximum net electric output of 30 GWe or less than the electric power capacity of California. This is far smaller than the 10 TWe output suggested by some OTEC advocates (see www.seasolarpower.com/).

The massive up-flow of cold, deep waters for a 20 TWe OTEC system will change the nutrients, gas content, and biota of the surface and deep waters. There is ~50 times more CO_2 in the ocean than the atmosphere (Herzog *et al.*, 2000). The ocean/atmospheric exchange of CO varies over a six year interval (Battle *et al.*, 2000). The effects of changing ocean circulation must be understood. It is not unreasonable to anticipate restricting the flow of cold, deep waters to the surface tropical waters to perhaps 20% of full potential flow. At 20 TWe, most of the mined waters will likely be pumped back to the depths. The pumping energy will reduce the OTEC's efficiency and warm the local deep waters. Over time the depth-to-surface temperature difference decreases. This reduces to ~1/5th the useful local inventory. The factors of 1%, 4/20ths, and 20% multiply to 0.04%. Thus, the ultimate inventory of energy may be reduced to the order of 0.04% of ultimately extractable energy at high pumping rates. This implies a useful inventory of ~90 TWt-y (or ≤6 TWe-y) that might be extracted from favorable locations. These crude estimates must be revised using detailed models of the ocean circulation through and about the most favorable sites. The United States National Renewable Energy Laboratories provides an extensive web site on OTEC and references (www.nrel.gov/otec/).

Geothermal (#10)

The thermal energy of the Earth is enormous but non-renewable. It originates from the in-fall energy of the materials that form the Earth and the ongoing decay of radioactive elements. This geothermal power flows from Earth at the rate of 0.06 W/ m^2 (Twidell and Weir, 1986: p. 378). Thus, Earth releases only 30 TWt or less than half that required for a 60 TWt global power system. Approximately 9×10^6 TWt-y of high temperature rock exists 1 to 7 km beneath the surface of the Earth. Only a tiny fraction of the energy is currently accessible at high temperatures at continental sites close to volcanic areas, hot springs, and geysers. However, in principle, these rocks can be drilled, water circulated between the rocks and turbines on the surface, and energy extracted. However, the costs are high (Nakicenovic *et al.*, 1998: 56). There is considerable uncertainty in maintaining re-circulating flow of water between hot deep rocks and the surface (Trinnaman and Clarke, 1998: 279). The useful global potential is likely less than 0.5 TWe.

Wind turbines (#11)

Winds near the surface of Earth transport ~300 TWm of mechanical power (Isaacs and Schmitt, 1980). The order of 100 TWm is potentially accessible over the continents, especially in coastal regions. Trinnaman and Clarke (1998: pp. 299–300) report the continental wind power resource to be ~100 times that of the hydro power resource, or approximately 160 TWm. Wind turbines (#11) are efficient and offer access to a major source of renewable power. Approximately 3 to 10 MWe (average) can be generated by a wind farm that occupies 1 square kilometer of favorable terrain. If wind farms occupy 2.4% to 7% of the continents then averaged output can be ~ 20 TWe. However, at the level of a commercial power system wind farms demonstrate a major limitation of all terrestrial renewable power sources.

Wind farms are NOT STAND-ALONE (NSA) power systems. For example, wind farms are now connected to power grids that take over power production when the wind is not adequate. Wind farms in California have supplied as much as 8% of system demand during off-peak hours. Research indicates that 50% penetration is feasible (Wan and Parsons, 1993). For these reasons the "feasible" power level is taken to be less than 6 TWe. This is consistent with *Case A2* of Nakicenovic *et al.* (1998: p. 69) in which wind farms supply all the renewable commercial power, or 23% of global power. Refer to Strickland (1996) for a less hopeful discussion of continental-scale use of wind power and other renewable systems that provide intermittent power.

It is necessary to examine the effect of large scale wind farms on the coupling of the Earth and the atmosphere. Such studies have not been done. A 20 TWe system of wind farms would extract approximately one-tenth of the global wind power. Climate changes comparable to mountain ranges might be induced by wind farming operating at a global level. It is known that winds and the rotation of the Earth couple through the friction of the winds moving over the land and oceans to produce a seismic hum within the free-oscillation of the Earth (Nishida *et al.*, 2000). What happens when larger coupling between the specific areas on Earth and the global winds is established?

9.2.3 Terrestrial solar power systems

The continents and the atmosphere above them intercept $\sim 50\,000$ TWs of solar power with a free-space intensity of 1.35 GWs/km^2 ("s" denotes solar power in space). However, due to the intermittent nature of solar power at the surface of the Earth, it is very difficult for a dedicated terrestrial solar power system, complete with power storage and regional power distribution, to output more than 1 to 3 MWe/km^2 when averaged over a year. Even very advanced technology will be unlikely to provide more than 20 MWe/km^2 (Criswell and Thompson, 1996; Hoffert and Potter, 1997). Terrestrial solar power systems (TSPS) are NOT STAND-ALONE sources of commercial solar power. For dependable system power, the solar installations, either thermal or photovoltaic, must be integrated into other dependable systems such as fossil fuel systems. Terrestrial solar photovoltaic systems have been growing in capacity at 15%/y. A doubling of integral world capacity is associated with a factor of 1.25 decrease in the cost of output energy. At this rate of growth and rate of cost decrease, TSPS energy may not be competitive commercially for another 50 years (Trinnaman and Clarke, 1998: p. 265).

Strickland (1996; in Glaser *et al.*, 1998: Ch. 2.5) examined both a regional TSPS and much larger systems distributed across the United States. Costs and the scale of engineering are very large compared to hydroelectric installations of similar capacity. Intercontinental solar power systems have been proposed. Klimke (1997) modeled a global system of photovoltaic arrays and intercontinental power grids scaled to provide Europe with ~ 0.5 TWe of averaged power. Capital costs for a larger 20 TWe global photovoltaic system that delivers 1000 TWe-y might exceed 10 000 trillion dollars, $= 1 \times 10^{16}$ dollars, and provide electric energy at a cost of $\sim 60\cent$/kWe-h (Criswell, 1999). The system could be shut down by bad weather over key arrays.

Even an intercontinental distribution of arrays does not eliminate the problems of clouds, smoke from large regional fires, or dust and gases from major volcanoes or small asteroids (<100 meters in diameter). Changes in regional and global climate could significantly degrade the output of regional arrays installed at enormous expense. It is impossible to predict the longest period of bad weather. Thus, it is impossible to engineer ancillary systems for the distribution of power and the storage of energy during the worst-case interruptions of solar power at the surface of Earth. In addition, large arrays are likely to be expensive to maintain and may induce changes in their local microclimates. In Table 9.4, the total power from both options #12 and #13 is taken to be ≤3.3 TWe or the equivalent of 10 TWt. This is 30% of the total global power, 33.3 TWt in 2050, for *Case A3* of Nakicenovic *et al.* (1998: p. 98).

9.2.4 Nuclear power systems

Nuclear fission (#14, 15, 16)

At the beginning of the 21st century, nuclear reactors output ~0.3 TWe and provide 17% of the world's electric power. By 1996 the world had accumulated ~8400 reactor-years of operating experience from 439 reactors. By 2010 nuclear operating capacity may be ~0.4 TWe. Adequate economically recoverable uranium and thorium exist on the continents to yield 270–430 TWt-y of energy, depending of the efficiency of fuel consumption. This corresponds to 4 to 7 years of production at 20 TWe. Nakicenovic *et al.* (1998: p. 52, p. 69 Case A1) estimate that nuclear systems may provide as much as 1.6 TWe by 2050. Krakowski and Wilson (2002) estimate that conventional nuclear plants may provide as much as 5 TWe by 2100. A major increase in commercial nuclear power requires the introduction of breeder reactors.

Breeder reactors potentially increase the energy output of burning a unit of uranium fuel by a factor of ~60 (Trinnaman and Clarke, 1998: Chaps. 5 & 6, back cover). Continental fuels could supply 20 TWe for ~500 years. The oceans contain 3.3 parts per billion by weight of uranium, primarily ^{238}U, for a total of 1.4×10^9 tons (see www.shef.ac.uk/chemistry/web-elements/fr-geol/U.html). Burned in breeder reactors this uranium can supply 20 TWe for ~ 300 000 years. There are wide-spread concerns and opposition to the development and use of breeder reactors. Concerns focus on proliferation of weapons-grade materials, "drastically improving operating and safety" features of reactors, and the disposal of spent fuels and components.

Wood *et al.* (1998) propose sealed reactors that utilize a "propagation and breeding" burning of as-mined actinide fuels and the depleted uranium already accumulated worldwide in the storage yards of uranium isotopic enrichments plants. These known fuels can provide ~1000 TWe-y and enable the transition to lower-grade resources. To provide ~1 kWe/person, this scheme requires ~10 000 operating reactors of ~2 GWt capacity each. Each sealed reactor would be buried hundreds of meters below the surface of the Earth and connected via a high pressure and high temperature helium gas loop to gas turbines and cooling systems at the surface. At the end of a reactor's operating life, ~30 years, the fuel/ash core would be extracted, reprocessed, and sealed into another new reactor. The used reactors, without cores, would remain buried. A 20 TWe world requires the construction and emplacement of ~2 reactors a day. Spent reactors accumulate at the rate of 20000 per century. A major increase in research, development, and demonstration activities is required to enable this option by 2050. Krakowski and Wilson (2002) do not envision breeder reactors as providing significant commercial power until the 22nd century. Perhaps electrodynamically accelerated nuclei can enable commercial fission with sub-critical masses of uranium, thorium, deuterium, and tritium. This could reduce proliferation problems and reduce the inventory of radioactive fuels within reactors (#17).

The nuclear power industry achieved ~2.4 TWe-y of output per major accident through the Chernobyl event. Shlyakhter *et al.* (1995) note that the current goal in the United States is to provide nuclear plants in which the probability of core melt-down is less than one per 10 TWe-y of power output. This corresponds to 10 TWe-y per core melt-down. Suppose a 20 TWe world is supplied exclusively by nuclear fission and the objective is to have no more than one major accident per Century. This implies 2000 TWe-y per major accident or a factor of ~200 increase in industry-wide safety over the minimum current safety standard for only core melt-downs. A 20 TWe nuclear industry would provide many other opportunities for serious health and economic accidents. Less severe accidents have destroyed the economic utility of more commercial reactors than have reactor failures. Many utilities are unwilling to order new nuclear plants due to financial risks. Also, political concerns have slowed down the use of nuclear power through the regulatory processes in several nations (Nakicenovic *et al.*, 1998: p. 84–87).

Nuclear fusion (#18, 19)

Practical power from controlled fusion installations for the industrial-scale burning of deuterium and tritium is still a distant goal. Europe, Japan, Russia, and the United States have decreased their funding for fusion research (Browne, 1999). It is highly unlikely that fusion systems will supply significant commercial power by 2050. Large-scale power output, ≥20 TWe, is further away. At this time the economics of commercial fusion power is unknown and in all probability cannot be modeled in a reasonable manner.

The fuel combination of deuterium and helium-3 (^{3}He) produces significantly fewer neutrons that damage the inner walls of a reactor chamber and make reactor components radioactive. However, this fusion process requires ten times higher energy to ignite than deuterium and tritium. Unfortunately helium-3 is not available on the Earth in significant quantities. Helium-3 is present at ~ 10 parts per billion by mass in lunar surface samples obtained during the Apollo missions. Kulcinski (NASA, 1988, 1989) first proposed mining ^{3}He and returning it to Earth for use in advanced fusion reactors. It is reasonable to anticipate that ^{3}He is present on most, if not all, of the surface lunar soils. The distribution of ^{3}He with depth is not known. The ultimately recoverable tonnage is not known. It is estimated that lunar ^{3}He might potentially provide between 100 and 1×10^5 TWt-y of fusion energy (Criswell and Waldron, 1990). Far larger resources of ^{3}He exist in the atmospheres of the outer planets and some of their moons (Lewis, 1991). Given the lack of deuterium–^{3}He reactors, and experience with massive mining operations on the Moon, it is unlikely that lunar ^{3}He fusion will be operating at a commercial level by 2050.

Three essential factors limit the large-scale development of nuclear power, fission and fusion, within the biosphere of Earth.

- The first factor is physical. To produce useful net energy the nuclear fuels must be concentrated within engineered structures (power plant and associated structures) by the order of 10^6 to 10^8 times their background in the natural environment of the continents, oceans, and ocean floor. A fundamental rule for safe systems is to minimize the stored energy (thermal, mechanical, electrical, etc.) that might drive an accident or be released in the event of an accident. Nuclear fission plants store the equivalent of several years of energy output within the reactor zone. In addition, the reactors become highly radioactive. Loss of control of the enormous stored energy can disrupt the reactor and distribute the radioactive materials to local regions of the biosphere, and even distant regions, at concentrations well above normal background. A 20 TWe fission world will possess ~ 60 to 600 TWt-y of fissionable materials in reactors and reprocessing units. Fusion may present relatively fewer problems than fission. However, even ^{3}He fusion will induce significant radioactivity in the reactor vessels.

- The concentration of fuels from the environment, maintenance of concentrated nuclear fuels and components, and long-term return of the concentrated radioactive materials to an acceptable background level present extremely difficult combinations of physical, technical, operational, economic, and human challenges. Nuclear materials and radioactive components of commercial operations must be isolated from the biosphere at levels now associated with separation procedures of an analytical chemistry laboratory (parts per billion or better). These levels of isolation must be maintained over 500 to 300000 years by an essential industry that operates globally on an enormous scale. Such isolation requires enormous and focused human skill, intelligence, and unstinting dedication. Research scientists, such as analytic chemists, temporarily focused on particular cutting-edge, research projects sometimes display this level of intelligence and dedication.

The energy output must be affordable. Thus, total costs must be contained. How can such human talent be kept focused on industrial commodity operations? Can automated control of the nuclear industry be extended from the microscopic details of mining operations to the level of international needs? Ironically, if this level of isolation can be achieved the nuclear fission industries will gradually reduce the level of natural background radiation from continental deposits of uranium and thorium.

- The last factor is more far ranging. Given the existence of the Sun and its contained fusion reactions, is it necessary to develop commercial nuclear power for operation within the biosphere of Earth? Are the nuclear materials of Earth and the solar system of much higher value in support of the future migration of human beings beyond the solar system? Once large human populations operate beyond the range of commercial solar power, it becomes imperative to have at least two sources of independent power (fission and fusion). Such mobile societies will likely require massive levels of power. Terrestrial and solar system nuclear fuels are best reserved for these longer-range uses.

9.2.5 Space lunar solar power systems

Introduction to solar electric power from space

It is extremely difficult to gather diffuse, irregular solar power on Earth and make it available as a dependable source of commercially competitive stand-alone power. The challenges increase as irregular terrestrial solar power becomes a larger fraction of total regional or global commercial electric power. Research indicates that terrestrial solar may provide 5% to 17% of renewable power to conventional small power grids. Fifty percent supply of power by ter-restrial solar, and wind, is conceivable. However, an increasing fraction of renewable power is limited by the higher cost of renewable sources, high costs of storage and transmission of renewable power, institutional resistance, and regulator effects (Wan and Parsons, 1993).

Conversely, above the atmosphere of Earth and beyond Earth's cone of shadow the sunlight is constant. In space, very thin structures that would be destroyed by water vapor, oxygen, winds, and other hostile elements of Earth's biosphere, can be deployed, collect the dependable sunlight (1.35 GWs/km^2 near Earth), and convert it to electric power. The electric power is then con-verted into microwave beams and directed to receivers on Earth at the relatively low intensity of ~ 0.2 GWe/km^2. Microwaves of ~ 12 cm wavelength, or ~ 2.45 GHz, are proposed because they travel with negligible attenuation through the

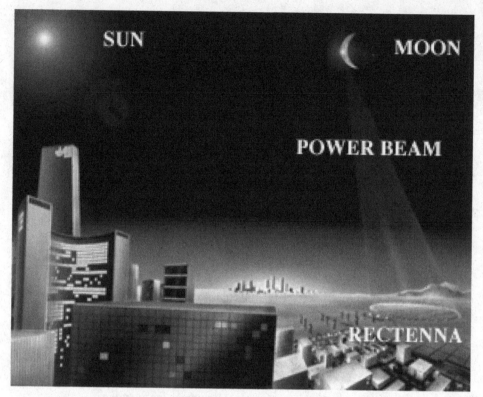

Figure 9.2   The lunar solar power system.

atmosphere and its water vapor, rain, dust, and smoke. Also, microwaves in this general frequency range can be received and rectified by the planar antennas, called rectennas (bottom right of Figure 9.2), into alternating electric currents at efficiencies in excess of 85%.

The beams will be 2 to 20 times more intense than recommended for continuous exposure by the general population. The beams will be directed to rectennas that are industrially zoned to exclude the general population. Microwave intensity under the rectenna will be reduced to far less than is permitted by continuous exposure of the general population through adsorption of the beam power by the rectenna and by secondary electrical shielding. The beams will be tightly focused. A few hundred meters beyond the beam, the intensity will be far below that permitted for continuous exposure of the general population. Humans flying through the beams in aircraft will be shielded by the metal skin of the aircraft, or by electrically conducting paint on composite aircraft. Of course aircraft can simply fly around the beams and the beams can be turned off or decreased in intensity to accommodate unusual conditions. The low-intensity beams do not pose a hazard to insects or birds flying directly through the beam. Active insects and birds will, in warm weather, tend to avoid the beams due to a slighter higher induced body temperature. See Glaser *et al.* (1998: Ch. 4.5).

The Earth can be supplied with 20 TWe by several thousand rectennas whose individual areas total to $\sim 10 \times 10^4$ km^2. Individual rectennas can, if the community desires, be located relatively close to major power users and thus minimize the need for long-distance power transmission lines. Individual rectennas can be as small as ~ 0.5 km in diameter and output ~ 40 MWe or as large in area as necessary for the desired electric power output. Note that existing thermal and electric power systems utilize far larger total areas and, in many cases, such as strip-mining or power line right-of-ways, degrade the land or preclude multiple uses of the land.

An "average" person can be provided with 2 kWe, for life, from ~ 10 m^2 of rectenna area or a section ~ 3 m, or 10 feet, on a side. This "per capita" section of the rectenna would have a mass of a few kilograms and be made primarily of aluminum, semi-conductors, glass, and plastics. This is a tremendous reduction in resources to supply each person with adequate commercial energy. In contrast, coal fired power systems will use $\sim 517\,000$ kilograms of coal to provide 2kWe to an energy-rich "average" person for a lifetime of 80 years. This is 160 kWe-y or $\sim 4\,210\,000$ kWt-h and is now done in developed nations.

Rectenna areas can be designed to reflect low-quality sunlight back into space and thereby balance out the net new energy the beams deliver to the biosphere. Space/lunar solar power systems introduce net new commercial energy into the biosphere that allows humankind to stop using the energy of the biosphere. Space/lunar solar power enables the production of net new wealth, both goods and services, without depleting terrestrial resources (Criswell, 1994, 1993).

Beams will be directed to commercially and industrially zoned areas that the public avoids. Power outside the tightly collimated beams will be orders of magnitude less than is permissible for continuous exposure of the general population. Considerable "knee-jerk" humor is directed at the concept of beaming microwave power to Earth. However, the essential microwave technologies, practices, environmental considerations, and economic benefits are understood. Microwaves are key to radio and television broadcasting, radar (air traffic control, weather, defense, imaging from Earth orbit), industrial microwave processing, home microwave ovens, cellular and cordless phones and other wireless technologies. Planetary radar is used to observe the Moon, asteroids, Venus, and other planets. It should be noted that medical diathermy and magnetic resonance imaging operate in the microwave. Medical practices and lightning associated with thunderstorms produce microwave intensities in excess of those proposed for beaming of commercial power.

   The core space/lunar solar technologies emerged from World War II research and development. These technologies are the drivers of economic growth in the developed nations. The space/lunar components are technologically similar to existing solar cells, commercial microwave sources (e.g., in cellular phones, microwave oven magnetrons, and klystrons), and solid-state phased-array radar systems. These are commercial and defense technologies that receive considerable research and development funding by commercial and government sources worldwide. The essential operating technologies for space/lunar solar power receive more R&D funding than is directed to commercial power systems and advanced systems such as fusion or nuclear breeder reactors.

   Space solar power systems output electric power on Earth without using terrestrial fuels. Few, if any, physical contaminants such as CO_2, NOx, methane, ash, dust, or radioactive materials are introduced into the biosphere. Space/lunar solar power enables the terrestrial economy to become fully electric while minimizing or eliminating most cost elements of conventional power systems (Nakicenovic *et al.*, 1998: p. 248, p. 103). Eventually, the cost of commercial space/lunar solar power should be very low. The commercial power industry and various governments are starting to acknowledge the potential role of commercial power from space and from installations on the Moon (Trinnaman and Clarke, 1998; Deschamps, 1991; Glaser *et al.*, 1998; ESA, 1995; Stafford, 1991; Moore, 2000; World Energy Council, 2000).

Geosynchronous space solar power satellites (SSPS) deployed from Earth (#20)

Following the "petroleum supply distribution" crises of the early 1970s, the United States government directed the new Department of Energy and NASA to conduct environmental impact studies and preliminary systems analyses of SSPS to supply electric power to Earth. The studies focused on construction of a fleet of 30 extremely large satellites deployed one a year over 30 years. Each satellite, once positioned in geosynchronous orbit, would provide 0.01 TWe of baseload power to a rectenna in the United States for 30 years (Glaser *et al.*, 1998).

   An SSPS reference design (Ref-SSPS) was developed for the 0.01 TWe satellite and used to conduct full life-cycle analyses of engineering, operations, and financing. Smaller 0.005 TWe electric SSPS were also studied. The 0.01 TWe Ref-SSPS was approximately 10 km by 20 km on a side, had a mass of ~100 000 tons, and required ~1000 tons/y of supplies, replacement components, and logistics support. In addition, a facility in geosynchronous orbit, with a mass of ~50 000 tons, was used to complete final assembly and testing. The one assembly facility and the 30 Ref-SSPS were to be partially constructed in orbit about Earth from components manufactured on Earth and shipped to space by very large reusable rockets. The assembly facility components and first Ref-SSPS self-deploy from low Earth orbit to geosynchronous orbit using solar power and ion propulsion. Ion propulsion requires a propellant mass in low Earth orbit of approximately 20% of the initial mass in low Earth orbit. These numbers allow an approximate estimate of the total mass, ~160 000 tons, required in low orbit about Earth to deploy one Ref-SSPS and maintain it for 30 years.

Weingartner and Blumenberg (1995) examined the energy inputs required for the construction and emplacement of a 0.005 TWe SSPS. They considered first the use of 50 micron (50×10^{-6} m) thick crystalline solar cells. The following comments assume they included the GEO construction facility, make-up mass, and reaction mass for the ion propulsion in their calculations. Specific energy of production of the satellite at geosynchronous orbit and its operation over 30 years was found to be 3044 kWh/kg. Details are given in the annotations to the reference.

One 0.01 TWe Ref-SSPS that outputs 0.3 TWe-y of base-load electric energy on Earth over 30 years delivers ~24 000 kWe-h per kilogram of SSPS in geosynchronous orbit. This is approximately the same as for a 0.005 TWe SSPS. The SSPS delivers a net energy of ~21 000 kWe-h/kg. In principle, the SSPS components can be refurbished on-orbit for many 30-year lifetimes using solar power. In this way the "effective energy yield" on Earth of a given SSPS can approach the ratio of energy delivered to Earth divided by the energy to supply station-keeping propellants, parts that cannot be repaired on orbit, and support of human and/or robotic assembly operations. Assuming a re-supply of 1% per year of Ref-SSPS mass from Earth, the asymptotic net energy payback for Earth is ~ 60 to 1 after several 30-year periods. Eventually, the refurbished SSPS might supply ~ 88 000 kWe-h of energy back to Earth per kilogram of materials launched from Earth. This high payback assumes that solar power in space is used to rebuild the solar arrays and other components.

For comparison, note that burning 1 kg of oil releases 12 kWt-h or ~4 kWe-h of electric energy. The first Ref-SSPS equipment has a potential "effective energy yield" ~5000 times that of an equal mass of oil burned in air. If the Ref-SSPS can be refurbished on-orbit with only 1000 tons/y of make-up mass (components, propellants, re-assembly support) then the Refurbished-SSPS yields ~ 22 000 times more energy per kilogram deployed from Earth than does a kilogram of oil. By comparison, the richest oil fields in the Middle East release ~20 000 tons of crude oil through the expenditure of 1 ton of oil for drilling and pumping the oil. However, one ton of oil is required to transport 10 to 50 tons of oil over long distances by ship, pipeline, or train (Smil,

1994: 13). Unlike the energy from burning oil, the SSPSs add high-quality industrially useful electric energy to the biosphere without the depletion of resources or the introduction of material waste products into the biosphere. These estimates must be tempered by the observation that practical terrestrial solar cells are the order of 500 microns in thickness and take considerably more energy to produce than is estimated above.

NASA-DoE developed life-cycle costs for a small fleet of 30 Ref-SSPSs of 0.3 TWe total fleet capacity. The calculations were done in 1977 US dollars. In the following cost estimates, 1990$/1977$ = 1.7 is assumed and all costs are adjusted to 1990$. NASA-DoE estimated that the power provided by the Ref-SSPS would cost approximately 0.10 $/kWe-h. This corresponds to ~1300 T$ to supply 1500 TWe-y. The National Research Council of the National Academy of Sciences reviewed the NASA-DoE program in the late 1970s and did concede that the basic technologies were available for the Ref-SSPS in the 1980s for both construction and operations (Criswell and Waldron, 1993 and references therein). However, the NRC projected energy and overall costs to be approximately a factor of 10 higher. In particular, solar arrays were estimated to be 50 times more expensive. The NASA-DoE estimated launch cost of 800 $/kg was increased three times to approximately 2400 $/kg. The NRC estimates of cost were consistent with ~13000 T$ to supply 1500 TWe-y. Significantly, the NRC accepted the estimated costs for constructing and maintaining the rectennas (~26 M$/GWe-y).

Row #20 of Table 9.6 summarizes the characteristics of a fleet of Geo-SSPS, located in geosynchronous orbit, to supply commercial electric power by the year 2050. A Geo-SSPS supplies baseload power. This power is supplied via one or multiple beams to one or a set of fixed rectennas that can be viewed by the SSPS from geosynchronous orbit. A geosynchronous SSPS will be eclipsed a total of 72 minutes a day for 44 day periods twice a year during the equinoxes. The eclipse occurs near local midnight for the rectennas. Adjacent un-eclipsed SSPSs might provide power to the rectennas normally serviced by the eclipsed SSPS.

Most rectennas will need to output a changing level of power over the course of the day and the year. "Stand-alone" SSPSs must be scaled to provide the highest power needed by a region. They will be more costly than is absolutely necessary. The alternatives are to:

- Employ a separate fleet of relay satellites that redistribute power around the globe and thus minimize the total installed capacity of SSPS in geosynchronous orbit;
- Construct and employ an extremely extensive and expensive set of power lines about the Earth, a global grid, to redistribute the space solar power;
- Provide expensive power storage and generation capacity at each rectenna;

- Provide expensive conventional power supplies that operate intermittently, on a rapid demand basis, as excess power is needed; or
- Provide a mixture of these systems and the SSPS fleet optimized for minimum cost and maximum reliability.

These trades have not been studied. A fleet of geosynchronous SSPS does not constitute a stand-alone power system. A 20 TWe SSPS system will either be over-designed in capacity to meet peak power needs or require a second set of power relay satellites. Alternatively, the order of 10 to 100 TWe-h of additional capacity will be supplied either through power storage, on-Earth power distribution, or other means of producing peak power.

As noted earlier, the rectennas will output the order of 200 We/m^2 of averaged power. This is 10 to 200 times more than the time-averaged output of a stand-alone array of terrestrial photovoltaics and associated power storage and distribution systems.

It is highly unlikely that Geo-SSPS can supply 20 TWe by the year 2050 or thereafter. Major issues include, but are not limited to, total system area and mass in orbit, debris production, low-cost transport to space, environmentally acceptable transport to space, and the installation rate. Extrapolating a fleet of Ref-SSPS to 20 TWe implies 220 000 km^2 of solar collectors and support structure, 3100 km^2 of transmitting aperture, and an on-orbit mass of 200 000 000 metric tons. If the 2000 to 3000 Ref-SSPS were co-located at geosynchronous altitude, they would collectively appear 1.7 to 2 times the diameter of the Moon. The individual satellites would be distributed along the geosynchronous arc with concentrations above Eurasia, North America, and South America. Few would be required over the Pacific Ocean. They would be highly visible, far brighter than any star under selected conditions, and sketch out the equatorial plane of Earth across the night sky.

Each of the 2000 to 3000 satellites would have to be actively managed, through rockets and light pressure, to avoid collisions with the others. If evenly distributed along geosynchronous orbit, they would be 80 to 130 km apart or separated by 4 to 7 times their own length. Allowing for clumping over Eurasia and North and South America, they would almost be touching (Criswell, 1997).

Micrometeorites will impact SSPSs and generate debris. Much of this debris will enter independent orbits about Earth and eventually impact the SSPSs. It is estimated that over a 30-year period a small fleet of 30 SSPS with 0.3 TWe capacity will convert 1% of the fleet mass into debris (Glaser *et al.*, 1998: p. 8). A 20 TWe fleet would eject ~6 × 10^5 tons/y of debris. By contrast, in 1995 the 478 satellite payloads in geosynchronous orbit had an estimated collective surface area of ~ 0.06 km^2 (Loftus, 1997). There were also 110 rocket bodies.

The estimated collision rate is $\sim 10^{-2}$ impacts/km^2-y (Yasaka *et al.*, 1996; Table 2). For the 20 TWe SPS fleet, a minimum initial rate of 2000 collisions/y is implied against existing manmade objects. Nature poses inescapable hazards. Meteor storms exist with fluxes 10^4 times nominal background. A large SSPS fleet in geosynchronous orbit may, under meteorite bombardment, release sufficient debris that the accumulating debris re-impacts the arrays and destroys the fleet. Special orbits about Earth that are located within the "stable plane" may minimize self-collisions of SPS debris (Kessler and Loftus, 1995). However, satellites in these orbits do not remain fixed in the sky as seen from Earth. Far more artificial debris is present in low Earth orbit. A major fleet of LEO-SPS could generate sufficient debris to make travel from Earth to deep space extremely hazardous, perhaps impossible.

Ref-SSPS in geosynchronous orbit, or lower, will be the dominant source of radio noise at the primary frequency of the microwave power beam and its harmonics (higher frequencies) and sub-harmonics (lower frequencies). The preferred 12 cm microwave wavelength, $\sim$2.45 GHz, for power beaming is inside the "industrial microwave band" that is set aside by most nations for industrial usage. Combinations of new active filtering techniques and reallocation of existing communications bands will be required for delivery of beamed power to rectennas on Earth. Neither national nor international agreements for the allocation of the industrial microwave band for power transmission are now in place. Personal communications and wireless data transmission systems are now being used without license in this frequency range.

Fleets of massive Earth-to-orbit rockets were proposed to deploy Ref-SSPS components and construction equipment to low orbit about Earth. Very large single-stage and two-stage-to-orbit launch systems were designed that could place $\sim$300 tons of payload into orbit. The objective was to reduce launch costs to low Earth orbit to $\sim$250 $ per pound ($\sim$500 $/kg). Analyses were restricted to hydrogen-oxygen launch vehicles. Launch noise would be a serious problem unless operations were moved from populated areas, such as the east coast of Florida, to remote areas. Also, the water vapor deposited in the upper atmosphere might deplete ozone and affect other aspects of atmospheric chemistry in the stratosphere and above.

Approximately one launch a day was required to deploy 0.01 TWe of electric capacity each year. This implies $\leq$0.4 TWe could be deployed between 2010 and 2050 for the scale of the industry and investments assumed for the Ref.-SSPS.

Freshlook study

In 1996 the United States Congress directed NASA to reexamine space solar power. Approximately 27 million dollars was expended through the year 2000.

A publicly available summary of the first part of the Freshlook Study is provided by Feingold *et al.* (1997) and NASA (1999). All resources continue to focus on versions of power satellites deployed from Earth to orbits about Earth. Contractor and community studies explored a wide range of low- and medium-altitude demonstration satellites and finally converged again on two designs for geosynchronous satellites – the solar "power tower" aligned along a radius to the Earth and the spinning "solar disk" that directly faces the Sun. The systems were projected to provide power at ~0.1 to 0.25 $/kWe-h. Costs are similar to those for the 1970s NASA-DoE Ref-SSPS. However, recent costing models are far more aggressive and project wholesale electricity cost as low as 5¢/kWe-h supplied to the top of the rectenna. Low projected beam costs are achieved through:

- Attainment of launch costs of ~120 – 150 $/kg, a factor of 3 to 5 lower than the 1970s Ref-SSPS studies and a factor of 100 lower than current practices;
- Avoiding the need for large assembly facilities in low or geosynchronous orbit through the use of SSPS components designed to "self-assemble" in low and geosynchronous orbits;
- Extensively utilizing "thin-film" components and minimal structural supports; and
- Assuming 40 years operational lifetime for satellites versus 30 years.

Costs for the complete system are not included. Estimates of major systems costs were reduced through:

- Minimizing up-front research and development through use of highly standardized components;
- Minimizing time between first deployment of a satellite and start of first power delivery;
- Providing power initially to countries that now use high cost power; and
- Other investors paying at least 50% of the costs of all ground facilities (launch facilities, rectennas, component manufacturing and testing, ground assembly and transportation, etc.).

The above conditions raise serious concerns. NASA, the US Air Force, and several major launch services companies have the goal of reducing launch costs to the order of 1000 $/lb. or approximately 2200 $/kg early in the 21st century. The "power tower" was projected to be ~20% more massive per TWe-output than the Ref-SSPS. A very simple model of SSPS mass and power output and launch costs can be adjusted for these two factors. For total electric cost to be 0.1 $/kWe-h, including the cost of rectennas, the mass of the "power tower" or "solar disk" and its make-up mass over 30 years has to decrease from ~160 000 tons to ~12 000 tons. The original SSPS and Freshlook designs pushed photoconversion, electrical, and structural limits. Another factor of ten reduction in mass per unit of power is extremely challenging and is likely to be physically impossible. Preliminary reports from the final "Freshlook" studies indicate that space solar power satellites deployed from Earth will not be competitive with conventional power systems (Macauley *et al.*, 2000)

Conversely, consider the challenge of deploying a space-based power system into orbit from Earth that delivers busbar electricity, 90 percent duty-cycle, at 1 ¢/kWe-h. Including all the mass elements associated with the Ref-SSP (satellite, make-up mass and components, assembly facility and supplies, ion-engine reaction mass), each kilogram of Ref-SSPS related mass launched to orbit is associated with the delivery to Earth of 17 000 kWe-h over 30 years. Selling the energy at 1 ¢/kWe-h yields ~165 $/kg. This return must cover launch costs and all other investments and expenditures on both the space components and the construction and operation of the rectennas on Earth. In this model the rectennas on Earth will be the dominant expense, ~60%, of a space power system that delivers inexpensive energy to Earth. It is necessary to invest *less than* 50 $/kg (@ 0.9 ¢/kWe-h) in the space components. Allowable space expenditures might increase to ≤170 $/kg for satellite systems that are ~3 times less massive per kWe than the Ref-SSPS. This is an extremely difficult target, probably impossible, with financing, the load-following SSPS is impossible 1 ¢/kWe.

LEO/MEO – solar power satellites (#21)

As an alternative to Geo-SSPS, several groups have proposed much smaller solar power satellites, 10 to 100s MWe. A wide range of orbital altitudes above Earth have been proposed, from low altitude (LEO <2000 km) to medium altitude (MEO ≤6000 km), and orbital inclinations ranging from equatorial to polar.

Communications satellites are the core of the most rapidly growing space industry. The satellites provide transmission of television and radio to Earth, and radiotelephony and data transmission between users across the globe. Hoffert and Potter (see Glaser *et al.*, 1998: Ch. 2.8) propose that LEO and MEO solar power satellites be designed to accommodate communications and direct transmission capabilities for the terrestrial markets. The primary power beam would be modulated to provide broadcast, telephony, and data transmissions to Earth. For efficient transmission of power from a satellite, the diameter of its transmitting antenna must increase with the square of the distance from the receiver on the Earth. Also, larger transmitting antennas are required on the satellite as the receivers on Earth decrease in diameter. Thus, attention is restricted to LEO and MEO orbits to enable efficient transmission of power to Earth. Otherwise, the power transmitter dominates the entire mass of the satellite and makes synergistic operation with communications functions far less attractive. Engineering and economics of these satellites will be generally similar to experimental LEO-SSPS units proposed in Japan.

The Japanese government, universities, and companies have sponsored modeling and experimental studies of commercial space solar power. These have focused on the proposed SPS 2000. SPS 2000 is seen by its developers as an experimental program to gain practical experience with power collection, transmission, delivery to Earth, and integration with small terrestrial power networks (Matsuoka, 1999; Glaser *et al.*, 1998: see Nagatomo, Ch. 3.3). This satellite is to be in equatorial orbit at an altitude of 1100 km above Earth. Studies indicate a satellite mass of ~200 tons. Power output on orbit is to be ~10 MWe. Approximately 0.3 MWe is delivered to a rectenna immediately under the equatorial ground path of the unit satellite. Power will be transmitted by the satellite to a given ground receiver 16 times a day for ~ 5 minutes. This implies a duty cycle (D) of the satellite and one rectenna to be ~ (1/12 hr) × 16/24 hr = 0.056 ~ 6%. Thus, ~18 (1/0.056) unit satellites would be required to provide continuous power to a given rectenna. Power users would be restricted to equatorial islands and continental sites. A given satellite would be over land and island rectennas no more than ~30% of its time per orbit. This reduces the effective duty cycle for power delivery to ~2%.

It is highly unlikely that LEO and MEO satellites can provide low-cost solar electric power to Earth. They essentially face the same burden of launch costs as described in the foregoing Ref-SSPS. However, the low duty cycle ($0.01 \leq D \leq 0.3$) increases the cost challenges by at least a factor of three to 100. In addition, orbital debris is far more of a concern in MEO and LEO orbits than in GEO. More debris is present. Relative orbital velocities are higher and collisions are more frequent.

The supply of 20 TWe from LEO and MEO is an unreasonable expectation. A factor of 10 increase in satellite area over GEO, due to a low duty cycle, implies >2 000 000 km² area of satellites close to Earth with a total mass >2 000 000 000 tons. The area would be noticeable. Collectively, it will be >20 times the angular area of the Moon. The components will pose physical threats to any craft in orbit about Earth. The heavy components will pose threats to Earth. For comparison, Skylab had a mass of ~80 tons. The International Space Station will have a mass of ~300 tons.

Space solar power satellites using non-terrestrial materials (#22 and #23)

O'Neill (1975; also see Glaser *et al.*, 1998, Ch. 4.10) proposed that SPS be built of materials gathered on the moon and transported to industrial facilities in deep space. These are termed LSPS. It was argued that without redesign at least 90% of the mass of an SSPS could be derived from common lunar soils. Transport costs from Earth would be reduced. Design, production, and construction could be optimized for zero gravity and vacuum. NASA funded

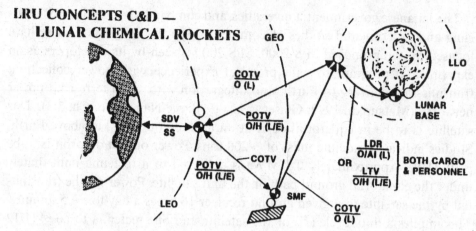

Figure 9.3   Facilities and transportation for construction of lunar-derived LSPS.

studies on the production of LSPS. MIT examined the production and design of LSPS and factories for LSPS in geosynchronous orbit (Miller, 1979). Prior to these studies a team at the Lunar and Planetary Institute examined the feasibility of producing engineering materials from lunar resources (Criswell *et al.*, 1979, 1980).

General Dynamics, under contract to the NASA-Johnson Space Center, developed systems-level engineering and cost models for the production of one 0.01 TWe LSPS per year over a period of 30 years (Bock, 1979). It was com- pared against a NASA reference model for a 0.01 TWe SSPS to be deployed from Earth that established the performance requirements and reference costs (Johnson Space Center, 1977, 1978). General Dynamics drew on the studies conducted at MIT, the Lunar and Planetary Institute, and others. The General Dynamics studies assumed there was no existing space program. New rockets and a spaceport were constructed. New space facilities were built in low orbit about the Earth and the Moon and in deep space. Note the annotations to the Bock (1979) reference. A ten-year period of R&D and deployment of assets to space and the Moon was assumed. The General Dynamics studies explicitly estimated costs of research and development, deployment over 30 years of a fleet of 30 LSPS with 0.3 TWe capacity, and operation of each LSPS for 30 years. They also included the establishment and operation of rectennas on Earth.

Figure 9.3 illustrates two of the three major facilities and transportation concepts (*C and D*) developed by General Dynamics for the systematic analy- sis of lunar production options. Study *Case D* assumed extensive production of chemical propellants (Al and O_2) derived from lunar materials. The lunar base was sized for the production of 90% of the LSPS components from lunar materials. Most of the components were made in deep space from raw and semi-processed materials transported to deep space by chemical rockets and electrically driven mass drivers. General Dynamics projected a base on the Moon with ~25 000 tons of initial equipment and facilities, 20 000 tons of propellants, and ~4500 people. Approximately 1000 people were directly involved in production of components for shipment to space. The rest supported logistics, upkeep, and human operations. People worked on the Moon and in space on six-month shifts. The space manufacturing facility (SMF) in GEO had a mass of ~50 000 tons and a crew of several hundred people. The lunar base and space manufacturing facility were deployed in three years. This fast deployment required a fleet of rockets similar to that required to deploy one 0.01 TWe Ref-SSPS per year from Earth, ~100 000 tons/year to LEO at a cost of ~500 $/kg. NASA-JSC managers required this similarly sized fleet to ease comparisons between Ref-SSPS systems deployed from Earth and those constructed primarily from lunar materials. Hundreds of people crewed the logistics facilities in low Earth orbit (40 000 tons) and tens of people the facility in lunar orbit (1000 tons).

General Dynamics concluded that LSPS would likely be the same or slightly less expensive than Ref-SSPS after production of 30 units. LSPS would require progressively smaller transport of mass to space than SPS after the completion of the second LSPS.

LSPS production could not be significantly increased without an expansion of the lunar base, the production facility in deep space, and the Earth-to-orbit fleet. In the context of the Ref-SSPS studies it is reasonable to anticipate by 2050 that total LSPS capacity would be no more than 1 TWe and likely far less.

These systems, engineering, and costs studies by General Dynamics provided the core relations used to model the Lunar Solar Power System. Thus, the LSP System studies, described in the following section, build directly on two million dollars of independent analyses that focused on utilization of the Moon and its resources.

Over the long term power satellites can be located beyond geosynchronous orbit, #23 in Table 9.6, where sunlight is never interrupted and SSPS power capacity can be increased indefinitely. The satellites will constitute no physical threat to Earth and appear small in the terrestrial sky. These remote SSPS will be beyond the intense radiation belts of Earth but still exposed to solar and galactic cosmic rays. Two favorable regions are along the orbit of the Moon in the gravitational potential wells located 60° before and after the Moon (L4 and L5). Power bases on the Moon and relays and/or LSPS at L4 and L5 can provide power continuously to most receivers on Earth. Advanced power satellites need not be restricted to the vicinity of Earth or even the Earth–Moon system. For example, there is a semi-stable region (L2 ~1.5 million kilometers toward the Sun from the Earth) where satellites can maintain their position with little or no use of reaction mass for propulsive station keeping. A power satellite located in this region continuously faces the Sun. The aft side continuously faces the Earth. It can continuously broadcast power directly back to Earth and to a fleet of relay satellites orbiting Earth. Such power satellites can be very simple mechanically and electrically (Landis, 1997). Asteroid and lunar materials might be used in their construction (Lewis, 1991a)

9.3 Lunar Solar Power (LSP) System

9.3.1 Overview of the LSP system

Figure 9.4 shows the general features of the LSP System. Pairs of power bases on opposite limbs of the moon convert dependable solar power to microwaves. The Earth stays in the same region of the sky as seen from a given power base on the moon. Thus, over the course of a lunar month, pairs of bases can continuously beam power toward collectors, called rectennas, on Earth (shown in the lower right of Figure 9.2). Rectennas are simply specialized types of centimeter-size TV antennas and electric rectifiers. They convert the microwave beam into electricity and output the pollution-free power to local electric distribution systems and regional grids. Rectennas are the major cost element of the LSP System. Figure 9.4 greatly exaggerates the size of the rectenna depicted by the circle in Brazil. The LSP system shown in Figure 9.4 may include three to five hours of power storage on the Moon or on Earth. This LSP system is projected to provide cheaper and more reliable power than can solar installations on Earth.

The LSP System (Ref) is a more advanced reference system that includes solar mirrors in orbit about the moon (LO). The LO mirrors are not shown in Figure 9.4. The LSP system (Ref) also includes microwave "orbital redirector or reflector" satellites in moderate altitude, high inclination orbits about Earth (EO) that are shown in Figure 9.4. EO relays will redirect LSP beams to rectennas on Earth that cannot directly view the power bases. The rectennas can be decreased in area compared to the LSP system described in the paragraph above. The LSP system (Ref) will very likely include sets of photovoltaics across the limb of the moon from each power base (*X-limb*). These X-limb stations are not shown in Figure 9.4.

The electric power capacity of LSP has been projected in terms of the key physical and engineering parameters and the level of technology. Refer for details to Table 4 of Criswell (1994) and discussion by Criswell and Waldron

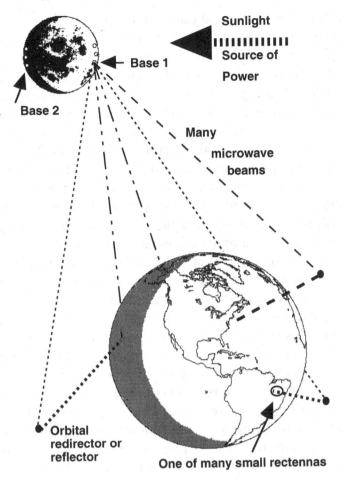

Figure 9.4   Schematic of the lunar solar power system.

(1993). Using 1980s technology, the LSP System can output 20 TWe by occupying ~25% of the lunar surface. Technologies likely to be available relatively early in the 21st century allow the LSP system to output 20 TWe while occupying only 0.16% of the lunar surface. Energy from LSP is projected to be less costly than energy from all other large-scale power systems at similar levels of power and total energy output. An electric energy cost of less than 1 ¢/kWe-h is projected for the mature system (Criswell and Waldron, 1990) and lower costs are conceivable. LSP with redirectors in Earth orbit can provide load-following power to rectennas located anywhere on Earth.

Technology base for operating system

The LSP System is an unconventional approach to supplying commercial power to Earth. However, the key operational technologies of the LSP have been demonstrated by NASA and others at a high technology readiness level (TRL ≥ 7). TRL = 7 denotes technology demonstrated at an appropriate scale in the appropriate environment (Criswell, 2000).

Power beams are considered esoteric and a technology of the distant future. However, Earth-to-Moon power beams of near-commercial intensity are an operational reality. Figure 9.5 is a picture of the South Pole of the Moon that was taken by the Arecibo radar in Puerto Rico. This technique is used for mapping the Moon, determining the electrical properties of the lunar surface, and even examining the polar regions for deposits of water ice (Margot *et al.*, 1999). The Arecibo beam passes through the upper atmosphere with an intensity of the order of 20–25 W/m^2. The LSP System is designed to provide power beams at Earth with intensities of less than 20% of noontime sunlight (≤230 W/m^2). Lower intensity beams are economically reasonable. The intensity of microwaves scattered from the beam will be orders of magnitude less than is allowed for continuous exposure of the general population.

Load-following electrical power, without expensive storage, is highly desirable. Earth orbiting satellites can redirect beams to rectennas that cannot view the Moon, and thus enable load-following power to rectennas located anywhere on Earth. Rectennas on Earth and the lunar transmitters can be sized to permit the use of Earth-orbiting redirectors that are 200 m to 1000 m in diameter. Redirected satellites can be reflectors or retransmitters. The technology is much more mature than is realized by the technology community at large.

Figure 9.6 is an artist's concept of the Thuraya-1 regional mobile communications satellite [operated by Thuraya Satellite Telecommunications Co. Ltd. of the United Arab Emirates which was placed in orbit in October 2000 (permission: Boeing Satellite Systems, Inc.)]. The circular reflector antenna is 12.25 m in diameter. C. Couvault (1997) reported that the US National Reconnaissance Office has deployed to geosynchronous orbit a similar, but much larger, "Trumpet" satellite. The Trumpet reflector is reported to be approximately 100 meters in diameter. The Trumpet reflector, only a few tons in mass, has a diameter within a factor of three of that necessary to redirect a low-power beam to a 1 km diameter or larger rectenna on Earth. Power beams and redirector satellites can minimize the need for long-distance power transmission lines, their associated systems, and power storage.

Alternatively, a relay satellite can receive a power beam from the Moon. The relay satellite then retransmits new beams to several rectennas on Earth. Unmanned and manned spacecraft have demonstrated the transmission of beams, with commercial-level intensity in low Earth orbit. Figure 9.7 illustrates the NASA Shuttle with a phased array radar. The radar fills the cargo bay of the shuttle, making a synthetic-aperture radar picture of the Earth. Near the

Figure 9.5   Arecibo radar picture of the Moon.

Figure 9.6   Concept of NRO trumpet satellite.

Shuttle, the beam has an intensity of the order of 150 W/m^2. This is well within the range for commercial transmission of power (Caro, 1996).

Approximately once a year the Earth will eclipse all the lunar power bases for up to three hours. This predictable outage can be accommodated by power storage of defined capacity or reserve generators on Earth. Alternatively, a fleet

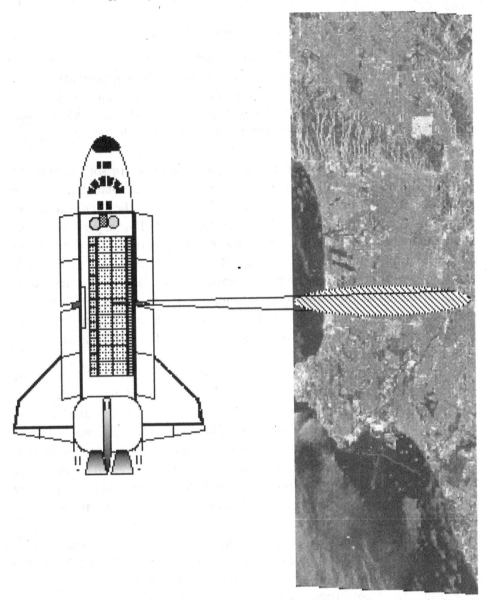

Figure 9.7   Shuttle synthetic aperture radar.

of solar mirrors in orbit about the Moon can reflect solar power to selected bases during eclipses and during sunrise and sunset. These solar reflectors, actually types of solar sails, will be far less expensive, per unit area, to build and operate than high-precision reflectors such as those in Figure 9.6.

9.3.2 LSP demonstration base

The lunar portion of an LSP System prototype Power Base is depicted in Figure 9.8. A Power Base is a fully segmented, multi-beam, phased array radar powered by solar energy. This demonstration Power Base consists of tens to hundreds of thousands of independent power plots. A demonstration power plot is depicted in the middle to lower right portion of the figure. A mature power plot emits multiple sub-beams.

A power plot consists of four elements. There are arrays of solar converters, shown here as north–south aligned rows of photovoltaics. Solar electric power is collected by a buried network of wires and delivered to the microwave transmitters. Power plots can utilize many different types of solar converters and many different types of electric-to-microwave converters. In this example the microwave transmitters are buried under the mound of lunar soil at the earthward end of the power plot. Each transmitter illuminates the microwave reflector located at the anti-earthward end of its power plot. The reflectors overlap, when viewed from Earth, to form a filled lens that can direct very narrow and well-defined power beams toward Earth. The Earth stays in the sky above the Power Base.

Extremely large microwave lenses, the circles on the Moon in figure 9.2, are required on the Moon to direct narrow beams to receivers (≥ 0.5 km diameter) on Earth. Large lenses are practical because of fortuitous natural conditions of the Moon. The same face of the Moon always faces Earth. Thus, the many small reflectors shown in Figure 9.8 can be arranged in an area on the limb of the moon so that, when viewed from Earth, they appear to form a single large aperture. The Moon has no atmosphere and is mechanically stable. There are virtually no moon quakes. Thus it is reasonable to construct the large lens from many small units.

Individually controllable sub-beams illuminate each small reflector. The sub-beams are correlated to combine coherently on their way toward Earth, to form one power beam. In the mature power base there can be hundreds to a few thousand sets of correlated microwave transmitters. These arrangements of multiple reflectors, likely including additional subreflectors or lenses in front of each main reflector, and transmitters form a fully segmented multibeam phased array radar.

9.3.3 LSP constructed of lunar materials on the Moon

To achieve low unit cost of energy, the lunar portions of the LSP System are made primarily of lunar-derived components (Criswell, 1996, 1995; Criswell

Figure 9.8   LSP system prototype Power Base and demonstration Power Plots.

and Waldron, 1993). Factories, fixed and mobile, are transported from the Earth to the Moon. High output of LSP components on the Moon greatly reduces the impact of high transportation costs of the factories from the Earth to the Moon. On the Moon the factories produce hundreds to thousands of times their own mass in LSP components. Construction and operation of the rectennas on Earth constitute greater than 90% of the engineering costs. Using lunar materials to make significant fractions of the machines of production and to support facilities on the Moon can reduce up-front costs (Criswell, 1995a). Personnel in virtual work places on Earth can control most aspects of manufacturing and operations on the Moon (Waldron and Criswell, 1995).

An LSP demonstration Power Base, scaled to deliver the order of 0.01 to 0.1 TWe, can cost as little as 20 billion dollars over ten years (Criswell and Waldron, 1993; Glaser *et al.*, 1998, Ch. 4.11). This assumes the establishment of a permanent base on the Moon by one or more national governments that is devoted to the industrial utilization of lunar resources for manufacturing and logistics. Such a base is the next logical step for the world space programs after completion of the International Space Station.

LSP is practical with 1980s technology and a low overall efficiency of conversion of sunlight to power output on Earth of ~0.15%. Higher system efficiency, ≥35%, is possible by 2020. An LSP System with 35% overall efficiency will occupy only 0.16% of the lunar surface and supply 20 TWe to Earth. Also, greater production efficiencies sharply reduce the scale of production processes and up-front costs.

There are no "magic" resources or technologies in Figure 9.8. Any handful of lunar dust and rocks contains at least 20% silicon, 40% oxygen, and 10% metals (iron, aluminum, etc.). Lunar dust can be used directly as thermal, electrical, and radiation shields, converted into glass, fiberglass, and ceramics, and processed chemically into its elements. Solar cells, electric wiring, some micro-circuitry components, and the reflector screens can be made from lunar materials. Soil handling and glass production are the primary industrial operations. Selected micro-circuitry can be supplied from Earth.

Unlike Earth, the Moon is the ideal environment for large-area solar converters. The solar flux to the lunar surface is predicable and dependable. There is no air or water to degrade large-area, thin-film devices. The Moon is extremely quiet mechanically. It is devoid of weather, significant seismic activity, and biological processes that degrade terrestrial equipment. Solar collectors can be made that are unaffected by decades of exposure to solar cosmic rays and the solar wind. Sensitive circuitry and wiring can be buried under a few to tens of centimeters of lunar soil and completely protected against solar radiation, temperature extremes, and micrometeorites.

The United States has sponsored over 500 million dollars of research on the lunar samples and geophysical data since the first lunar landing in 1969. This knowledge is more than adequate to begin designing and demonstrating on Earth the key lunar components and production processes. Lunar exploration is continuing. The DoD Clementine probe and the Lunar Prospector (lunar.arc.nasa.gov; *Science, 266*: 1835–1861, December 16; Binder, 1998) have extended the Apollo-era surveys to the entire Moon.

9.4 LSP System versus Other Power System Options at 20 TWe

Table 9.7 summarizes the materials and manufacturing scales of major options to provide 600 to 960 TWe-y of electric power (Criswell and Waldron, 1990). The second column indicates the fuel that would be used over the seventy-year period by the conventional systems in rows 1, 2, and 3. The third column indicates the scale of machinery to produce and maintain the power plants and provide the fuel. The rightmost column shows the total tonnage of equipment needed to produce a TWe-y of power (specific mass). The higher the specific mass the more effort is required to build and maintain the system and the greater the opportunity for environmental modification of the biosphere of Earth. Notice that the LSP units on the Moon are approximately 600 000 times more mass efficient in the production of power than a hydroelectric dam and 3000 times more mass efficient than a coal system. Estimates for rectennas, row #4, assume 1970s technology of small metallic dipoles supported by large aluminum back planes and concrete stands. Rectennas can now be incorporated into integrated circuits on plastic or can employ low mass reflectors to concentrate the incoming microwaves. Specific mass of the rectennas can likely be reduced by an order of magnitude (Waldron and Criswell, 1998)

LSP does not have the mechanical directness of SSPS. To achieve the lowest cost of energy the LSP System needs microwave orbital redirectors about the Earth. Compared to an SSPS the specific mass of beam redirectors can be very low for the power they project to the rectenna. This is because the LSP orbital redirectors can achieve far higher efficiency in retransmitting or reflecting microwaves than can an SSPS in converting sunlight into microwaves. Also, the LSP microwave reflectors can be much smaller in area than an SSPS that transmits an equal level of power. This is because the LSP orbital unit can be illuminated by microwave beams in space that are more intense than solar intensity. LSP requires the smallest amount of terrestrial equipment and final materials of any of the power systems, as can be seen from line 7 of Table 9.7.

Engineers would not have built the large hydroelectric dams on Earth if it had been necessary to excavate the catchment areas and river valleys first. The

Table 9.7 *Terrestrial fuel and equipment tonnage & energy output of 20 TWe power systems*

Terrestrial Systems	Fuel(70 y) (T)	Equip. & plant (T)	Total energy (TWe-y)	Specific mass (T/TWe-y)
1. Hydro & TSP (without storage)	9×10^{16}	8×10^{10}	900	9×10^8
2. Nuclear fission (non-breeder)	6×10^7	2×10^{10}	600	3×10^7
3. Coal plants, mines, & trains	3×10^{12}	6×10^9	600	1×10^7
4. Rectenna Pedestals (SPS & LSP)	—	4×10^9	—	4×10^6
(Electronic elements)	—	2×10^7	—	2×10^4

Space Systems (Mass shipped from Earth)	First year equip. (T)	Total equip. (T)	Total energy (TWe-y)	Specific mass (T/TWe-y)
5. SSPS made on Earth (@10 T/MWe)	2×10^6	3×10^8	600	5×10^5
6. LSPS from lunar materials	2×10^7	5×10^7	600	8×10^4
7. LSP (Ref)	3×10^4	3×10^6	960	3×10^3

water and geography were gifts of nature that minimize the amount of earth that must be moved and enable the smallest possible dams. The Moon provides the solid state equivalent of a "natural watershed" for the 21st century. It is there, correctly positioned, composed of the materials needed, and lacking the environment of Earth that is so damaging to thin-film solid-state devices.

The General Dynamics–Convair models for construction of space solar power satellites from lunar materials (refer to section 9.2 and Figure 9.3) were adapted to modeling the construction of the Lunar Solar Power System (Criswell and Waldron, 1990). Use of the Moon eliminates the need to build extremely large platforms in space. LSP components can be manufactured directly from the lunar materials and then immediately placed on site. This eliminates most of the packaging, transport, and reassembly of components delivered from Earth or the Moon to deep space. There is no need for a large manufacturing facility in deep space. This more focused industrial process reduces the fleet of rockets necessary to transport components, manufacturing facilities, and people from Earth to space and the Moon, compared to the General Dynamics–Convair model. If the LSPS and LSP use similar technologies for deployment, manufacturing, and operations in space and on the Moon, LSP power capacity can be installed at ~50 times the rate of LSPS for similar levels of expenditures over similar times. Higher LSP System emplacement rates are conceivable with future operating technologies, higher levels of automation of the production process, and use of lunar materials to build the more massive elements of production machinery (Criswell and Waldron, 1990; Criswell, 1998c). The higher production efficiencies and lower cost of LSP support the assertion that the LSP System is the only likely means to provide 20 TWe of affordable electric power to Earth by 2050 (Table 9.6, Row 23).

Figure 9.9 illustrates the growth of power transmission capacity on the Moon (power units). The installation units, which are the mobile units in Figure 9.8, are initially transported to the Moon and produce and emplace the power plots that are termed power units in Figure 9.8. As experience is gained, an increasing fraction of installation units can be made on the Moon by manufacturing units that primarily use lunar resources. Manufacturing units might be contained in modules deployed from Earth such as are depicted in the middle-left of Figure 9.8. Power unit production can be increased without significantly increasing the transportation of materials from Earth (Criswell, 1998c). The long life and increasing production of power units enable an "exponential" growth in LSP transmission capacity and similar growth in the delivery of net new energy to Earth. Development of manufacturing units and the extensive use of lunar materials is considered to be a reasonable goal (Bekey *et al.*, 2000).

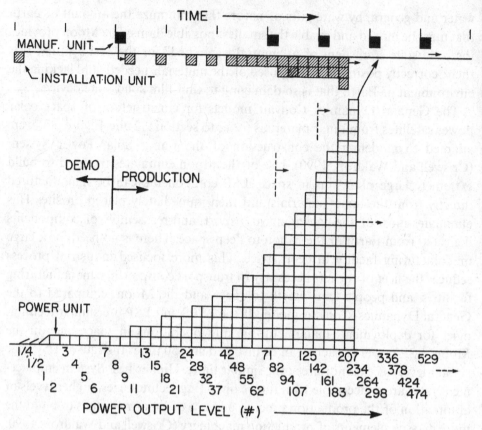

TIME ──────▶

MANUF. UNIT

INSTALLATION UNIT

DEMO ◀── | ──▶ PRODUCTION

POWER UNIT

1/4		3		7		13		24		42		71		125		207		336		529
	1/2		4		8		15		28		48		82		142		234		378	
	1		5		9		18		32		55		94		161		264		424	---
2		6		11		21		37		62		107		183		298		474	

POWER OUTPUT LEVEL (#) ──────▶

Figure 9.9   Exponential growth of LSP System power output from the Moon.

Table 9.8 compares estimated "median" life-cycle costs of five power systems scaled to provide 1500 TWe-y of energy. The costs are given in trillions $(1 \text{ T} = 1 \times 10^{12})$ of US dollars. The estimates are based on studies of systems utilizing 1990s levels of technology (Criswell, 1997, 1997a, 1997b; Criswell and Thompson, 1996). The major cost categories are capital, labor, fuel, and waste handling and mitigation. Nominal costs for labor, capital, and fuel are taken from 1980s studies of advanced coal and nuclear plants (Criswell and Thompson, 1996).

Costs of the coal and fission plants in Table 9.8 are consistent with the esti-mated cost of a "mixed" power system as described by *Case A2* of Nakicenovic *et al.* (1998) in Section 2 and summarized in Table 9.1. Thirty percent of the costs of the coal and fission systems are for regional power distribution systems. Nakicenovic *et al.* (1998) project a capital cost of 120 T$ for a mixed system that consumes 3000 TWt-y of energy by 2100. This corresponds to ~140 T$ for 4500 TWt-y. Table 9.8 assumes that all the thermal energy is con-

Table 9.8 *Nominal costs (T$) of power systems to deliver 1500 TWe-y*

	Coal	Fission	TTSP	TPSP	LSP(ref)	LSP(X-limb)	LSP(no EO)
Labor	20	60	113	233	2	3	7
Capital	570	713	1340	2166	63	105	286
Fuel	243	0	0	0	0	0	0
Wastes	914	3000	0	0	0	0	0
Total	1746	3773	1452	2399	64	108	293

Notes:
TTSP – Terrestrial thermal solar power.
TSPS – Terrestrial solar photovoltaic power.
LSP(Ref) – Lunar Solar Power System with beam redirectors in orbit about Earth
   and solar reflectors in orbit about the Moon.
LSP(X-limb) – LSP System with fields of photovoltaics across the lunar limb from
   each power base, no solar reflectors in orbit about the Moon, and three hours of
   electric storage capacity on Earth.
LSP(no EO) – Similar to LSP(X-limb) but no redirectors in orbit about Earth and at
   least 18 hours of electric power storage capacity on Earth.

verted to electric energy. To a first approximation, this doubles the cost of the
capital equipment to 380 T$ for 4500 TWt-y. The Table 9.8 estimate of capital
costs of ~570 T$ (for fossil only) or 713 T$ (for fission only) assumed 1980s
levels of technology and no technology advancement during the 21st century.
Case A2 assumes far higher costs for fossil fuels, ~830 T$, than the ~243 T$
assumed for the nominal case in Table 9.8.

Externality cost, corresponding to wastes costs in Table 9.8, of ~913 T$ for
coal are comparable to the ~800 T$ in *Case A2* where externality cost is
assumed to be proportional to the price of the fossil fuel. The costs of coal and
fission plants are dominated by "waste" handling, which includes estimates of
damage to human health and to the environment of the entire fuel process
(mining to burning to disposal).

The cost estimates in Table 9.8 assume that TTSP(thermal) and
TPSP(photovoltaic) systems are scaled to storage energy for only 1 day of local
operations. These systems must be scaled up considerably to feed power to a
global network of power lines. Analyses provided by Klimke (1997) indicate
that the delivery 1500 TWe-y would cost the order of 10000 T$. Power storage
is not included in the estimated cost of the global system.

The LSP (Ref) System is the lowest in cost. It uses reflectors in orbit about
the Moon to illuminate the power bases during eclipses of the Moon by the
Earth. It also uses redirectors in orbit about Earth to provide load-following

power to any rectenna on Earth. The LSP(Ref) provides 1500 TWe-y at 1/27th the cost of the coal-fired system.

The LSP(X-limb) System, column #7 of Table 9.8, does not use solar reflectors in orbit about the Moon. Rather, each power base is provided with a field of photovoltaics across the limb of the Moon. Power lines connect the power base and extra photovoltaics. Energy storage is provided on Earth for 3 TWe-h, and power bases and rectennas are slightly scaled up. The LSP(no EO) is the LSP(X-limb) without redirectors in orbit about the Earth. Approximately 18 TWe-h of power storage is supplied on Earth. Deep-pumped hydro is assumed. The power bases and rectennas are scaled up over the LSP(X-limb) case. Even the LSP(no EO) provides 1500 TWe-y of energy at less cost than conventional coal or fission. The LSP Systems can provide savings of the order of 1000 trillion dollars for the delivery of 1500 TWe-y over coal and the order of 9000 trillion dollars in savings over global solar photovoltaics.

Criswell and Thompson (1996) analyzed the effects of changes of a factor of 10 in "waste" handling costs and reasonable variations in the costs of labor, capital, and fuel for coal, advanced nuclear fission, terrestrial solar thermal and photovoltaic, and the LSP(Ref) systems. This prototype analysis found the LSP(Ref) system to be 10 to 16 times less expensive in the delivery of end-user electric power than the closest competitor, coal.

New types of mixed systems may enable substantial reductions in costs. Berry (1998: *Target Scenario*) proposes a power system scaled to supply the United States with ~1.3 TWe. Approximately 45% is used directly by end-users. The majority (55%) is used to make hydrogen for transportation vehicles. The primary power sources are wind (0.85 TWe), solar thermal (0.85 TWe), hydro-electric and nuclear (0.15 TWe), and distributed photovoltaics. Energy is stored as hydrogen to be used in load leveling and peaking. See Bockris (1980) for an early but extensive discussion of a hydrogen economy and its technology.

A novel aspect of the Target Scenario is to store the hydrogen in the fuel tanks of cars, vans, trucks, building power supplies, and similar power users (Appleby, 1999; Lloyd, 1999). The vehicles are assumed to be connected to the national power grid when not in use. Thus, the capital associated with the trans-portation fleet and end-use production is also used to provide "prepaid" energy storage facilities. Assuming very low-cost advanced technologies, the simula-tions project ~0.4 T$/TWe-y for the cost of delivering end-use and transpor-tation energy. One thousand TWe-y would cost ~400 T$. This is ~50 times less costly than a stand alone global system of photovoltaics and power distribu-tion. Given reasonable costs for reversible fuel cells, the primary concern becomes accurate knowledge of the longest period of unsuitable weather (cloudy or smoky skies, low winds, etc.) in a region or globally. However, given the complexity of the biosphere, this "longest" duration is essentially unknowable. The LSP System provides a power source that is decoupled from the biosphere and can provide the power as needed through all conditions of fog, clouds, rain, dust, and smoke.

Berry's mixed system, or Target Scenario, is ~20 times more costly than the least expensive option of the Lunar Solar Power System. The LSP System, and other systems, could also use the vehicular and other energy storage units of the Target Scenario to reduce costs.

Total costs of power for an energy-prosperous world are so enormous that it is difficult to understand their scale and significance. One method is to calculate the simple sum of gross world product (GWP) from 2000 to 2100 under the population assumptions in Table 9.1. For ten years after the oil disruptions of the 1970s GWP/person was ~4,000 $/person-y. Assume an energy constrained 21st century and a constant GWP/person. This sums to 3900 T$. This projected "poor" world simply cannot afford to build and operate terrestrial solar power systems necessary to provide 2 kWe/person by 2050 or 1500 TWe-y by 2100. The world economy will be extremely hard pressed to provide the energy by means of coal and fission as noted in Tables 9.2–9.6.

For *Case A*, Nakicenovic *et al.* (1998: p. 6), project a global economy that sums to ~12000 T$ by 2100. Per capita income is ~4300 $/person-y in 2000 and ~30000 $/person-y in 2100. Case A assumes no major "waste" costs for the large-scale use of fossil and nuclear energy. They project ~8800 T$ for *Case C* (mostly renewable energy) between 2000 and 2100. Note that approximately 10% of GWP is now expended on the production and consumption of commercial energy. This corresponds to 240 to 540 trillion dollars between 2000 and 2100. These sums are much smaller than the costs of conventional power systems to supply adequate power. But they are larger than projected for the LSP System. A poor world must remain energy poor if it uses only conventional power systems. However, the less costly LSP System electricity can save money, minimize or even eliminate the pollution associated with energy production, and accelerate the generation of wealth.

Why is the LSP so attractive as a large-scale power system? The Sun is a completely dependable fusion reactor that supplies free and ashless high-quality energy at high concentrations within the inner solar system, where we live. The LSP primarily handles this free solar power in the form of photons. Photons weigh nothing and travel at the speed of light. Thus, passive and low mass equipment (thin-films, diodes, reflectors, and rectennas) can collect and channel enormous flows of energy over great distances, without physical connections, to end uses when the energy is needed. The LSP is a distributed system that can be operated continuously while being repaired and evolving. All other power systems require massive components to contain and handle matter under intense conditions, or require massive facilities to store energy. Low mass and passive equipment in space and on the Moon will be less expensive per unit of delivered energy to make, maintain, decommission, and recycle at the end of its useful life than massive and possibly contaminated components on Earth.

The Moon is a uniquely suitable and available natural platform for use as a power station. It has the right materials, environment, mechanical stability, and orientation and remoteness with respect to Earth. The major non-terrestrial components of LSP can be made of lunar materials and the large arrays can be sited on the Moon.

The rectennas on Earth are simple and can be constructed as needed and begin to produce net revenue at a small size. The LSP can be far less intrusive, both in the physical and electromagnetic sense, than any other large power system. Most of the power can be delivered close to where it is needed. LSP can power its own net growth and establish new space and Earth industries. Finally, all of this can be done with known technologies within the period of time that the people of Earth need a new, clean, and dependable source of power that will generate new net wealth.

9.5 Implications of the Lunar Solar Power System

Between 1960 and 1986, the total electric energy, Ee, (y), used every year, measured in TWe-y, was an excellent index of the annual GWP in trillions of dollars ($T\$_e(y)$) in a given year "$y$" (Starr, 1990; Criswell, 1997). Equation (1) presents this empirical relation. Equation (1) includes the annual increase in productivity of energy ($Eff(y) = 1\%/y$). The maximum cost of 1500 TWe-y of energy delivered by the LSP System between 2000 and 2100 is estimated to be 300 T\$.

$$T\$e\ (y) = 2.2\ T\$ + [10.5\ T\$/TWe\text{-}y] \times Ee(y) \times Eff(y) - 300T\$/(100\ y) \quad (1)$$

A new electric power system is initiated in 2000. Capacity builds to 20 TWe by 2050 and then remains at 20 TWe capacity until 2100. The new power system delivers 1510 TWe-y over the 21st century. Applying Equation (1) to this profile predicts an integral net GWP ~25 800 T\$ by 2100. Assume the growth in world population presented in Table 9.1. These relations predict a global per capita income of ~30 000 \$/y-person in 2050 as a result of the acceleration of global electrification. By 2100, global per capita income is ~38 000 \$/y-person because of the 1%/y growth in economic productivity of a unit of electric energy. The electric power capacity of the new system, and the net new wealth it produces, could be further increased for users on the Earth and in space.

These gains are enormous in total GWP compared to *Case A2* of Nakicenovic *et al.* (1998). Refer to Table 9.1. The all-electric world supplied by the LSP system has ~2.5 times greater economic gain and retains enormous reserves of fossil and nuclear fuels. Also, there is no additional contamination of the atmosphere or Earth.

Case A2 assumes that aggressive use of oil, natural gas, and especially coal will not degrade the environment and that costs of environmental remediation, health effects, and pollution control will all be low. However, it is not obvious this should be so. During the 1990s the world per capita income remained near 4000 $/y-person. There was little growth in the Developing Countries because of increases in population and recessions. Without a major new source of clean and lower-cost commercial energy it will be very difficult to increase per capita income in the Developing Nations. Suppose per capita income remains at 4000 $/y-person throughout the 21st century. The integral of gross world product will be ~4000 T$ or only 2.2 times the total energy costs for *Case A2* in Table 9.1. Over the 21st century the LSP System offers the possibility of economic gains ~80 to 900 times energy costs.

Enormous attention is directed to discovering and promoting "sustainable" sources of energy and seeking more efficient means of utilizing conventional commercial and renewable energy. However, there are clear limits to the conventional options. Over 4 billion of Earth's nearly 6 billion people are poor in both wealth and energy. Their existence depends primarily on new net energy taken from the biosphere. This energy is harvested as wood, grass, grain, live stock from the land, fish from the seas, and in many other direct and indirect products. The biosphere incorporates each year approximately 100 TWt-y of solar energy in the form of new net plant mass (algae, trees, grass, etc.). It is estimated that humanity now directly extracts ~5% of that new energy and disturbs a much greater fraction of the natural cycles of power through the biosphere. People divert almost 50% of the new solar photosynthetic energy from its natural cycles through the biosphere. Humankind now collects and uses approximately 50% of all the rainwater that falls on accessible regions of the continents. Given the continuing growth of human population, most of the fresh water used by humans will be obtained through desalination (Ehrlich and Roughgarden, 1987; Rees and Wachernagel, 1994).

Human economic prosperity is possibly now using 6 kWt/person. In the next century an LSP supplying ~2 to 3 kWe/person will enable at least an equal level of prosperity with no major use of biosphere resources. For a population of 10 billion people this corresponds to 2000 to 3000 TWe-y, of electric energy per century (Goeller and Weinberg, 1976; Criswell, 1998, 1994, 1993). Much more energy might be desirable and can be made available.

It is widely recognized that the lack of affordable and environmentally benign commercial energy limits the wealth available to the majority of the human population (WEC, 1993, 1998, 2000). However, there is almost no discussion of how to provide the enormous quantities of quality commercial energy needed for an "energy-rich" world population. The carbon curve of Figure 9.1 depicts the cumulative depletion of terrestrial fossil thermal energy by a prosperous human population in terawatt-y of thermal energy. There is approximately 4000 to 6000 TWt-y of economically accessible fossil fuels. Thus, the fossil energy use stops around 2100 when the prosperous world consumes the fossil fuels. Economically available uranium and thorium can provide only the order of 250 TWt-y of energy. Fission breeder reactors would provide adequate energy for centuries once seawater is tapped for uranium and thorium. However, given the political opposition, health and safety risks, and economic uncertainty of nuclear power at the end of the 20th century, it is unlikely that nuclear fission will become the dominant source of power within the biosphere by 2050.

The LSP System is recommended for consideration by technical, national, and international panels and scientists active in lunar research (NASA, 1989; Stafford, 1991; ESA, 1995; ILEWG, 1997; Sullivan and McKay, 1991; Spudis, 1996). An LSP System scaled to enable global energy prosperity by 2050 can, between 2050 and 2070, stop the depletion of terrestrial resources and bring net new non-polluting energy into the biosphere. People can become independent of the biosphere for material needs and have excess energy to nurture the biosphere. The boundaries of routine human activities will be extended beyond the Earth to the Moon, and a two-planet economy will be established.

Acknowledgements

It is a pleasure to acknowledge the reviews, comments, and challenges provided on this chapter by Dr. Robert D. Waldron (retired – Boeing/Rockwell International, Canogo Park, CA), Professor John Lewis (Planetary Research Inst., University of Arizona), Professor Gerald Kulcinski (Fusion Power Institute, University of Wisconsin), and Professor Martin Hoffert (Physics Department, New York University). Special thanks is extended to Prof. Robert Watts (Tulane University, New Orleans) and my wife Ms. Paula Criswell (Schlumberger – Geoquest, Houston) for reviewing the chapter and providing extensive editorial suggestions. Appreciation is certainly extended to Dr. John Katzenberger (Director – Aspen Global Change Institute) and Ms. Susan Joy Hassol (Director of Communication – Aspen Global Change Institute) for organizing, participating in, and administrating the 1998 summer workshop on

Innovative Energy Strategies for CO₂ Stabilization and documenting the results in the AGCI publication *Elements of Change 1998*.

Definitions of Special Terms

Al = aluminum

1 bbl = one barrel of oil = 42 US gallons ~ 159 liters

1 billion = 1×10^9 (also = 1 giga = 1 G)

C = carbon

°C = temperature measured in degrees Celsius

D = duty cycle (fraction of a complete cycle in which action occurs)

e = electric (e.g. 1 We = 1 Watt of electric power)

EO = Earth orbit, an orbit about the Earth

GDP = gross domestic product

Geo = geosynchronous orbit about Earth (satellite stays fixed in sky directly above the equator of Earth)

1 GHz = 1×10^9 cycle per second

1 GWs = one gigawatt of solar energy in free space (above the atmosphere of Earth)

1 GtC = one gigaton of carbon

1 GTce = 1 billion or gigaton of coal

1 GTce = Energy released by burning one billion or one gigaton of coal (~0.93 TWt-y = 2.93×10^{19} Joules)

1 GToe = Energy released by burning one gigaton of oil (~1.33 TWt-y = 4.2×10^{19} Joules)

He = helium

1 J = 1 Joule = 1 Newton of force acting through 1 meter (m) of length (a measure of energy)

1 k = 1 kilo = 1×10^3

1 kg = one kilogram of mass (1 kg exerts 1 Newton of force, = 1 kg-m/sec², under 9.8 m/sec² acceleration; 1 Newton of force ~0.225 pounds of force)

1 km = one kilometer = 1000 meters (measure of length)

1 km² = one square kilometer of area (= 1×10^6 m²)

1 kWe = 1 kilowatt of electric power (functionally equivalent to ~ 3 kWt)

1 kWt = 1 kilowatt of thermal power

LEO = low Earth orbit (an object in low altitude orbit about the Earth, ≤1000 km altitude)

LO = lunar orbit, an orbit about the Moon

LSP = lunar solar power (system)

1 meg = 1 M = 1×10^6

David R. Criswell

MEO = medium Earth orbit (an object in medium altitude orbit about the Earth, ≤10000 km altitude)

1 m = one meter (measure of length)

1 m^2 = one square meter (measure of area)

1 MWe = one megawatt of electric power ($= 1 \times 10^6$ watts of electric power)

NA = Not applicable

NSA = Not stand alone (a power system, such as wind, that must be attached to other power systems, such as coal or oil, to provide dependable power)

N = nitrogen

O = oxygen

OECD = Organization of Economic Cooperation and Development: Australia, Austria, Belgium, Canada, Czech Republic, Denmark, Finland, France, Germany, Greece, Hungary, Iceland, Ireland, Italy, Japan, Korea, Luxembourg, Netherlands, New Zealand, Norway, Poland, Portugal, Spain, Sweden, Switzerland, Turkey, United Kingdom, United States

1 lb = 1 pound

SSPS = space solar power satellite

t = thermal (Wt = Watt of thermal power)

1 tera = 1 T = 1×10^{12}

1 ton = one tonne ($= 1 \times 10^3$ kg) (usually 1 ton = 2000 lbs in US units)

TSPS = terrestrially based solar power system

TPSP = terrestrially based solar power system using photoconversion devices (e.g., photovoltaics)

TTSP = terrestrially based solar power system using concentrated solar thermal power (e.g., solar power tower surrounded by fields of mirrors)

1 TW = 1 terawatt = 1×10^{12} watts

1 TWe = 1 terawatt of electric power

1 TWe-y = one terawatt-year of electric energy = 3.156×10^{19} Joules of electric energy but often functionally equivalent at end use to $\sim 9.5 \times 10^{19}$ Joules of input thermal energy

1 TWm = one terawatt of mechanical power

1 TWt = 1 terawatt of thermal power

1 TWt-y = one terawatt-year of thermal energy = 3.156×10^{19} Joules

1 T$ = 1×10^{12} dollars

1 watt = 1 Joule/sec (measure of power)

1 y = 1 year

1 $ = 1 United States dollar (usually 1990 value)

1 ¢ = 0.01 $

References and notes

Appleby, A. J. (1999, July) The electrochemical engine for vehicles, *Scientific American*, 74–79.

Avery, W. H. and Wu, C. (1994), *Renewable Energy from the Ocean. A guide to OTEC*, Oxford University Press, New York, 446 pp.

Battle, M., Bender, M. L., Tans, P. P., White, J. W. C., Ellis, J. T., Conway, T., and Francey, R. J. (2000), Global carbon sinks and their variability inferred from atmospheric O_2 and $\partial^{13}C$, *Science*, **287**: 2467–2470.

Bekey, G., Bekey, I., Criswell, D. R., Friedman, G., Greenwood, D., Miller, D., and Will, P. (2000, April 4–7), NSF-NASA Workshop on autonomous construction and manufacturing for space electrical power systems, Chapter 5, Arlington, VI (robot.usc.edu/spacesolarpower/presentations/4–7_criswell.ppt).

Berry, G. D. (1998), *Coupling hydrogen fuel and carbonless utilities*, position paper, Energy Program, Lawrence Livermore National Laboratory, Livermore, CA: 10pp.

Binder, A. B. (1998) Lunar prospector: overview, *Science*, **281**: 1475–1480: September 4. (See also 1480–1500).

Blunier, T. (2000) Frozen methane from the sea floor, *Science*, **288**: 68–69.

Bock, E. (1979), *Lunar Resources Utilization for Space Construction*, Contract NAS9–15560, DRL Number T-1451, General Dynamics – Convair Division, San Diego, CA.

    a. Final Presentation (21 February 1979), Line Item 3, DRD Number DM253T, ID# 21029135, 171 pp.

    b. Final Report (30 April 1979) Volume II, Study Results, DRD No. MA-677T, Line Item 4, eight chapters, approximately 500 pp.

    Figure 9.3 acronyms (working from left to right): SDV = shuttle derived vehicle, SS = space station, LEO = low earth orbit, POTV = personnel orbital transfer vehicle, O = oxygen, H = hydrogen propellant, L = propellant supplied from Moon, E = propellant supplied from Earth, COTV = cargo orbital transfer vehicle (solar electric powered), SMF = space manufacturing facility, GEO = geosynchronous orbit, LTV = Lunar transfer vehicle, LDR = lunar mass driver delivered materials/cargo, LUNAR BASE, LLO = low lunar orbit. The open circles represent transfer/logistic facilities.

Bockris, J. O'M. (1980), *Energy Options: Real Economics and the Solar-Hydrogen System*, Australia & New Zealand Book Co., 441pp.

Broecker, W. S. (1997, 28 Sept.), Thermohaline circulation, the Achilles Heel of our climate system: will man-made CO_2 upset the current balance, *Science*, **278**: 1582–1589.

Browne, M. W. (1999, 8 June), Reviving quest to tame energy of stars, *The New York Times: Science Times*, Y, D1 – D2.

Caro E. (1996, 25 September), personal communication (NASA/JPL SAR Prog. Eng.).

Couvault, C. (1997, 8 December), NRO radar, sigint launches readied, *Av. Week & Space Technology*, 22–24. Also, "Boeing's Secret", (1 September, 1998): 21.

Criswell, D. R. (1993), Lunar solar power system and world economic development, Chapter 2.5.2 , 10 pp., of *Solar Energy and Space Report, in World Solar Summit*, UNESCO. Paris.

Criswell, D. R. (1994, October), Net growth in the two planet economy (invited), *45th Congress of the International Astronautical Federation*: Session: a comprehensive rationale for astronautics, Jerusalem, IAF-94-IAA.8.1.704, 10 pp.

Criswell, D. R. (1995), Lunar solar power: scale and cost versus technology level, boot-strapping, and cost of Earth-to-orbit transport, *46th Congress of the International Astronautical Federation*, IAF-95-R.2.02, 7 pp., Oslo.

Criswell, D. R. (1995a, July 31–August 4), Lunar solar power system: systems options, costs, and benefits to Earth, IECEC Paper No. AP-23, *Proc. 30th Intersociety Energy Conversion Engineering Conf.*, 1: 595–600.

Criswell, D. R. (1996, April/May), Lunar-solar power system: Needs, concept, challenges, pay-offs, *IEEE Potentials*: 4–7.

Criswell, D. R. (1997), Challenges of commercial space solar power, *48th Congress of the International Astronautical Federation*, IAA-97-R.2.04, 7 pp., Turin, Italy.

Criswell, D. R. (1997a), Lunar-based solar power and world economic development, *48th Congress of the International Astronautical Federation*, IAA-97-IAA.8.1.04, 6 pp., Turin, Italy.

Criswell, D. R. (1997b), Twenty-first century power needs, challenges, and supply options, *Proc. SPS '97 Space Solar Power Conf.*, 6 pp. (August 24–28).

Criswell, D. R. (1998), Lunar solar power for energy prosperity within the 21st century, *Proc. 17th Congress of the World Energy Council*, Div. 4: The Global Energy Sector: Concepts for a sustainable future: 277–289, Houston, (September 13–17).

Criswell, D. R. (1998b), Solar power system based on the Moon, in *Solar Power Satellites: A Space Energy System for Earth*: 599–621, Wiley-Praxis, Chichester, UK.

Criswell, D. R. (1998c), Lunar Solar Power: Lunar unit processes, scales, and challenges, 6 pp. (ms), *ExploSpace: Workshop on Space Exploration and Resources Exploitation*, European Space Agency, Cagliari, Sardinia (October 20–22).

Criswell, D. R. (1999), Commercial lunar solar power and sustainable growth of the two-planet economy. Special issue on the exploration and utilization of the Moon (Proc. 3rd International Lunar Conference), invited editor E. M. Galimov, *Solar System Research*, **33**, No. 5, 356–362.

Criswell, D. R. (2000), Lunar solar power: review of the technology readiness base of an LSP system, *Acta Astronautica*, **46**, 8, 531–540.

Criswell, D. R. and Thompson, R. G. (1996), Data envelopment analysis of space and terrestrial-based large scale commercial power systems for Earth: a prototype analysis of their relative economic advantages, *Solar Energy*, **56**, No. 1: 119–131.

Criswell, D. R. and Waldron, R. D. (1990), Lunar system to supply electric power to Earth, *Proc. 25th Intersociety Energy Conversion Engineering Conf.*, 1: 61–70.

Criswell, D. R. and Waldron, R. D. (1991), Results of analysis of a lunar-based power system to supply Earth with 20,000 GW of electric power, *Proc. SPS'91 Power from Space: 2nd Int. Symp.*: 186–193. Also in *A Global Warming Forum: Scientific, Economic, and Legal Overview*, Geyer, R. A., (editor) CRC Press, Inc., 638pp., Chapter 5: 111–124.

Criswell, D. R. and Waldron, R. D. (1993), International lunar base and the lunar-based power system to supply Earth with electric power, *Acta Astronautica*, **29**, No. 6: 469–480.

Criswell, D. R., Waldron, R. D., and Erstfeld, T. (1979), *Extraterrestrial Materials*

Processing and Construction, available on microfiche, National Technical Information Service, 450 pp.

Criswell, D. R., Waldron, R. D., and Erstfeld, T. (1980), *Extraterrestrial Materials Processing and Construction,* available on microfiche, National Technical Information Service, 500pp.

Deschamps, L. (editor) (1991), *SPS 91 Power from Space: Second International Symposium,* Société des Électriciens et des Électroniciens, Société des Ingénieurs et Scientifiques de France, Paris/Gif-Sur-Yvete (August 27–30), 641pp.

Dickens, G. R., (1999), The blast in the past, *Nature* **401**: 752–753.

Egbert, G. D. and Ray, R. D. (2000), Significant dissipation of tidal energy in the deep ocean inferred from satellite altimeter data, *Nature,* **405**: 775–778.

Ehrlich, P. R. and Roughgarden, J. (1987), *The Science of Ecology,* see Table 23–1 and p. 524–525, Macmillan Pub. Co.

ESA (European Space Agency) (1995), *Rendezvous with the new millennium: the report of ESA's Long-term Space Policy Committee,* (38–45), SP-1187 Annex, 108pp.

Feingold, H., Stancati, M., Friedlander, A., Jacobs, M., Comstock, D., Christensen, C., Maryniak, G., Rix, S., and Mankins, J. C. (1997), *Space solar power: a fresh look at the feasibility of generating solar power in space for use on Earth,* SAIC-97/1005, 321 pp.

Glaser, P., Davidson, F. P., and Csigi, K. (editors) (1998: rev. 2), *Solar Power Satellites,* Wiley, 654 pp.

Goeller, H. E. and Weinberg, A. M. (1976), The age of substitutability, *Science,* **191**: 683–689.

Haq, B. U. (1999), Methane in the deep blue sea, *Science,* **285**: 543–544.

Herzog, H., Eliasson, B., and Kaarstad, O. (2000, February), Capturing greenhouse gases, *Scientific American*: 72–79.

Hoffert, M. I. and Potter, S. D. (1997), Energy supply, Ch. 4: 205–259, in *Engineering Response to Global Climate Change,* (ed. R. G. Watts), CRC Press LCC.

Hoffert, M. I., Caldeira, K., Jain, A. K., Haites, E. F., Harvey, L. D., Potter, S. D., Schlesinger, M. E., Schneiders, S. H., Watts, R. G., Wigley, T. M., and Wuebbles, D. J., (1998), Energy implications of future stabilization of atmospheric CO_2 content, *Nature,* **395**: 881–884, (October 29).

ILEWG (1997), *Proc. 2nd International Lunar Workshop,* organized by: International Lunar Exploration Working Group, Inst. Space and Astronautical Science, and National Space Development Agency of Japan, Kyoto, Japan, (October 14–17), 89pp.

Isaacs, J. D. and Schmitt, W. R. (1980, 18 January), Ocean Energy: forms and prospects, *Science,* **207**, #4428: 265–273.

Johnson Space Center (1977), *Satellite Power System (SPS) Concept Evaluation Program,* NASA Johnson Space Center, (July). (See also General Dynamics 6th monthly Lunar Resources Utilization progress report, 1978).

Johnson Space Center (1978), *A recommended preliminary baseline concept,* SPS concept evaluation program, (January 25). (See also *Boeing SPS System Definition Study, Part II.* Report No. D180–22876 (December, 1977).

Kennett, J. P., Cannariato, K. G., Hendy, I. L., and Behl, R. J. (2000), Carbon isotopic evidence for methane hydrate instability during Quaternary Interstadials, *Science,* **288**, 128–133.

Kerr, R. A. (2000), Globe's "missing warming" found in the ocean, *Science,* **287**: 2126–2127.

Kerr, R. A. (2000a), Draft report affirms human influence, *Science*, **288**: 589–590.

Kerr, R. A. (2000b), Missing mixing found in the deep sea, Science, **288**: 1947–1949.

Kessler, D. and Loftus, Jr., J. P. (1995), Orbital debris as an energy management problem, *Adv. Space Research*, **16**: 39–44.

Klimke, M. (1997), New concepts of terrestrial and orbital solar power plants for future European power supply, in *SPS'97: Conference on Space Power Systems, Energy and Space for Mankind*, 341 pp., Canadian Aeronautics and Space Inst. (Montreal) and Société des Electriciens et Electroniciens (France), pp. 67–72.

Krakowski, R. A. and Wilson R. (2002), What nuclear power can accomplish to reduce CO_2 emissions, Chapter 8 in *Proc. Innovative Energy Systems and CO_2 Stabilization: Aspen Global Climate Change Workshop* (editor R. Watts), (July 12–24 1998).

Landis, G. (1997), A Supersynchronous Solar Power Satellite, *SPS-97: Space and Electric Power for Humanity*: 327–328, Montreal, Canada, (24–28 August).

Levitus, S., Antonov, J. I., Boyer, T. P., and Stephens, C. (2000, March 24), Warming of the world ocean, *Science*, **287**: 2225–2229.

Lewis, J. (1991), Extraterrestrial sources for ^{3}He for fusion power, *Space Power*, **10**: 363–372.

Lewis, J. (1991a), Construction materials for an SPS constellation in highly eccentric Earth orbit, *Space Power*, **10**: 353–362.

Lloyd, A. C. (1999, July), The power plant in your basement, *Scientific American*, 80–86.

Loftus, J. P. (1997), personal communication, NASA-Johnson Space Center.

Macauley, M. K., Darmstadter, J., Fini, J. N., Greenberg, J. S., Maulbetsch, J. S., Schaal, A. M., Styles, G. S. W., and Vedda, J. A. (2000), *Can power from space compete: a discussion paper*, 32 pp., Resources for the Future, Washington, D.C.

Margot, J. L., Campbell, D. B., Jurgens, R. F., and Slade, M. A. (1999), Topography of the lunar poles from radar interferometry: a survey of cold trap locations, *Science*, **284**: 1658–1660.

Matsuoka, H. (1999), Global environmental issues and space solar power generation: promoting the SPS 2000 project in Japan, *Technology in Society*, 21: 1–7.

Miller. R. (1979), *Extraterrestrial Materials Processing and Construction of Large Space Structures*, NASA Contract NAS 8–32935, NASA CR-161293, Space Systems Lab., MIT, 3 volumes.

Moore, T. (2000, Spring), Renewed interest in space solar power, *EPRI Journal*, pp. 6–17.

Nakicenovic, N., Grubler, A, and McDonald, A. (editors) (1998), *Global Energy Perspectives*, 299 pp., Cambridge University Press.

NASA (1988), *Lunar Helium-3 and Fusion Power*, NASA Conf. Pub. 10018, 241 pp., (25–26 April).

NASA (1989), *Lunar Energy Enterprise Case Study Task Force*, NASA TM-101652.

NASA (1999, 23 April), *Space Solar Power: Exploratory Research and Technology (SERT) Offerors Briefing NRA 8–32*, nais.msfc.nasa.gov/home.html, NASA Marshall Space Flight Center, 77 pp.

Nishida, K., Kobayashi, N., and Fukao, Y. (2000), Resonant oscillations between the solid Earth and the Atmosphere, *Science*, **287**: 2244–2246.

Norris, R. D. and Röhl, U. (1999), Carbon cycling and chronology of climate warming during the Palaeocene/Eocene transition, *Nature*, **402**: 775–782.

O'Neil, G. K. (1975), Space Colonies and energy supply to the Earth, *Science*, **190**: 943–947.

Postel, S. L., Daily, G. C., and Ehrlich, P. R. (1996), Human appropriation of renewable fresh water, *Science*, 271: 785–788.

Rees, W. E. and Wachernagel, M. (1994), Appropriated carrying capacity: Measuring the natural capital requirements of the human economy, in *Investing in Natural Capital: The Ecological Economic Approach to Sustainability*, A. M. Jansson, M. Hammer, C. Folke, and R. Costanza (eds.) Island Press, Washington, D.C. pp. 362–390.

Shlyakhter, A., Stadie, K., and Wilson, R. (1995), *Constraints Limiting the Expansion of Nuclear Energy*, United States Global Strategy Council, Washington, D.C., 41 pp.

Smil, V. (1994), *Energy in World History*, Westview Press, Boulder, CO., 300 pp.

Spudis, P. D. (1996), *The Once and Future Moon*, 308 pp., Smithsonian Inst. Press.

Stafford, T. (1991), *America at the Threshold: Report of the Synthesis Group on America's Space Exploration Initiative*, 181 pp., Washington, D.C., Government Printing Office.

Starr, C. (1990), Implications of continuing electrification, (p. 52–71), in *Energy: Production, Consumption, and Consequences*, National Academy Press, Washington, D.C., 296 pp.

Stevens, W. K. (1999), Lessons from ancient heat surge, *New York Times*, 23 November, D3.

Strickland, J. K. (1996), Advantages of solar power satellites for base load supply compared to ground solar power, *Solar Energy*, 56, No. 1: 23–40. (See also Glaser *et al.*, 1998, Ch. 2.5).

Sullivan, T. A. and McKay, D. S. (1991), *Using Space Resources*, pp. 11–12, NASA Johnson Space Center, 27 pp.

Trinnaman, J. and Clarke, A. (editors) (1998), *Survey of Energy Resources 1998*, World Energy Council, London, 337 pp.

Twidell, J. and Weir, T. (1986), *Renewable Energy Resources*, 439 pp., Spon, London. OTEC Note 1: Under an ideal Carnot cycle, for Delta $T = 20\,°C$, 5.7×10^{12} tons of deep cold water will extract 20 TWt-y of energy from the warm surface waters. Real engines extract less than 25% of this energy. Losses in pumping sea water ($\sim$25%), turbine losses ($\sim$50%) to convert mechanical to electrical power, and systems losses ($\geq$25%) decrease efficiency further. To a first approximation 20 TWe will require $\geq 1 \times 10^{15}$ tons/y of deep waters.

Waldron, R. D. and Criswell, D. R. (1995), Overview of Lunar Industrial Operations, *Proc. of the 12th Symposium on Space Nuclear Power and Propulsion: The Conference on Alternative Power from Space*, AIP Conf. Proc. 324, Part Two: 965–971.

Waldron, R. D. and Criswell, D. R. (1998), *Costs of space power and rectennas on Earth*, IAF-98-R.4.03: 5, Melbourne, Australia, (2 October).

Wan, Y.-H. and Parsons, B. K. (1993, August), *Factors relevant to utility integration of intermittent renewable technologies*, NREL/TP-463–4953, National Renewable Energy Laboratory, 106pp.

Watts, R. G., (1985), Global climate variation due to fluctuations in the rate of deep water formation, *J. Geophysical Res.*, 95, No. D5: 8067 – 8070, (August 20).

WEC (World Energy Council) (1993), *Energy for Tomorrow's World*, St. Martin's Press, 320 pp.

WEC (World Energy Council) (1998), *17th Congress of the World Energy Council – Roundup*, Houston, TX, (September 10–17).

WEC (World Energy Council) (2000), see www.wec.co.uk/wec-geis/publications/

Weingartner, S. and Blumenberg, J. (1995), Solar power satellite-life-cycle energy recovery considerations, *Acta Astronautica*, **35**, No. 9–11: 591–599.

For the 0.005 TWe SSPS that utilizes crystalline solar cells (50 micron thickness) the breakdown of energy inputs are: PV production = 1628 kW-h/kg; other SSPS components = 531 kW-h/kg; on-orbit installation = 177 kW-h/kg; and transport to space = 708 kW-h/kg. The theoretical minimum for transport from the Earth to LEO ($\sim$1000 km altitude) is $\sim$10 kWt-h/kg. Rockets place approximately 5% of their propellant energy into the payload on orbit. This implies $\sim$200 kWt-h/kg of launch energy. Total energy input for this 0.05 TWe satellite is estimated to be $\sim$177TW-h and for the 0.05 TWe rectenna $\sim$25 TW-h. A 0.005 TWe satellite using amorphous silicon solar cells requires less input energy. An SSPS designed for GaAlAs photocells requires an input of 50 TW-h. Weingartner and Blumenberg also estimate the energy input for a terrestrial array of photovoltaics that feed power, when produced, into an existing grid. An energy payback time of 42 to 86 months is predicted. However, a stand alone terrestrial array to supply a region would be far larger and likely never pay back its energy of production and maintenance. See Strickland (1996) and Hoffert and Potter (1997).

Wood, L., Ishikawa, M., and Hyde, R. (1998), Global warming and nuclear power, UCRL-JSC131306 (preprint), Workshop on Innovative Energy Systems and CO_2 stabilization, Aspen, (July 14–24), 20 pp.

Wunsch, C. (2000), Moon, tides and climate, *Nature*, **405**: 743–744, (15 June).

Yasaka, T., Hanada., T., and Matsuoka, T. (1996), Model of the geosynchronous debris environment, *47th Cong. Intern. Astron. Fed.*, IAF-96–IAA.6.3.08, 9 pp.

10

Geoengineering the Climate: History and Prospect[1]

10.1 Introduction

The possibility of using geoengineering – the deliberate manipulation of the planetary environment – to counteract anthropogenic climate change is deeply controversial. At least in name, geoengineering has largely been ignored in recent climate assessments (Bruce, Lee *et al.*, 1996; Watson, Zinyowera *et al.*, 1996). Under close examination, however, the distinction between geoengineering and other responses to the CO_2-climate problem proves to be fuzzy. Use of the term geoengineering is shifting, as advocates of response strategies that were formerly labeled geoengineering now seek to avoid the term. Section 10.2 elaborates a definition of geoengineering; assessment of the implications of its shifting meaning are deferred to the concluding discussion.

Historical perspective is vital to understanding the role of geoengineering in human choices about climate. The historical background sketched in Section 10.3 shows that proposals to engineer the climate are deeply woven into the history of the CO_2-climate problem. The focus is on the postwar rise of weather and climate modification and the interweaving of its decline with rising concern about inadvertent climate modification. The evolving status of geoengineering as a response to anthropogenic climate change is examined through a review of US climate assessments and the IPCC assessment reports.

Section 10.4 reviews proposals to geoengineer the climate. Structure for the review is provided by a taxonomy of anthropogenic climate modification that includes geoengineering to counter anthropogenic climate forcing as a special case. Whereas the structure is broad, treatment of detailed proposals focuses on recent work that was not covered by previous reviews of geoengineering (Keith and Dowlatabadi, 1992; Flannery, Kheshgi *et al.*, 1997; Kitzinger and

[1] This article originally appeared under the same title in *Annual Reviews of Energy and Environment*, **25**: 245–284 (2000). The two versions are identical excepting minor editorial corrections.

411

Frankel, 1998; Michaelson, 1998; Allenby, 1999). Recent developments include analysis of very low-mass scattering systems for altering planetary albedo (Teller, Wood *et al.*, 1997), climate model simulation of the effect of albedo geoengineering, improved scientific understanding of the role of iron as a limiting nutrient in oceanic ecosystems (Cooper, Watson *et al.*, 1996; Behrenfeld and Kolber, 1999), and speculation about the use of genetically modified organisms to enhance carbon sinks (Reichle, Houghton *et al.*, 1999).

Frameworks for assessing geoengineering are surveyed in Section 10.5; they include economics, risk and uncertainty, politics and law, and environmental ethics. Finally, the concluding section suggests that the fuzziness of the boundary demarcating geoengineering from conventional mitigation arises from deep uncertainties about the appropriate extent of deliberate human management of global biogeochemical systems. Although most geoengineering methods may reasonably be viewed as marginal to the debate about climate change, the failure of modern assessments to consider the implications of geoengineering has encouraged avoidance of questions about the appropriate extent of deliberate planetary management – questions that warrant serious debate.

10.2 Defining Geoengineering

10.2.1 Etymology and definition

In this review *geoengineering* is defined as intentional large-scale manipulation of the environment. Scale and intent play central roles in the definition. For an action to be geoengineering, the environmental change must be the primary goal rather than a side effect, and the intent and effect of the manipulation must be large in scale; e.g., continental to global. Two examples serve to demonstrate the roles of scale and intent. First, intent without scale. Ornamental gardening is the intentional manipulation of the environment to suit human desires, yet it is not geoengineering because neither the intended nor realized effect is large-scale. Second, scale without intent. The modification of global climate due to increasing atmospheric CO_2 has global effect, yet it is not geoengineering because it is a side effect resulting from combustion of fossil fuels with the aim of providing energy services.

Manipulations need not be aimed at changing the environment, but rather may aim to maintain a desired environmental state against perturbations – either natural or anthropogenic. Indeed, the term *geoengineering* has usually been applied to proposals to manipulate the environment with the goal of reducing undesired climate change caused by human influences. The focus of this review is likewise on climatic geoengineering, primarily – but not exclusively – to counter CO_2-induced climate change. In this usage geoengineering implies a countervailing measure or a "technical fix". As we will see, the definition of geoengineering is ambiguous, and the distinction between geoengineering and other responses to climate change is of degree, not of kind. Three core attributes will serve as markers of geoengineering: scale and intent, plus the degree to which the action is a countervailing measure.

The first use of *geoengineering* in approximately the sense defined above was by Marchetti in the early 1970s to describe the mitigation of the climatic impact of fossil fuel combustion by the injection of CO_2 into the deep ocean (Marchetti, 1977). The term entered the mainstream of debate about climate change during the last decade, particularly with publication of the 1992 NAS assessment (see Table 10.1).

Geoengineering is not found in standard dictionaries. In technical usage it has at least one other not wholly unrelated meaning, as a contraction of geotechnical engineering: the "science that deals with the application of geology to engineering". If the definition above is accepted, a fitting etymology is readily constructed: *geoengineering* as *geo-* from the Greek root *ge* meaning earth and *engineering* meaning "the application of science to the optimum conversion of the resources of nature to the uses of humankind".

10.2.2 Geoengineering and carbon management

The long-term use of fossil energy without emissions of CO_2 is an energy path that may substantially lower the economic cost of mitigating anthropogenic climate change. I call the required technologies Industrial Carbon Management (ICM), defined as the linked processes of capturing the carbon content of fossil fuels while generating carbon-free energy products such as electricity and hydrogen and sequestering the resulting CO_2.

The distinction between ICM and geoengineering is both imprecise and interesting. In drawing the distinction we may first consider climatic geoengineering as a category of response to the CO_2-climate problem. Figure 10.1 shows a simple schematic of the climate problem for which the response strategies are mitigation, geoengineering, or adaptation. In this scheme geoengineering is any manipulation of the climate system that alters its response to anthropogenic forcing; and the status of ICM is unclear because it resembles both conventional mitigation and geoengineering.

The definition adopted here emerges from an elaboration of the three-part schematic. It permits a clear distinction between mitigation of fossil fuel consumption and mitigation of CO_2 emissions, and it draws the line between ICM and geoengineering at emission of CO_2 to the active biosphere. Figure 10.2

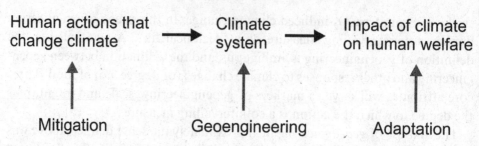

Figure 10.1   Three-part schema of the climate problem. The black arrows in the top row show the causal chain in this version of the anthropogenic climate problem. The gray arrows and the second row define the modes of intervention.

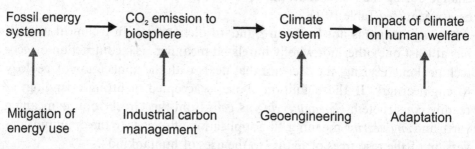

Figure 10.2   Four-part schema of the climate problem. The interpretation follows that of Figure 10.1. Note the distinction between mitigation of fossil energy use, carbon management and geoengineering that illustrated the definition described in Section 10.2.

shows a four-part schematic that illustrates the definition. It focuses on CO_2, ignoring other anthropogenic climate forcings, and distinguishes between control of CO_2 emissions to the active biosphere (ICM) and control of atmospheric CO_2 post-emission (geoengineering). The implications of this distinction are discussed in the concluding section of the review.

10.3 History

10.3.1 Introduction

While the term *geoengineering* is an invention of the last few decades, explicit consideration of intentional large-scale manipulation of the environment has a history measured in centuries. This review focuses on the post-World War II history of weather and climate modification as a direct precursor to current thinking about geoengineering. Modern understanding of the CO_2-climate problem emerged at a time when climate and weather modification was an important focus of science policy. Our aim is to explore the implications of this background for the treatment of proposals to employ countervailing measures in response to the CO_2-climate problem.

While the focus here is post-World War II, the link between scientific under-standing of the CO_2-climate connection and proposals for its manipulation extends to the beginning of the twentieth century. Writing around 1905, Arrhenius speculated about a virtuous circle in which CO_2 emissions from a growing fossil-fueled civilization would warm the climate, pushing back the northern limits of agriculture and so enhancing agricultural productivity as required to sustain the growth in population (Arrhenius, 1908). Similarly, Eckholm discussed the beneficial effects of elevated CO_2, including effects on both climate and on plant growth, and speculated about the possibility of climate modification via engineered enhancements of CO_2 emission (Ekholm, 1901).

The historical sketch presented here is necessarily incomplete, and its weak-nesses highlight the absence of a thorough historical treatment of deliberate climate modification. While there are modern intellectual histories of climate change (Feming, 1998), and treatments of climate and weather modification that date from the 1970s (Taubenfeld, 1970; Green, 1977), there is little modern analysis that explores the links between weather and climate modification and current concerns about climate change.

As we will see, "weather and climate modification" or "weather control" was a centerpiece of research in the atmospheric sciences during the 1950s and 60s, and was viewed as a priority by the governments of the United States and the USSR. In that context what are now called climate impacts was then called inadvertent climate modification; and, what is now called geoengineering bears a strong similarity to what was then called weather and climate modification.

We may ask, what degree of continuity exists between the older concerns about deliberate and inadvertent climate modification and current concerns about climate impacts and geoengineering? With respect to inadvertent climate modification the case for continuity is strong. Consider, for example, the NAS66 report titled *Weather and Climate Modification* (see Table 10.1 for definition of the NASxx style mnemonics). The report contains an excellent summary of the CO_2-climate problem in a chapter titled "Inadvertent Modification of Atmospheric Processes". This is the first extensive treatment of the climate problem in an NAS document, and it shares language and authorship with *Restoring the Quality of our Environment* (PSAC65), an early and influential assessment of the CO_2-climate problem.

The correspondence between the 1960s concern with weather and climate modification and current discussion of geoengineering is less precise in that the aim of weather and climate modification was "improvement" of the natural state or mitigation of natural hazards, whereas the aim of recent geoengineering proposals is the mitigation of anthropogenic hazards. Weather and climate modification therefore had two of the three defining attributes (Section 10.2.1) of geoengineering – scale and intent – but not the third, as it was not a countervailing measure. The case for continuity rests on the similarity of proposed technical methods, the continuity of citations to earlier work, a similarity of debate about legal and political problems, and finally, the strong resemblance of climate and weather modification to geoengineering as defined here.

10.3.2 USSR

In the USSR, sustained interest in weather modification predated WWII. Beginning with the establishment of Leningrad's Institute of Rainmaking in 1932, work on cloud modification moved outside the laboratory, with airborne cloud seeding experiments using calcium chloride beginning as early as 1934 and continuing until 1939 (Zikeev and Doumani, 1967). Work resumed immediately after the war with tests of cloud seeding using dry ice (1947) and silver iodide (1949). In the 1950s and early 1960s Soviet interest in climate and weather modification reached its zenith. A single experiment during the winter of 1960–61, for example, is reported to have cleared clouds over an area of 20000 km^2.

In the United States, despite common use of the phrase "weather and climate modification", the emphasis was almost entirely on weather control, particularly on the enhancement of precipitation. In contrast, in the USSR there was sustained interest in climate modification, although the bulk of the effort was likewise devoted to weather modification. Climate modification appears to have attracted significant government interest and research funding. In 1961, for example, the 22nd Congress of the Soviet Communist Party listed the development of climate-control methods among the most urgent problems of Soviet science (Fletcher, 1968).

Taking the 51 abstracts on climate modification cataloged by Zikeev as a guide (Zikeev and Doumani, 1967), we find that most of the work during this period addressed the possibility of climate change owing to hydrological modifications such as the construction of large reservoirs and the physical or chemical control of evaporation. There was also persistent interest in the grand project of removing the arctic sea ice to warm Russia. The analysis of the day showed that "the annihilation of the ice cover of the Arctic would be permanent: once destroyed it would never be re-established". (Joint Publications Research Service, 1963, p. 7).

Plans for global climate modification attracted occasional interest, perhaps the most extravagant being the proposals to place aerosol "Saturn rings" in earth orbit to heat and illuminate the polar regions. Independent proposals in 1958 and 1960 called for the injection of metallic aerosols into near-earth orbit to form rings that would supply heat and light to northern Russia or would shadow equatorial regions to provide their inhabitants with the supposed benefits of a temperate climate (Rusin and Flit, 1960).

The triumphant tone of Soviet thinking during the period is well captured in the concluding paragraph of *Man Versus Climate* (Rusin and Flit, 1960). "Our little book is now at an end. We have described those mysteries of nature already penetrated by science, the daring projects put forward for transforming our planet, and the fantastic dreams to be realized in the future. Today we are merely on the threshold of the conquest of nature. But if, on turning the last page, the reader is convinced that man can really be the master of this planet and that the future is in his hands, then the authors will consider that they have fulfilled their purpose."

In the absence of a thorough historical study, one may speculate about the roots of post-war Soviet interest in climate modification. Three preconditions seem relevant: (a) a social climate in which demonstration of technological power expressed in rapid industrial expansion and in the "space race" was central to state ideology, (b) a natural climate that is harsh by European standards, and finally, (c) the existence of relevant scientific expertise.

Discussions of inadvertent climate modification, and of the potentially harmful side effects of deliberate modifications, punctuate the Soviet literature on climate and weather modification as they did in the United States. For example, perhaps the earliest proposal to engineer a cooling to counter the climatic warming caused by industrial progress was made in 1964 (Zikeev and Doumani, 1967), roughly coincident with similar proposals in the United States.

10.3.3 USA

The 1946 discovery of cloud seeding by Schaefer and Langmuir (Schaefer, 1946) at the General Electric research labs ignited a commercial boom in weather modification.[2] Within five years private cloud seeding ventures had total annual receipts of $3–5 million, and in 1951 had targeted an area equal to 14% of the landmass of the lower 48 states (ACWC57, see Table 10.1). The boom rapidly attracted government attention with the first court case involving

---

[2] Contemporary documents, and more recent historical summaries, ignore prior work in the USSR.

liability for cloud seeding occurring in 1950, the first senate hearings in 1951, and the formation by congress of the Advisory Commission on Weather Control in 1953.

In the late 1950s weather modification became entangled in the politics of the cold war. Instead of regulating a growing industry, the focus became national security, and during the next decade the issue moved to the top drawer of national science politics. Apparently central to this transformation was growing knowledge of the Soviet effort in the area combined with concern about the possibility of superior Soviet scientific accomplishment marked by the launch of Sputnik in 1957.

Henry Houghton, the chair of the MIT meteorology department, summarized these fears in an influential 1957 address, "Man's material success has been due in large degree to his ability to utilize and control his physical environment. ... As our civilization steadily becomes more mechanized and as our population density grows, the impact of weather will become ever more serious. ... The solution lies in ... intelligent use of more precise weather forecasts and, ideally, by taking the offensive through control of weather." Of Soviet effort he said, "I shudder to think of the consequences of a prior Russian discovery of a feasible method for weather control. Fortunately for us and the world we were first to develop nuclear weapons . . . International control of weather modification will be as essential to the safety of the world as control of nuclear energy is now.",He concluded "Basic research in meteorology can be justified solely on the economic importance of improved weather forecasting but the possibility of weather control makes it mandatory." (Orville, 1957, Vol. II, p. 286).

During the 1960s federal support for weather and climate modification grew rapidly, reaching ~$10 million by the decade's end. A series of NAS and NSF reports echoed – and occasionally quoted – Houghton's claims, confirming the central importance of the topic in the atmospheric sciences and repeating concerns about Soviet leadership in the area (e.g., NAS66, see Table 10.1).

In the United States the focus was on weather, with large-scale climate modification receiving distinctly less attention than it did in the USSR. Occasional counter examples stand out, for example, in a 1958 paper in *Science*, the head of meteorological research at the United States weather bureau speculated about the use of nuclear explosives to warm the arctic climate via the creation of infrared reflecting ice clouds (Wexler, 1958).

By 1966 theoretical speculation about use of environmental modification as a tool of warfare (MacDonald, 1968) became realized as the United States began a campaign of cloud seeding in Vietnam that ultimately flew more than 2600 sorties and had a budget of ~3.6 $m/yr. Public exposure of the program in 1972 generated a rapid and negative public reaction, and lead to an international treaty, the "Convention on the Prohibition of Military or Any Other Hostile Use of Environmental Modification Techniques" (1976).

The gradual demise of weather modification after the mid-1970s may, arguably, be attributed to three forces: (a) backlash against the use of weather modification by the US military, (b) the growing environmental movement, and (c) the growing realization of the lack of efficacy of cloud seeding.

Beginning in the early 1960s, concerns about CO_2-induced climate change and other forms of inadvertent climate modification become interwoven with work on climate and weather modification. The gradual shift in concern is evident in National Academy documents charged with planning the research agenda for the atmospheric sciences and in the history of climate assessments that is the topic of Section 10.3.5.

10.3.4 Terraforming

Terraforming is "planetary engineering specifically directed at enhancing the capacity of an extraterrestrial planetary environment to support life" (Fogg, 1995). The topic is relevant to the assessment of geoengineering because the terraforming literature is remarkably broad. In addition to technical papers in mainstream scientific publications (Sagan, 1961; Sagan, 1973; McKay, Toon *et al.*, 1991), it includes popular fiction and work by environmental philosophers that examines the moral implications of planetary engineering (Hargrove, 1986). Though fragmentary, this work compliments the geoengineering literature, which is almost exclusively technical. They are linked by commonality of proposed technologies, ethical concerns, and by their ambiguous position between the realms of science fiction and reasoned debate about human use of technology.

Speculation about geoengineering – in the form of climate and weather control – and about terraforming both emerged in the 1950s during an era of technological optimism. The history of terraforming is well summarized by Fogg (1995). Both the concept of terraforming and the term itself originated in science fiction of the 1940s and 1950s. In 1961 a paper by Sagan in *Science* momentarily brought speculation about terraforming into the "respectable" scientific literature, with a suggestion that "planetary engineering" of Venus could be accomplished by seeding its clouds with photosynthetic microbes to liberate O_2 from CO_2 (Sagan, 1961). Another paper by Sagan in 1973 considered terraforming Mars via alteration of the polar cap albedo using dark dust or living organisms (Sagan, 1973). Beginning in the mid 1970s, a small community of research on and advocacy of terraforming grew around a nucleus of professional planetary scientists. While clearly at the margins of the scientific mainstream, the terraforming community has nevertheless been able to generate a remarkable continuity of dialogue.

Interestingly, the terraforming community has generated a more robust debate about ethical concerns than exists for geoengineering. Rolston and Callicott, for example, have separately attempted to integrate a value for extraterrestrial life into their separate conceptions of a terrestrial environmental ethic (Hargrove, 1986).

10.3.5 *Geoengineering in assessments*

Arguably the first high-level government policy assessment that stated the CO_2-climate problem in modern terms[3] was *Restoring the Quality of Our Environment*, issued in 1965 by Johnson's Science Advisory Committee (PSAC65). In concluding the section of the report devoted to climate, the sole suggested response to the "deleterious" impact of CO_2-induced climate change is geoengineering: "The possibilities of deliberately bringing about countervailing climatic changes therefore need to be thoroughly explored." The report continues with analysis of a scheme to modify the albedo by dispersal of buoyant reflective particles on the sea surface, concluding, "A 1% change in reflectivity might be brought about for about $500 million a year. Considering the extraordinary economic and human importance of climate, costs of this magnitude do not seem excessive." The possibility of reducing fossil fuel use is not mentioned.

It is noteworthy that the NAS report on climate and weather modification (NAS66), though it was written contemporaneously with PSAC65, does not suggest use of climate modification to counteract human impacts, although it does contain a fair summary of the CO_2-climate problem in its chapter on "Inadvertent Modification of Atmospheric Processes".

The *Study of Critical Environmental Problems* (SCEP70) and the subsequent *Study of Man's Impact on Climate* (SMIC71) (see Table 10.1 for references), both led by MIT during 1970–71, reflect a sharp break with the tone of optimism about technology that marks the meteorology assessments of the 1960s. Both reports include broad statements that exemplify the emerging environmental consciousness. SMIC, for example, notes the increasing demands

---

[3] The report combines analysis of atmospheric CO_2 content based on the then ~6 year record of accurate measurements with estimates of global fossil fuel combustion to estimate future concentrations. It then combines concentration estimates with early radiative convective models to estimate temperature change, and then compares that estimate to observed changes with consideration given to intrinsic climate variability. Finally, it speculates about possible impacts beyond temperature, e.g., CO_2 fertilization of plant growth.

"man" places on "fragile biological systems" and asks "How much can we push against the balance of nature before it is seriously upset?" Neither report devotes significant attention to possible responses to the CO_2-climate problem, although SCEP70 does note that reduction in fossil fuel consumption is the only solution and cites nuclear energy as the sole alternative. Neither report suggests countervailing measures (geoengineering). SMIC71 explicitly considers weather and climate modification as a potential environmental threat, noting that "like so many human endeavors, cloud seeding is showing evidence of unexpected side effects", and recommending "that an international agreement be sought to prevent large-scale[4] experiments in persistent or long-term climate modification".

The release of the NAS report *Energy and Climate* in (NAS77) coincided with an increasing federal research and assessment effort on the CO_2-climate issue centered at the Department of Energy. It marks the beginning of a continuing chain of NAS reports on the topic that are linked by shared authorship, and explicit cross-references (e.g., NAS79, NAS83, NAS92). Like PSAC65, the report linked projections of fossil fuel consumption with models of the carbon cycle and the climate to estimate future climate change. In contrast to SCEP70 and SMIC71, and like PSAC65, geoengineering was again on the agenda. The fourth of four "crucial" questions listed in the introduction to NAS77 is "What, if any, countervailing human actions could diminish the climatic changes or mitigate their consequences?" Several possibilities were examined, including fertilization of the ocean surface with phosphorus, engineered increases in planetary albedo (citing PSAC65), and massive afforestation with subsequent preservation of woody biomass. However, the report is less optimistic than PSAC65 about countervailing measures and concludes that mitigation via "increased reliance on renewable resources . . . will emerge as a more practical alternative". Though not given prominence, the report concludes its introductory statement of the "Nature of the Problem" with an implicit taxonomy of responses that presages the formal taxonomy in NAS92 seen in Figure 10.2: "If the potential for climate change . . . is further substantiated then it may be necessary to *(a)* reverse the trend in the consumption of fossil fuels. Alternatively, *(b)* carbon dioxide emissions will somehow have to be controlled or *(c)* compensated for." (Geophysics Study Committee 1977, p. 3).

Geoengineering in its most recent incarnation, as a means of counteracting CO_2-induced climate change, receives its most serious airing in the NAS reports of 1982 and 1992.

NAS83 articulated a general framework for understanding the implications

---

[4] Large-scale was specified as $>10^6$ km^2.

of climate change. The explicit aim of the framework was to broaden the debate beyond CO_2, to examine the spatial and temporal inequalities in the distribution of impacts, and finally to consider the problem dynamically over an extended timescale. The report considered measures of CO_2 control separately from countervailing climate modification. With respect to CO_2 control NAS82 notes the importance of the distinction between pre- and post-emission CO_2 control and discusses post-emission sequestration in terrestrial and oceanic ecosystems, including the burial of trees at sea to effect more permanent sequestration. With respect to countervailing measures NAS83 notes that "in principle weather and climate modification are feasible; the question is only what kinds of advances . . . will emerge over the coming century", and adds that "interest in CO_2 may generate or reinforce a lasting interest in national or international means of climate and weather modification; once generated, that interest may flourish independent of whatever is done about CO_2." Finally, NAS83 speculated about the political consequences arising from the possibility of unilateral action to engineer the climate.

In contrast to the NAS83 report, NAS92 made less effort in the direction of an overarching framework. Rather, it focused on detailed technical analysis and included a chapter titled *Geoengineering* that included detailed analysis of a diverse array of options. NAS92 did contain a brief three part taxonomy of response strategies like that presented in Figure 10.1, in which CO_2 capture from the atmosphere (post-emission) is considered geoengineering and in which sequestration of CO_2 from industrial systems is grouped with other methods of reducing emissions from the energy system. In a synthesis chapter, NAS92 heroically attempted a uniform comparison of the cost effectiveness of all mitigation options, including geoengineering and presented aggregate mitigation supply curves for many options – a comparison that has not since been repeated.

In the chapter titled "Geoengineering", NAS92 analyzed four options: reforestation, ocean fertilization, albedo modification, and removal of atmospheric chlorofluorocarbons. Multiple cost estimates were presented for reforestation, oceanic fertilization with iron, albedo modification with space-based mirrors or with aerosols in the stratosphere and troposphere. The chapter's introduction included a discussion of predictability and risk assessment, comparing the risk of geoengineering to the risk of inadvertent climate modification. A summary of steps toward further assessment suggested small-scale experiments and the study of side effects including consideration of reversibility and predictability. The chapter ends by observing that "perhaps one of the surprises of this analysis is the relatively low costs at which some of the geoengineering options might be implemented" and concluded that "this analysis does suggest that further inquiry is appropriate".

Beginning in the late 1980s the trickle of climate assessments became a flood.

A selected set of assessments is summarized in Table 10.1; most mention geo-engineering peripherally or not at all. I conclude this survey of geoengineering in assessments with a summary of first and second assessment reports[5] (FAR and SAR) of the IPCC.

The FAR dealt with mitigation in the report of Working Group III (IPCC, 1990) whose sole charge was to "formulate response strategies". The report adopts an abstract tone and contains little detailed economic or technical analysis. Neither the FAR nor SAR include a general framework for categorizing of response strategies as was done in the NAS studies of 1977, 1982, and 1992. The FAR mentions the possibility of "CO_2 separation and geological or marine disposal" as a long-term option but does not describe the possibility further. Enhancement of natural carbon sinks is discussed only for forestry, and as an aside to a more detailed discussion of preventing further emissions by slowing deforestation.

Working Groups were reorganized for the SAR, with WGII charged with scientific and technical analysis of mitigation. WGII treated enhancement of terrestrial sinks in separate chapters devoted to forests and agricultural lands, and covered capture and sequestration of industrial carbon emission in the chapter on mitigation in the energy sector. The WGII report included a three-page section (~0.3% of the report) on "Concepts for Counterbalancing Climate Change". The text is primarily descriptive, presenting a taxonomy[6] and review of geoengineering methods including enhancements to the oceanic carbon sink, alteration of albedo, and manipulation of feedback mechanisms. The SAR nowhere addresses the question of why enhancement of terrestrial carbon sinks is treated as mitigation while enhancement of oceanic sinks is treated as geoengineering. In contrast to NAS92 there is no attempt at cost estimation nor is there mention of broad ethical implications of geoengineering. Risks and uncertainties are stressed, but again in contrast to NAS92, no general heuristics for assessing risk are mentioned.[7] Despite the absence of any cost calculations or attempts at risk assessment, the WGII report and the SAR "Summary for Policy Makers" concludes that geoengineering is "likely to be ineffective, expensive to sustain and/or to have serious environmental and other effects that are in many cases poorly understood".

For the SAR, WGIII was charged with assessing the socio-economic

---

[5] The author is a contributor to the section of the Third Assessment Report that is provisionally titled "Biological uptake in oceans and fresh-water reservoirs; and geo-engineering". While technical details of some of the new work described here in Section 10.3.5 will be included, the report will not significantly improve on the SAR with respect to assessment of geoengineering.

[6] The first two elements of the SAR's four-part taxonomy are identical to "albedo" and "emissivity" categories used here (Figure 10.3). The third element SAR's taxonomy covers all of the "energy transport" category. The fourth element, "counteracting the harmful effects of changes that do occur" represents a different view of the problem from that presented here.

[7] For example, comparison of the magnitude of natural to engineered effect.

Table 10.1 *Selected climate assessments. Note the definition of the NASxx style mnemonics used in the text. The notes are focused on the treatment of geoengineering*

Mnemonic	Title	Notes
ACWC57	Advisory Committee on Weather Control (Orville, 1957)	Efficacy of weather control; legal implications; peripheral mention of deliberate and inadvertent climate modification.
NAS66	Weather and Climate Modification: problems and prospects (Committee on Atmospheric Sciences, 1966)	Focus on weather, but extended discussion of deliberate and inadvertent climate modification including the CO_2-climate problem.
PSAC65	Restoring the quality of our environment (President's Science Advisory Committee, 1965)	Seminal modern statement of the CO_2-climate problem. Countervailing measures were the *only* mitigation method considered.
SCEP70	Study of Critical Environmental Problems (Study of Critical Environmental Problems, 1970)	Detailed examination of CO_2-climate problem as one of a small set of critical environmental problems.
SMIC71	Study of Man's Impact on Climate: Inadvertent Climate Modification (Study of Critical Environmental Problems, 1970)	Very little on mitigation. Concern for the impacts of weather modification.
NAS77	Energy and Climate (Geophysics Study Committee, 1977)	Stressed importance of limiting fossil emissions and of understanding countervailing measures.
NAS79	Carbon Dioxide and Climate: A Scientific Assessment (National Research Council, 1979)	Focus on estimating climate sensitivity; mitigation was not addressed.
NAS83	Changing Climate (Carbon Dioxide Assessment Committee, 1983)	General framework of responses to climate change includes countervailing measures and CO_2 control. General discussion of methods; little technical analysis.
EPA83	Can we delay a greenhouse warming? (Seidel and Keyes, 1983)	Focus on mitigation in energy sector. Terrestrial sequestration, countervailing measures, and ocean CO_2 injection covered as "Nonenergy options".

Table 10.1 (*cont.*)

Mnemonic	Title	Notes
NAS92	Policy Implications of Greenhouse Warming (Panel on Policy Implications of Greenhouse Warming, 1992)	Included a chapter titled *geoengineering* that considered many options and attempted to estimate marginal CO_2-equivalent mitigation costs.
EPA90	Policy Options for Stabilizing Global Climate (Lashof and Tirpak, 1990)	Only mitigation in energy sector considered.
IPCC90	The IPCC Response Strategies (Intergovernmental Panel on Climate Change, 1991)	No general framework for response strategies; CO_2 capture and enhancing forest sinks get minor mentions.
OTA91	Changing by Degrees: Steps to Reduce Greenhouse Gases (Office of Technology Assessment, 1991)	Focus on mitigation in energy sector. Analysis of carbon sequestration by afforestation.
IPCC95	Climate Change 1995: Impacts, Adaptation, and Mitigation of Climate Change: Scientific-Technical Analysis (Watson, Zinyowera *et al.*, 1996)	No general framework for response strategies; terrestrial sinks covered extensively, oceanic sinks and geoengineering mentioned peripherally.

dimensions of climate change and was specifically instructed to "be comprehensive, cover(*ing*) all relevant sources, sinks and reservoirs of greenhouse gases". The report, however, contains no analysis of geoengineering *per se*. It briefly mentions sequestration of carbon from industrial sources, but does not address any socio-economic implications of issues raised by those technologies, such as the gradual re-release of sequestered carbon. The possible enhancement of ocean carbon sinks is not addressed, and while enhancement of terrestrial sinks is considered, little discussion of the social, economic, and biological consequences of the enhancement is presented.

10.4 Taxonomy and Review of Proposals to Manipulate the Climate

10.4.1 Taxonomy

The myriad proposals to geoengineer the climate may usefully be classified by their mode of action (Figure 10.3). The root division is between alteration of

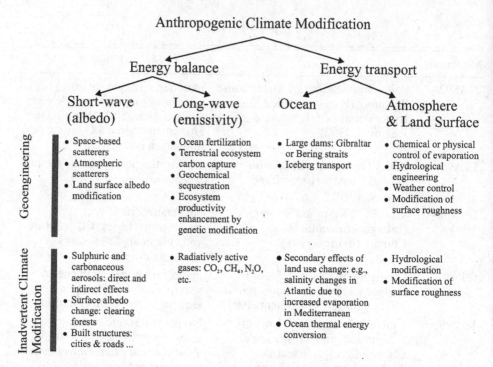

Anthropogenic Climate Modification

Figure 10.3   Taxonomy of climate modification. The taxonomy organizes the modes of climate modification – equivalently, possibilities for anthropogenic forcing of climate – both deliberate and inadvertent. The modes of climate modification listed as geoengineering have been proposed with the primary aim of climate modification. Note that some modes appear both as geoengineering and as inadvertent climate modification.

radiative fluxes to regulate the net thermal energy input and alteration of the internal dynamics that redistribute energy in the climate system.[8] The overwhelming majority of geoengineering proposals aim to alter radiative energy fluxes, either by increasing the amount of outgoing infrared radiation through reduction of atmospheric CO_2, or by decreasing the amount of absorbed solar radiation through an increase in albedo. With more generality we subdivide alteration of radiative energy fluxes into alteration of thermal (long-wave) radiation or solar (short-wave) radiation. Proposals to alter internal dynamics have focused on the oceans or on surface/atmosphere interaction, and are subdivided accordingly in Figure 10.3. Here we focus on the means of climate modification in general – either deliberate or inadvertent – whereas the categor-

---

[8] Some treatments use a forcing/feedback division in place of the energy-inputs/internal-dynamics division used here (Watson, Zinyowera *et al.*, 1996), however this is not as precise since internal feedbacks (e.g., ice-albedo feedback) modify the energy input.

ization illustrated in Figure 10.2 describes responses to anthropogenic climate change. Figure 10.3 emphasizes this point by including a classification of human impacts on climate to stress the strong relationship between impacts and geoengineering.

With respect to geoengineering aimed at countering CO_2-induced global change, there is a fundamental difference between controlling CO_2 and controlling its effects. Albedo modification schemes aim to offset the effect of increasing CO_2 on the global radiative balance, and thus on average surface temperatures; climatic change may, however, still occur owing to changed vertical and latitudinal distributions of atmospheric heating (Section 10.4.2.1). Moreover, increased CO_2 has substantial effects on plant growth independent of its effect on climate – an effect that cannot be offset by an increase in albedo. In addition, modification of albedo using shields in space or in the stratosphere would reduce the sunlight incident on the surface. The possible effects of this reduction on ecosystem productivity have not been examined.

10.4.2 Energy balance: albedo

10.4.2.1 Aim and effect of albedo modification

It has long been suggested that albedo geoengineering aimed at countering the climatic effects of increased CO_2 would produce significant alterations in climate even if perfect control of mean surface temperature were achieved (Keith and Dowlatabadi, 1992; Dickinson, 1996; Schneider, 1996). A recent numerical experiment, however, has demonstrated that modification of albedo can compensate for increased CO_2 with remarkable fidelity.

Govindasamy and Caldeira (2000) tested the effects of albedo geoengineering using a high quality model known to do a good job of simulating the global radiative balance.[9] They compared a control case with two tests cases, one with $2 \times CO_2$ and the other a "geoengineering" case with $2 \times CO_2$ and a reduction of solar constant[10] by 1.8%. By design, the geoengineering case had (almost) the same mean surface temperature as the control. Surprisingly, the spatial pattern of surface temperature showed little change despite the changed vertical and latitudinal distributions of atmospheric heating. Compared to the

---

[9] The model was version three of the Community Climate Model at a horizontal resolution of T42 with 18 vertical layers, run with interactive sea ice coupled to a slab ocean. The $2 \times CO_2$ climate sensitivity was 1.75°C in this configuration. The statistics presented below are derived from the last 15 years of a 40 year model run.

[10] Uniform modification of planetary albedo accomplished using scattering systems in space or in the stratosphere would produce a climatic effect equivalent to an alteration of the solar constant (the solar flux at the top of the atmosphere).

control, the geoengineering case produced statistically significant temperature changes over only 15% of the globe as compared to 97% for the $2 \times CO_2$ case. Contrary to expectations, there was very little change in the amplitude of the diurnal and seasonal cycles in the geoengineering case.

Although a single numerical experiment does not prove the case, it neverthe-less suggests that the climate is less sensitive to changes in the meridional dis-tribution of heating than is often assumed, and therefore the assumption that albedo geoengineering could not do an effective job of countering CO_2-induced climate change must be reexamined.

10.4.2.2 Atmospheric aerosols

Aerosols can increase albedo either directly by optical scattering or indirectly by acting as cloud condensation nuclei that increase the albedo and lifetime of clouds by decreasing the mean droplet size. The modification of climate via alteration of cloud and aerosol properties was first proposed in the 1950s (Section 10.3). The most famous early proposal was by Budyko who suggested increasing the albedo to counter CO_2-induced climate change by injecting SO_2 into the stratosphere where it would mimic the action of large volcanoes on the climate (Budyko, 1982). He calculated that injection of about 10^7 t per annum into the stratosphere would roughly counter the effect of doubled CO_2 on the global radiative balance. The NAS92 study showed that several technologically straightforward alternatives exist for injecting the required mass into the stratosphere at low cost.

As with other geoengineering proposals, deliberate and inadvertent climate modification are closely linked: anthropogenic sulfate aerosols in the tropo-sphere currently influence the global radiation budget by ~ 1 Wm^{-2} – enough to counter much of the effect of current anthropogenic CO_2.

Addition of aerosol to the stratosphere could have serious impacts, most notably, depletion of stratospheric ozone. Recent polar ozone depletions have clearly demonstrated the complexity of chemical dynamics in the stratosphere and the resulting susceptibility of ozone concentrations to aerosols. Although the possibility of this side effect has long been noted (Budyko, 1982; MacCracken, 1991; Keith and Dowlatabadi, 1992), no serious analysis has been published. In addition, depending on the size of particles used, the aerosol layer might cause significant whitening of the daytime sky. This side effect raises one of the many interesting valuation problems posed by geoengi-neering: How much is a blue sky worth?

Recent work by Teller et al. (Teller, Wood et al., 1997; Teller, Caldeira et al., 1999) has reexamined albedo geoengineering. In agreement with NAS92, Teller et al. found that 10^7 t of dielectric aerosols of ~ 100 nm diameter are sufficient to increase the albedo by ~1%, and suggested that use of alumina particles could minimize the impact on ozone chemistry. In addition, Teller *et al.* demonstrated that use of metallic or optically resonant scatterers could greatly reduce the total mass of scatterers required. Two configurations of metallic scatterers were analyzed: mesh microstructures and micro-balloons. Conductive metal mesh is the most mass efficient configuration.[11] In principle, only ~10^5 t of such mesh structures is required to achieve the benchmark 1% albedo increase. The proposed metal balloons have diameters of ~4 mm, are hydrogen filled, and are designed to float at altitudes of ~25 km. The required mass is ~10^6 t. Because of the much longer stratospheric residence time of the balloon system, the required mass flux (t/yr) to sustain the two systems is comparable. Finally, Teller *et al.* show that either system, if fabricated with aluminum, can be designed to have long stratospheric lifetimes yet oxidizes rapidly in the troposphere, ensuring that few particles are deposited on the surface.

It is unclear whether the cost of the novel scattering systems will be less than that of the older proposals, as is claimed by Teller *et al.*, because although the system mass is less, the scatterers will be more costly to fabricate. However, it is unlikely that cost would play any significant role in a decision to deploy stratospheric scatterers, because the cost of any such system is trivial compared to the cost of other mitigation options. The importance of the novel scattering systems is not in minimizing cost, but in their potential to minimize risk. Two of the key problems with earlier proposals were the potential impact on atmospheric chemistry, and the change in the ratio of direct to diffuse solar radiation with the associated visual whitening of the sky. The new proposals suggest that the location, scattering properties, and chemical reactivity of the scatterers could, in principle, be tuned to minimize both of these impacts.

10.4.2.3 Planetary engineering from space

Proposals to modify the climate using space-based technology reflect an extreme of confidence in human technological prowess. Fittingly, some of the grandest and earliest such proposals arose in the USSR immediately following the launch of Sputnik (Section 10.3.2). During the 1970s, proposals to generate solar power in space and beam it to terrestrial receivers generated substantial interest at NASA and among space technology advocates. Interest in the technology waned under the light of realistic cost estimates, such as the 1981 NAS analysis (Committee on Satellite Power Systems, 1981).

In principle, the use of space-based solar shields has significant advantages

---

[11] The thickness of the mesh wires is determined by the skin-depth of optical radiation in the metal (about 20 nm). The spacing of wires (~300 nm) must be ~1/2 the wavelength of scattered light.

over other geoengineering options. Because solar shields effect a "clean" alteration of the solar constant, their side effects would be both less significant and
more predictable than for other albedo modification schemes. For most plausible shield geometries, the effect could be eliminated at will. Additionally,
steerable shields might be used to direct radiation at specific areas, offering the
possibility of weather control.

The obvious geometry is a fleet of shields in low-earth orbit (NAS92).
However, solar shields act as solar sails and would be pushed out of orbit by
the sunlight they were designed to block. The problem gets worse as the mass
density is decreased in order to reduce launch costs. A series of studies published in 1989–92 proposed locating the shield(s) just sunward of the L1
Lagrange point between the Earth and the Sun where they would be stable with
weak active control (Early, 1989; Seifritz, 1989).

Teller *et al.* (Teller, Wood *et al.*, 1997) note that a scattering system at the L1
point need only deflect light through the small angle required for it to miss the
earth, about 0.01 rad as compared to ~1 rad for scatterers in near earth orbit
or in the stratosphere. For appropriately designed scattering systems, such as
the metal mesh described above, the reduced angular deflection requirement
allows the mass of the system to decrease by the same ratio. Thus, while a shield
at the L1 point requires roughly the same area as a near-earth shield, its mass
can be ~10^2 times smaller. Teller *et al.* estimate the required mass at ~3×10^3
t. The quantitative decrease in mass requirement suggested by this proposal is
sufficient to warrant a qualitative change in assessments of the economic feasibility of space-based albedo modification.

The cost of this proposal has not been seriously analyzed. An optimistic
order-of-magnitude estimate is 50–500 billion dollars.[12] Arguably, the assumptions about space technology that underlie this estimate could also make space
solar power competitive.

10.4.2.4 Surface albedo

The most persistent modern proposals for large-scale engineering of surface
albedo were the proposals to create an ice free Arctic Ocean to the supposed
benefit of northern nations (Section 10.3.2).

Modification of surface albedo was among the first geoengineering meas-

---

[12] Cost assessment is heavily dependent on expectations about the future launch costs. Current costs for
large payloads to low earth orbit (LEO) are just under 10 k$/kg. Saturn V launches (the largest launcher
ever used successfully) cost 6 k$/kg. The stated goal of NASA's current access to space efforts is to lower
costs to 2 k$/kg by 2010. This proposal requires ~30 launches of a Saturn V – approximately equal to the
cumulative total of payload lifted to LEO since Sputnik. We assume (i) that the transit to L1 can be
accomplished without large mass penalty (perhaps by solar sailing), and (ii) that average cost of hardware
is less than 10k$/kg.

ures proposed to counter CO_2-induced warming. For example, the possibility for increasing the oceanic albedo was considered in a series of US assessments (PSAC65, NAS77, and NAS92). Proposals typically involved floating reflective objects, however, "the disadvantages of such a scheme are obvious" (NAS77). They include contamination of shorelines, damage to fisheries, and serious aesthetic impacts.

Local modification of surface albedo accomplished by whitening of urban areas, can however, play an important role in reducing energy used by air conditioning and in adapting to warmer conditions.

10.4.3 Energy balance: emissivity

Control of long-lived radiatively active gases is the only important means of controlling emissivity.[13] We focus here on CO_2. Following the discussion in Section 10.2.2 above, we may usefully distinguish between (i) reduction in fossil fuel use, (ii) reduction in CO_2 emission per unit of fossil carbon used, and (iii) control of CO_2 by removal from the atmosphere. Following the discussion above (Section 10.2.2) we refer to these as conventional mitigation, carbon management, and geoengineering, respectively.

The distinction is sometimes made between technical and biological sequestration where the former is intended to label pre-emission sequestration and the latter post-emission. This labeling is imprecise, however, because there are proposals for non-biological capture from the atmosphere, and for pre-emission biological capture in engineered systems (Reichle, Houghton *et al.*, 1999).

10.4.3.1 Carbon capture in terrestrial ecosystems

The use of intensive forestry to capture carbon as a tool to moderate anthropogenic climate forcing was first proposed in the late 1970s (Dyson, 1977). It is now a centerpiece of proposals to control CO_2 concentrations under the Framework Convention on Climate Change, particularly under the Clean Development Mechanism. The focus of interest has moved beyond forests to other managed ecosystems such as croplands. There is an extensive literature on both the science and economics of such capture; the summary below aims to frame the issue with reference to geoengineering.

Four alternatives are considered for disposition of the carbon once captured. It may be (a) sequestered *in situ* either in soil or in standing biomass, (b)

---

[13] There is little opportunity to modify surface emissivity (typically values are 85–95% in the mid IR) and in any case modification has little effect since only a small fraction of surface radiation is transmitted to space. The main gas controlling atmospheric emissivity is water, but no direct means for controlling it have been proposed.

harvested and separately sequestered, (c) harvested and burned as fuel, or (d) harvested and burned as fuel with sequestration of the resulting CO_2.

In situ sequestration has been the focus of most of the FCCC-related analysis (Bruce, Lee *et al.*, 1996; Watson, Zinyowera *et al.*, 1996; Rosenberg, Izaurralde *et al.*, 1998). Uncertainty about the duration of sequestration is crucial. For example, recent analysis has demonstrated that changes in management of cropland, such as use of zero-tillage farming, can capture significant carbon fluxes in soils at low cost, but continued active management is required to prevent the return of carbon to the atmosphere by oxidation (Rosenberg, Izaurralde *et al.*, 1998). For both forest and cropland, uncertainty about the dynamics of carbon in these ecosystems limits our ability to predict their response to changed management practices or to climatic change, and thus adds to uncertainty about the duration of sequestration.

Sequestration of harvested biomass was considered in early analyses but has received little attention in recent work, perhaps because use of biomass as a fuel is a more economically efficient means to retard the increase in concentrations than is sequestration of biomass to offset fossil carbon emissions. Finally, biomass could be used to produce carbon-free energy (H_2 or electricity) with sequestration of the resulting CO_2 (IPCC95). This process illustrates the complexities of the definitions described above, because it combines pre- and post-emission capture and combines biological and technical methods.

Recent studies of carbon capture in cropland have identified the possible contributions of genetically modified organisms to achieving increases in carbon capture, and have stressed the importance of further research (Rosenberg, Izaurralde *et al.*, 1998). The US DOE research effort on sequestration currently includes genomic science as an important part of the sequestration research portfolio for both terrestrial and oceanic ecosystems (Reichle, Houghton *et al.*, 1999).

Use of terrestrial ecosystems to supply energy needs with minimal net carbon emissions – via any combination of sequestration to offset use of fossil fuels or via the use of biomass energy – will demand a substantial increase in the intensity and/or areal extent of land use. Whether captured by silviculture or agriculture, areal carbon fluxes are or order 1–10 tC/ha-yr. If the resulting biomass were used as fuel the equivalent energy flux would be 0.2–2 W/m², where the lower end of each range is for lightly managed forests and the upper end for intensive agriculture. Mean per-capita energy use in the wealthy industrialized world is ~5 kW. Thus about 1 hectare per capita would be required for an energy system based entirely on terrestrial carbon fixation, roughly equivalent to the current use of cropland and managed forest combined.

Is management of terrestrial ecosystems for carbon capture geoengineering?

As discussed in the concluding section, the ambiguity of the answer provides insight into shifting standards regarding the appropriate level of human intervention in global biogeochemical systems. Considering the defining attributes of *geoengineering* described in Section 10.2.1, we can describe a land management continuum in which, for example, land management that considers *in situ* carbon sequestration as one element in a balanced set of goals forms one pole of the continuum, and the large-scale extraction and separate sequestration of carbon from intensively irrigated and fertilized genetically modified crops forms the opposite pole. The land-use requirements discussed above suggest that manipulation of carbon fluxes at a level sufficient to significantly retard the growth of CO_2 concentrations would entail a substantial increase in the deliberate manipulation of terrestrial ecosystems. Put simply, enhancement of terrestrial carbon sinks with sufficient vigor to aid in solving the CO_2-climate problem is plausibly a form of geoengineering.

10.4.3.2 Carbon capture in oceanic ecosystems

Carbon can be removed from the atmosphere by fertilizing the "biological pump" which maintains the disequilibrium in CO_2 concentration between the atmosphere and deep ocean. The net effect of biological activity in the ocean surface is to bind phosphorus, nitrogen, and carbon into organic detritus in a ratio of $\sim 1:15:130$ until all of the limiting nutrient is exhausted. The detritus then falls to the deep ocean providing the "pumping" effect. Thus the addition of one mole of phosphate can, in principle, remove ~ 130 moles of carbon.[14]

The possibility of fertilizing the biological pump to regulate atmospheric CO_2 was discussed as early as the NAS77 assessment. At first, suggestions focused on adding phosphate or nitrate. Over the last decade it has become evident that iron may be the limiting nutrient over substantial oceanic areas (Watson, 1997; Behrenfeld and Kolber, 1999). The molar ratio Fe:C in detritus is $\sim 1:10000$, implying that iron can be a very efficient fertilizer of ocean-surface biota. Motivated in part by interest in deliberate enhancement of the oceanic carbon sink, two field experiments have tested iron fertilization *in situ*, and have demonstrated dramatic productivity enhancements over the short duration of the experiments (Martin, Coale *et al.*, 1994; Monastersky, 1995; Coale, Johnson *et al.*, 1998). However, it is not clear that sustained carbon removal is realizable (Peng and Broecker, 1991).

Ocean fertilization is now moving beyond theory. Recently, a commercial

---

[14] This ratio includes the carbon removed as $CaCO_3$ due to alkalinity compensation. This first order model of the biology ignores the phosphate–nitrate balance. Much of the ocean is nitrate limited. Adding phosphate to the system will only enhance productivity if the ecosystem shifted to favor nitrogen fixers. In many cases, nitrogen fixation may be limited by iron and other trace metals.

venture, Ocean Farming Incorporated, has announced plans to fertilize the ocean for the purpose of increasing fish yields and perhaps to claim a carbon sequestration credit under the emerging FCCC framework (Kitzinger and Frankel, 1998, p. 121–129).

Ocean fertilization may have significant side effects. For example, it might decrease dissolved oxygen with consequent increased emissions of methane – a greenhouse gas. In addition, any significant enhancement of microbiological productivity would be expected to alter the distribution and abundance of oceanic macro-fauna. These side effects are as yet unexamined.

10.4.3.3 Geochemical sequestration

On the longest timescales, atmospheric CO_2 concentrations are controlled by the weathering of magnesium and calcium silicates that ultimately react to form carbonate deposits on the ocean floor, removing the carbon from shorter timescale biogeochemical cycling. In principle, this carbon sink could be accelerated, for example, by addition of calcite to the oceans (Kheshgi, 1995) or by an industrial process that could efficiently form carbonates by reaction with atmospheric CO_2. I call this geochemical sequestration.

In either case, the quantity (in moles) of the required alkaline minerals is comparable to the amount of carbon removed. The quantities of material processing required make these proposals expensive compared to other means of removing atmospheric CO_2.

The most plausible application of geochemical sequestration is as a means to permanently immobilize carbon captured from fossil fuel combustion. Integrated power plant designs have been proposed, in which a fossil fuel input would be converted to carbon-free power (electricity or hydrogen) with simultaneous reaction of the CO_2 with serpentine rock (magnesium silicate) to form carbonates. Carbonate formation is exothermic; thus, in principle, the reaction requires no input energy. Ample reserves of the required serpentine rocks exist at high purity. The size of the mining activities required to extract the serpentine rock and dispose of the carbonate are small compared to the mining activity needed to extract the corresponding quantity of coal. The difficulty is in devising an inexpensive and environmentally sound industrial process to perform the reaction.

The importance of geochemical sequestration lies in the permanence with which it removes CO_2 from the biosphere. Unlike carbon that is sequestered in organic matter or in geological formations, once carbonate is formed the carbon is permanently removed. The only important route for it to return to active biogeochemical cycling is by thermal dissociation following the subduction of the carbonate-laden oceanic crust beneath the continents, a process with a time-scale of $>10^7$ years.

10.4.3.4 Capture and sequestration of carbon from fossil fuels

The CO_2 generated from oxidation of fossil fuels can be captured by separating CO_2 from products of combustion or by reforming the fuel to yield a hydrogen-enriched fuel stream for combustion and a carbon-enriched stream for sequestration. In either case, the net effect is an industrial system that produces carbon-free energy and CO_2 – separating the energy and carbon content of fossil fuels. The CO_2 may then be sequestered in geological formations or in the ocean.

Because the status of carbon management as geoengineering is ambiguous, and because there is now a large and rapidly growing literature on the subject (Herzog, Drake *et al.*, 1997; Parson and Keith, 1998), only a brief summary is included here despite its growing importance. Our focus will be on oceanic sequestration because it most clearly constitutes geoengineering (Section 10.2.2).

One may view CO_2-induced climate change as a problem of mismatched timescales. The problem is due to the rate at which combustion of fossil fuels is transferring carbon from ancient terrestrial reservoirs into the comparatively small atmospheric reservoir. When CO_2 is emitted to the atmosphere, atmosphere–ocean equilibration transfers ~80% of it to the oceans with an exponential timescale of ~300 years (Archer, Kheshgi *et al.*, 1997). The remaining CO_2 is removed with much longer timescales. Injecting CO_2 into the deep ocean accelerates this equilibration, reducing peak atmospheric concentrations. Marchetti used similar arguments in coining the term geoengineering in the early 1970s to denote his suggestion that CO_2 from combustion could be disposed of in the ocean (Marchetti, 1977). Oceanic sequestration is a technical fix for the problem of rising CO_2 concentrations; it is a deliberate planetary-scale intervention in the carbon cycle. It thus fits the general definition of geoengineering given above (Section 10.2) as well as the original meaning of the term.

The efficiency with which injected CO_2 equilibrates with oceanic carbon depends on the location and depth of injection. For example, injection at ~700 m depth into the Kuroshio current off Japan would result in much of the CO_2 being returned to the atmosphere in ~100 years, whereas injections that formed "lakes" of CO_2 in ocean trenches might more efficiently accelerate equilibration of the CO_2 with the deep-sea carbonates.

The dynamic nature of the marine carbon cycle precludes defining a unique static capacity, as may be done for geological sequestration. Depending on the increase in mean ocean acidity that is presumed acceptable, the capacity is of order $\sim 10^3$–10^4 gigatons of carbon (GtC), much larger than current anthropogenic emissions of ~ 6 GtC per year.

In considering the implications of oceanic sequestration one must note that – depending on the injection site – about 20% of the carbon returns to the atmosphere on the ~ 300 yr time-scale. Supplying the energy required for separating, compressing, and injecting the CO_2 would require more fossil fuel than would be needed if the CO_2 was vented to the atmosphere. Thus, while oceanic sequestration can reduce the peak atmospheric concentration of CO_2 caused by the use of a given amount of fossil-derived energy, it may increase the resulting atmospheric concentrations on timescales greater than the ~ 300 yr equilibration time.

10.4.4 Energy transport

The primary means by which humanity alters energy transport is by alteration of land surface properties. The most important influence is on hydrological properties, particularly through changes to surface hydrological properties and the rates of evapo-transpiration, but additionally via dams that create large reservoirs or redirect rivers. A secondary influence is on surface roughness via alteration of land use.

Inadvertent alteration of local and regional climate has already occurred due to alteration of land surface properties via either the means mentioned above or by alteration of albedo (Section 10.4.2.4). In addition, deliberate alteration of local microclimates is a common feature of human land management. Despite the long record of speculation about the alteration of surface properties with the intention of altering regional or global climate, it seems highly unlikely that geoengineering will ever play an important role in land management given the manifold demands on land use and the difficulty of achieving such large-scale alterations.

Other means of altering energy transport are more speculative. Examples include weather modification and redirection of ocean currents using giant dams. In principle, the direct application of mechanical work to alter atmospheric motions offers an energetically efficient means of weather modification, however, no practical means of applying such forces is known. Alternatively, weather modification may be accomplished by cloud seeding. Despite very large cumulative research expenditure over its long history, cloud seeding has demonstrated only marginal effectiveness. Accurate knowledge of the atmospheric state and its stability could permit leverage of small, targeted perturbations to effect proportionately larger alterations of the atmospheric dynamics.

The small perturbations could be effected by cloud seeding or direct application of thermal or mechanical energy. The increasing quality of analysis/forecast systems and the development of effective adjoint models that allow accurate identification of dynamic instabilities suggest that the relevant predictive capability is emerging.

10.5 Evaluating Geoengineering

Geoengineering is not now seriously incorporated into formal assessments of anthropogenic climate change (e.g., the IPCC). More specifically, (a) the word geoengineering is rarely used, (b) the methods defined here as geoengineering are generally not discussed (with the salient exception of biological sequestration), and finally, (c) the implications of deliberate planetary-scale environmental management are not seriously addressed. Where geoengineering is discussed, the focus is typically technical with scant consideration of broader implications such as the appropriate legal and political norms for decision-making and the distribution of risks and benefits. Here I review various framings for the assessment of geoengineering, but make no attempt at an overall synthesis.

10.5.1 Economics

The simplest economic metric for geoengineering is the cost of mitigation (COM), where mitigation of any kind is measured as an equivalent quantity of carbon removed from the atmosphere. Table 10.2 comprises a summary of COM for geoengineering computed in accord with the NAS92 methodology. The costs are highly uncertain. For albedo modification schemes additional uncertainty is introduced by the somewhat arbitrary conversion from albedo change to equivalent reduction in CO_2, which depends on assumptions about the climate's sensitivity to increased CO_2 and on the atmospheric lifetime of CO_2. The estimated COM varies by more than two orders of magnitude between various geoengineering methods. It is noteworthy that, for some methods, particularly albedo modification the costs are very low compared to emissions abatement.[15]

In principle, the COM permits a direct comparison among geoengineering methods and between geoengineering and abatement. In practice, differences in the distribution of costs and benefits as well as the non-monetary aspects of

---

[15] While they vary greatly, conventional economic models generally put the marginal COM between 100 and 500 $/tC for abatement beyond 50% of current emissions (Panel on Policy Implications of Greenhouse Warming, 1992; Watson, Zinyowera, *et al.*, 1996).

Table 10.2 *Summary comparison of geoengineering options*

Geoengineering Method	COM*	Technical Uncertainties	Risk of Side Effects	Non-technical Issues
Injection of CO_2 into the ocean.	50–150	Costs are much better known than for other geoengineering schemes. Moderate uncertainty about fate of CO_2 in ocean.	Low risk. Possibility of damage to local benthic community.	Like abatement this scheme is local with costs associated with each source. Potential legal and political concerns over oceanic disposal.
Injection of CO_2 underground.	50–150	Costs are known as for CO_2 in ocean; less uncertainty about geologic than oceanic storage.	Low risk.	Is geologic disposal of CO_2 geoengineering or a method of emissions abatement?
Ocean fertilization with phosphate	3–10	Uncertain biology: can ecosystem change its P:N utilization ratio? Is there significant long-term carbon captive?	Moderate risk. Possible oxygen depletion may cause methane release. Changed mix of ocean biota.	Legal concerns: Law of the Sea, Antarctic Treaty. Liability concerns arising from affect on fisheries; N.B. fisheries might be improved.
Ocean fertilization with iron	1–10	Uncertain biology: when is iron really limiting? Is there significant long-term carbon captive?	Moderate risk. Possible oxygen depletion may cause methane release. Changed mix of ocean biota.	Legal concerns: Law of the Sea, Antarctic Treaty. Liability concerns arising from effect on fisheries; N.B. fisheries might be improved
Intensive forestry to capture carbon in harvested trees.	10–100	Uncertainty about rate of carbon accumulation, particularly under changing climatic conditions.	Low risk. Intensive cultivation will impact soils and biodiversity.	Political questions: how to divide costs? Whose land is used?

	COM*			
Solar shields to generate an increase in the Earth's albedo.	0.05–0.5	Costs and technical feasibility are uncertain. Uncertainty dominated by launch costs.	Low risk. However, albedo increase does not exactly counter the effect of increased CO_2.	Security, equity and liability if system used for weather control.
Stratospheric SO_2 to increase albedo by direct optical scattering.	$\ll 1$	Uncertain lifetime of stratospheric aerosols.	High risk. Effect on ozone depletion uncertain. Albedo increase is not equivalent to CO_2 mitigation.	Liability: ozone destruction.
Tropospheric SO_2 to increase albedo by direct and indirect effects.	<1	Substantial uncertainties regarding aerosol transport and its effect on cloud optical properties.	Moderate risk: unintentional mitigation of the effect of CO_2 already in progress.	Liability and sovereignty because the distribution of tropospheric aerosols strongly effects regional climate.

Note:
* Cost of mitigation (COM) is in dollars per ton of CO_2 emissions mitigated. While based on current literature, the estimates of risk and cost are the author's alone.

geoengineering render such cost comparison largely irrelevant to real decisions about abatement.

Examination of the shape of the marginal COM functions provides an insightful comparison between geoengineering and abatement. Although the COM is uncertain, there is much less doubt about how the COM scales with the amount of mitigation required. First, consider conventional mitigation. Whereas econometric and technical methods for estimating the cost of moderate abatement differ, both agree that costs will rise steeply if we want to abate emissions by more than about 50% (between 100 and 500 $/tC). (Watson, Zinyowera *et al.*, 1996). In sharp contrast, geoengineering the planetary albedo has marginal costs that, although highly uncertain, are roughly independent of, and probably decrease with, the amount of mitigation required.[16] In particular, the COM for albedo modification will not rise steeply as one demands 100% abatement because the process of albedo modification has no intrinsic link to the scale of current anthropogenic climate forcing. One could, in principle, engineer an albedo increase several times larger than the equivalent anthropogenic forcing and thus cool the climate. These relationships are illustrated in Figure 10.4a.

Next, consider industrial carbon management (ICM) defined in the restrictive sense as including pre-emissions controls only (Section 10.2.2). For low levels of mitigation, the COM for ICM is higher than for conventional mitigation, but the marginal cost of carbon management is expected to rise more slowly. The addition of ICM to conventional mitigation is thus expected to substantially lower the cost of large emission reductions, as is shown schematically in Figure 10.4a. However, no matter how optimistically one assesses ICM technologies, the marginal COM will still rise steeply as one approaches 100% mitigation due to the difficulty of wringing the last high-marginal-cost emissions from the energy system.

Finally, consider geoengineering of CO_2 by enhancement of biological sinks or by physical/chemical methods. As with albedo modification, there is no link to the scale of current anthropogenic emissions. Rather, each kind of sink will have its own intrinsic scale determined by the relevant biogeochemistry. Marginal COM for each sink will rise as one demands an amount of mitigation beyond its intrinsic scale. Figure 10.4b shows examples of plausible marginal cost functions.

Examination of the marginal COM functions illuminates the question of whether enhancement of biological carbon sinks is a form of geoengineering. Industrial carbon management is like conventional mitigation in that it is tied

---

[16] While space-based albedo modification is much more expensive, both stratospheric and space based albedo modification have large initial fixed costs and likely decreasing marginal costs.

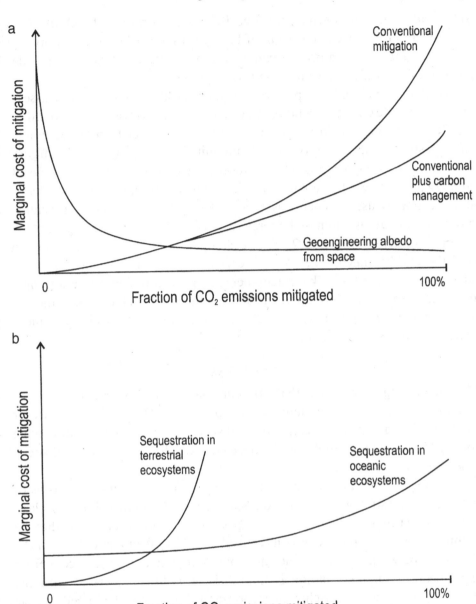

Figure 10.4   Schematic comparison between modes of mitigation. (a) Conventional mitigation means any method other than geoengineering or carbon management; e.g., conservation or use of non-fossil energy. The addition of carbon management lowers the cost of emissions mitigation, however, costs will still rise steeply as one tries to eliminate all emissions. Conversely, albedo modification from space has a very high initial capital costs, but can provide essentially unlimited effect low marginal cost. (b) Sequestration based on ecosystem modification will have costs that rise steeply at mitigation amount (carbon flux) set by the internal dynamics of the respective systems.

to the scale of anthropogenic emissions. In contrast, removal of CO_2 from the atmosphere, either by enhancement of biological sinks or by other methods, is like geoengineering of albedo because as a countervailing measure it is independent of the scale of anthropogenic emissions.

Geoengineering might, in principle, be incorporated into integrated assessments of climate change as a fallback strategy that supplies an upper bound on the COM. In this context a fallback strategy must either be more certain of effect, faster to implement, or provide unlimited mitigation at fixed marginal cost. Various geoengineering schemes meet each of these criteria. The *fallback strategy* defined here for integrated assessment is a generalization of a backstop technology used in energy system modeling, where it denotes a technology that can supply unlimited energy at fixed (usually high) marginal cost. Fallback strategies will enter if climate change is more severe than we expect or if the COM is much larger than we expect (Keith and Dowlatabadi, 1992). The existence of a fallback strategy permits more confidence in adopting a moderate response to the climate problem: without fallback options a moderate response is risky given the possibility of a strong climatic response to moderate levels of fossil-fuel combustion.

10.5.2 Risk

Geoengineering poses risks that combine natural and social aspects. For example, will stratospheric aerosols destroy ozone? Will the availability or implementation of geoengineering prevent sustained action to mitigate climate forcing? Here we focus on the technical risks, and defer consideration of social risks to the following section.

The biogeochemical risks differ markedly for the two principal classes of geoengineering strategy – albedo modification and CO_2 control. For each class, risks may be roughly divided into two types: risk of side effect and risk that the manipulation will fail to achieve its central aim. For albedo modification, the division is between side effects such as ozone depletion, that arise directly from the albedo-modifying technology, and risk of failure associated with the difficulty of predicting the climatic response to changes in albedo. Side effects of CO_2 control include loss of biodiversity or loss of aesthetic value that may arise from manipulating ecosystems to capture carbon, and risk of failure is associated with unexpectedly quick re-release of sequestered carbon.

The risks posed by geoengineering are sufficiently novel that, in general, the relevant biological and geophysical science is too uncertain to allow quantitative assessment of risk. Absent quantitative assessment, various avenues remain for robust qualitative risk assessment, for example, if a geoengineering scheme works by imitating a natural process we can compare the magnitude of the engineered effect with the magnitude and variability of the natural process, and then assume that similar perturbations entail similar results (Keith and Dowlatabadi, 1992; Michaelson, 1998). For example, the amount of sulfate released into the stratosphere as part of a geoengineering scheme and the amount released by a large volcanic eruption are similar. We may estimate the magnitude of stratospheric ozone loss by analogy.

In decisions about implementation, judgment about the risks of geoengineering would depend on the scalability and reversibility of the project. Can the project be tested at small scale, and can the project be readily reversed if it goes awry? These attributes are vital to enabling management of risk through some form of global-scale adaptive ecological management (Gunderson, Holling *et al.*, 1995; Allenby, 1999). Even crude qualitative estimates of risk can give insight into the relative merits of various geoengineering methods when considered in conjunction with other variables (Keith and Dowlatabadi, 1992).

We have examined the risk of geoengineering in isolation. More relevant to real choices about planetary management is a comparison of the risks and benefits of geoengineering with those of other response strategies. Here we are in unexplored territory as the literature has largely avoided this question. Without attempting such a comparison, we note that it would have to be explicit about the goals; i.e., is geoengineering a substitute for abatement, an addition to abatement, or a fallback strategy? Also, it would have to assess the risks of abatement or adaptation *per se*.

10.5.3 Politics and law

The politics of geoengineering rests on two central themes: the first emerges from the fact that many geoengineering schemes are amenable to implementation by independent action, whereas the second relates to geoengineering's status as a form of moral hazard. First consider independent action. Unlike other responses to climate change (e.g., abatement or adaptation), geoengineering could be implemented by one or a few countries acting alone. Various political concerns arise from this fact with respect to security, sovereignty, and liability; they are briefly summarized below.

Some geoengineering schemes raise direct security concerns; solar shields, for example, might be used as offensive weapons. A subtler, but perhaps more important security concern arises from the growing links between environmental change and security. Whether or not they were actually responsible, the operators of a geoengineering project could be blamed for harmful climatic events that could plausibly be attributed by an aggrieved party to the project. Given the current political disputes arising from issues such as the depletion of fisheries and aquifers, it seems plausible that a unilateral geoengineering project could lead to significant political tension.

International law would bear on these security and liability concerns. Bodansky (1996) points out that existing laws may cover several specific proposals; for example, the fertilization of Antarctic waters would fall under the Antarctic Treaty System, and the use of space-based shields would fall under the Outer Space Treaty of 1967. In addition, the IPCC95 report argues that many geoengineering methods might be covered by the 1977 treaty prohibiting the hostile use of environmental modification.

As in the current negotiations under the FCCC, geoengineering would raise questions of equity. Schelling (1996) has argued that in this case geoengineering might simplify the politics; geoengineering ". . . totally transforms the greenhouse issue from an exceedingly complicated regulatory regime to a simple – not necessarily easy but simple – problem in international cost sharing".

One must note that not all geoengineering methods are amenable to centralized implementation, in particular, most albedo modification methods are, while control of greenhouse gases generally is not.

Separate from the possibility of independent action is the concern that geoengineering may present a moral hazard. The root problem is simple: would mere knowledge of a geoengineering method that was demonstrably low in cost and risk weaken the political will to mitigate anthropogenic climate forcing? Knowledge of geoengineering has been characterized as an insurance strategy; in analogy with the moral hazard posed by collective insurance schemes, which encourage behavior that is individually advantageous but not socially optimal, we may ascribe an analogous hazard to geoengineering if it encourages suboptimal investment in mitigation. As the following examples demonstrate, geoengineering may pose a moral hazard whether or not its implementation is in fact a socially optimal strategy.

If the existence of low-cost biological sinks encourages postponement of effective action on emissions mitigation, and if such sinks prove leaky then the existence of these sinks poses a moral hazard.

To illustrate that geoengineering may be optimal yet still present a moral hazard, suppose that two or three decades hence real collective action is underway to reduce CO_2 emissions under a binding agreement that limits peak atmospheric CO_2 concentrations to 600 ppmv and which mandates that concentrations will be reduced to less than 450 ppmv by some fixed date. Suppose further that both the cost of mitigation and the climate sensitivity turn out to be higher than we now anticipate and that the political coalition supporting the agreement is just strong enough to sustain the actions necessary to meet the concentration targets, but is not strong enough to support lowering of the targets. Finally, suppose that a temporary space-based albedo modification system is proposed that will limit climate impacts during the period of peak CO_2 concentrations. Even if strong arguments can be made that the albedo modification is truly a socially optimal strategy, it may still present a moral hazard if its implementation encourages a retreat from agreed stringent action on mitigation.

The status of geoengineering as a moral hazard may partially explain the paucity of serious analysis on the topic. Within the policy analysis community, for example, there has been vigorous debate about whether discussion of geo-engineering should be included in public reports that outline possible responses to climate change, with fears voiced that its inclusion could influence policy makers to take it too seriously and perhaps defer action on abatement, given knowledge of geoengineering as an alternative (Schneider, 1996; Watson, Zinyowera *et al.*, 1996).

10.5.4 *Environmental ethics*

Discussion of the advisability of geoengineering has been almost exclusively limited to statements about risk and cost. Although ethics is often mentioned, the arguments actually advanced have focused on risk and uncertainty; serious ethical arguments about geoengineering are almost nonexistent. Many of the objections to geoengineering that are cited as ethical have an essentially prag-matic basis. Three common ones are:

*   *The slippery slope argument.* If we choose geoengineering solutions to counter anthropogenic climate change, we open the door to future efforts to systematically alter the global environment to suit humans. This is a pragmatic argument, unless one can define why such large-scale environmental manipulation is bad, and how it differs from what humanity is already doing.
*   *The technical fix argument.* Geoengineering is a "technical fix", "kluge", or "end-of-pipe solution". Rather than attacking the problems caused by fossil fuel com-bustion at their source, geoengineering aims to add new technology to counter their side effects. Such solutions are commonly viewed as inherently undesirable – but not for ethical reasons.
*   *The unpredictability argument.* Geoengineering entails "messing with" a complex, poorly understood system; because we cannot reliably predict results it is unethi-cal to geoengineer. Because we are already perturbing the climate system willy-nilly with consequences that are unpredictable, this argument depends on the notion that intentional manipulation is inherently worse than manipulation that occurs as a side effect.

These concerns are undoubtedly substantive, yet they do not exhaust the underlying feeling of abhorrence that many people feel for geoengineering. As a first step toward discussion of the underlying objections one may analyze geoengineering using common ethical norms; for example, one could consider the effects of geoengineering on intergenerational equity or on the rights of minorities. Such an analysis, however, can say nothing unique about geoengineering because other responses to the CO_2-climate problem entail similar effects. I sketch two modes of analysis that might be extended to address some of the underlying concerns about geoengineering. The first concerns the eroding distinction between natural and artificial and the second the possibility of an integrative environmental ethic.

The deliberate management of the environment on a global scale would, at least in part, force us to view the biosphere as an artifact. It would force a re-examination of the distinction between natural science and what Simon (1996) called "the sciences of the artificial" – that is, engineering and the social sciences. The inadvertent impact of human technology has already made the distinction between managed and natural ecosystems more one of degree than of kind, but in the absence of planetary geoengineering it is still possible to imagine them as clearly distinct (Smil, 1985; Allenby, 1999). The importance of, and the need for, a sharp distinction between natural and artificial, between humanity and our technology was described by Tribe in analyzing concerns about the creation of artificial environments to substitute for natural ones (Tribe, 1973; Tribe, 1974).

The simplest formulations of environmental ethics proceed by extension of common ethical principles that apply between humans. A result is "animal rights" (Singer, 1990) or one of its variants (Regan, 1983). Such formulations locate "rights" or "moral value" in individuals. When applied to a large-scale problem such as the choice to geoengineer, an ethical analysis based on individuals reduces to a problem of weighing conflicting rights or utility. As with analyses that are based on more traditional ethical norms, such analysis has no specific bearing on geoengineering.

In order to directly address the ethical consequences of geoengineering one might desire an integrative formulation of environmental ethics that located moral value at a level beyond the individual, a theory that ascribed value to collective entities such as a species or a biotic community. Several authors have attempted to construct integrative formulations of environmental ethics (Taylor, 1986; Norton, 1987; Callicott, 1989), but it is problematic to build such a theory while maintaining an individualistic conception of human ethics (Callicott, 1989), and no widely accepted formulation has yet emerged.

10.6 Summary and Implications

A casual look at the last few decades of debate about the CO_2-climate problem might lead one to view geoengineering as a passing aberration; an idea that originated with a few speculative papers in the 1970s, that reach a peak of public exposure with the NAS92 assessment and the contemporaneous American Geophysical Union and American Association for the Advancement of Science colloquia of the early 1990s, an idea that is now fading from view as international commitment to substantive action on climate grows ever stronger. The absence of debate about geoengineering in the analysis and negotiations surrounding the FCCC supports this interpretation. However, I argue that this view is far too simplistic. First, consider that scientific understanding of climate has co-evolved with knowledge of anthropogenic climate impacts, with speculation about the means to manipulate climate, and with growing technological power that grants the ability to put speculation into practice. The history of this co-evolution runs through the century, from Eckholm's speculation about the benefits of accelerated fossil fuel use, to our growing knowledge about the importance of iron as a limiting factor in ocean ecosystem productivity.

This view of climate history is in accord with current understanding of the history of science that sees the drive to manipulate nature to suit human ends as integral to the process by which knowledge is accumulated. In this view, the drive to impose human rationality on the disorder of nature by technological means constitutes a central element of the modernist program. This link between understanding and manipulation is clearly evident in the work of Francis Bacon that is often cited as a signal of the rise of modernism in the seventeenth century.

Moreover, the disappearance of the term geoengineering from the mainstream of debate, as represented by the FCCC and IPCC processes, does not signal the disappearance of the issue. The converse is closer to the truth: use of the term has waned as some technologies that were formerly called geoengineering have gained acceptance.

To illustrate the point, consider the shifting meaning of carbon management. The recent Department of Energy "roadmap", an important agency-wide study of "Carbon Sequestration Research and Development"(Reichle, Houghton *et al.*, 1999) serves as an example. The report uses a very broad definition of carbon management that includes (a) demand-side regulation through improved energy efficiency, (b) decarbonization via use of low-carbon and carbon-free fuels or nonfossil energy, and (c) carbon sequestration by any means, including not only carbon capture and sequestration prior to atmospheric emission, but all means by which carbon may be captured from the atmosphere. Although the report avoids a single use of the word *geoengineering* in the body of the text, one may argue from its broad definition of carbon management that the authors implicitly adopted a definition of *geoengineering* that is restricted to modifications to the climate system by any means other than manipulation of CO_2 concentration.

In this review, in contrast, I have drawn the line between geoengineering and industrial carbon management at the emission of CO_2 to the active biosphere. Three lines of argument support this definition. First, and most importantly, the capture of CO_2 from the atmosphere is a countervailing measure, one of the three hallmarks of geoengineering identified in Section 10.2.1. It is an effort to counteract emissions, and thus to control CO_2 concentrations, through enhancement of ecosystem productivity or through the creation of new industrial processes. These methods are unrelated to the use of fossil energy except in that they aim to counter its effects (Section 10.5.1). The second argument is from historical usage (Section 10.3.5); the capture of CO_2 from the atmosphere has been treated explicitly as geoengineering (MacCracken, 1991; Keith and Dowlatabadi, 1992; Watson, Zinyowera et al., 1996; Flannery, Kheshgi et al., 1997; Michaelson, 1998) or has been classified separately from emissions abatement and grouped with methods that are now called geoengineering. Finally, the distinction between pre- and post-emission control of CO_2 makes sense because it will play a central role in both the technical and political details of implementation.

As a purely semantic debate, these distinctions are of little relevance. Rather, their import is the recognition that there is a continuum of human responses to the climate problem that vary in resemblance to hard geoengineering schemes such as spaced-based mirrors. The *de facto* redefinition of geoengineering to exclude the response modes that currently seem worthy of serious consideration, and to include only the most objectionable proposals, suggests that we are moving down the continuum toward acceptance of actions that have the character of geoengineering (as defined here) though they no longer bear the name. The disappearance of geoengineering thus signals a lamentable absence of debate about the appropriate extent of human intervention in the management of planetary systems, rather than a rejection of such intervention.

Consider, for example, the perceived merits of industrial and biological sequestration. In the environmental community (as represented by environmental nongovernment agencies) biological sequestration is widely accepted as a response to the CO_2-climate problem. It has been praised for its multiple benefits such as forest preservation and the possible enrichment of poor nations via the Clean Development Mechanism of the FCCC. Conversely, industrial sequestration has been viewed more skeptically as an end-of-pipe solution that avoids the root problems. Yet, I have argued here that biological sequestration – if adopted on a scale sufficient for it to play an important role – resembles geoengineering more than does industrial sequestration. Whereas industrial sequestration is an end-of-pipe solution, biological sequestration might reasonably be called a beyond-the-pipe solution. Such analysis cannot settle the question; it merely highlights the importance of explicit debate about the implications of countervailing measures.

Looking farther ahead, I speculate that views of the CO_2-climate problem may shift from the current conception in which CO_2 emission is seen as a pollutant to be eliminated, albeit a pollutant with millennial timescale and global impact, toward a conception in which CO_2 concentration and climate are seen as elements of the Earth system to be actively managed. In concluding the introduction to the 1977 NAS assessment, the authors speculated on this question, asking "In the light of a rapidly expanding knowledge and interest in natural climatic change, perhaps the question that should be addressed soon is 'What should the atmospheric carbon dioxide content be over the next century or two to achieve an optimum global climate?' Sooner or later, we are likely to be confronted by that issue." (NAS77, p. ix).

Allenby argues that we ought to begin such active management (Allenby, 1999). Moreover, he argues that failure to engage in explicit "Earth system engineering and management" will impair the effectiveness of our environmental problem solving. If we take this step, then the upshot will be that predicted in NAS83: "Interest in CO_2 may generate or reinforce a lasting interest in national or international means of climate and weather modification; once generated, that interest may flourish independent of whatever is done about CO_2".

Although the need for improved environmental problem solving is undeniable, I judge that great caution is warranted. Humanity may inevitably grow into active planetary management, yet we would be wise to begin with a renewed commitment to reduce our interference in natural systems rather than to act by balancing one interference with another.

Acknowledgments

I thank Alex Farrell, Tim Johnson, Anthony Keith, Granger Morgan, Ted Parson, Peter Reinelt and Robert Socolow for their useful comments. The work was supported in part by NSF grant SES-9022738.

References

Allenby, B. (1999). Earth systems engineering: the role of industrial ecology in an engineered world. *J. Industrial Ecology* **2**: 73–93.

Archer, D., H. Kheshgi, and E. Maier-Reimer (1997). Multiple timescales for neutralization of fossil fuel CO_2. *Geophys. Res. Let.* **24**: 405–408.

Arrhenius, S. (1908). *Worlds in The Making: The Evolution of the Universe*. New York, Harper & Brothers.

Behrenfeld, M. J. and Z. S. Kolber (1999). Widespread iron limitation of phytoplankton in the South Pacific Ocean. *Science* **283**: 840–843.

Bodansky, D. (1996). May we engineer the climate?. *Climatic Change* **33**: 309–321.

Bruce, J. P., H. Lee, and E. F. Haites, eds. (1996). *Climate change 1995: economic and social dimensions of climate change*. Cambridge, Cambridge University Press.

Budyko, M. I. (1982). *The Earth's climate, past and future*. New York, Academic Press.

Callicott, J. B. (1989). *In defense of the land ethic: essays in environmental philosophy*. Albany, NY, SUNY press.

Carbon Dioxide Assessment Committee (1983). *Changing climate*. Washington, DC, National Academy Press.

Coale, K. H., K. S. Johnson, S. E. Fitzwater, S. D. G. Blain, T. P. Stanton, and T. L. Coley (1998). IronEx-I, an in situ iron-enrichment experiment: experimental design, implementation and results. *Deep-Sea Research* **45**: 919–945.

Committee on Atmospheric Sciences (1962). *The atmospheric sciences 1961–1971*. Washington, DC, National Academy of Sciences.

Committee on Atmospheric Sciences (1966). *Weather and climate modification problems and prospects; final report of the Panel on Weather and Climate Modification*. Washington, DC, National Academy of Sciences.

Committee on Atmospheric Sciences (1971). *The atmospheric sciences and man's needs; priorities for the future*. Washington, DC, National Academy of Sciences.

Committee on Satellite Power Systems (1981). *Electric power from orbit: a critique of a satellite power system: a report*. Washington, DC, National Academy Press.

Cooper, D. J., A. J. Watson, and D. D. Nightingale (1996). Large decrease in ocean-surface CO_2 fugacity in response to in situ iron fertilization. *Nature* **383**: 511–513.

DeLapp, R. A. (1997). *The politics of weather modification: shifting coalitions*. Department of Political Science, Fort Collins, CO, Colorado State University.

Dickinson, R. E. (1996). Climate engineering: a review of aerosol approaches to changing the global energy balance. *Climatic Change* **33**: 279–290.

Dyson, F. J. (1977). Can we control the carbon dioxide in the atmosphere? *Energy* **2**: 287–291.

Early, J. T. (1989). Space-based solar shield to offset greenhouse effect. *J. Brit. Interplanet. Soc.* **42**: 567–569.

Ekholm, N. (1901). On the variations of the climate of the geological and historical past and their causes. *Quart. J. Roy. Meteorol. Soc.* **27**: 1–61.

Flannery, B. P., H. Kheshgi, G. Marland, and M. C. MacCracken (1997). Geoengineering climate. In *Engineering Response to Global Climate Change*. R. G. Watts (ed.). Boca Raton, Lewis: 379–427.

Fleming, J. R. (1998). *Historical Perspectives on Climate Change*. New York, Oxford University Press.

Fletcher, J. O. (1968). *Changing Climate*. Santa Monica, CA, Rand Corp.: 27.

Fogg, M. J. (1995). *Terraforming: Engineering Planetary Environments*. Warrendale, PA, Society of Automotive Engineers.

Geophysics Study Committee (1977). *Energy and Climate*. Washington, D.C., National Academy of Sciences.

Gove, P. B., ed. (1986). *Webster's Third New International Dictionary of the English Language Unabridged*. Springfield, MA, Merriam-Webster.

Govindasamy, B. and K. Caldeira (2000). Geoengineering Earth's radiation balance to mitigate CO_2-induced climate change. *Geophys. Res. Let.* **27**: 2141–2144.

Green, F. (1977). *A Change in the Weather*. New York, Norton.

Gunderson, L. H., C. S. Holling, and S. S. Light, eds. (1995). *Barriers and Bridges to the Renewal of Ecosystems and Institutions*. New York, NY, Columbia University Press.

Hargrove, E. C. (1986). *Beyond Spaceship Earth: Environmental Ethics and the Solar System*. San Francisco, CA, Sierra Club Books.

Herzog, H., E. Drake, and E. Adams (1997). *CO_2 Capture, Reuse, and Storage Technologies for Mitigating Global Climate Change*. Washington, DC, Department of Energy.

Intergovernmental Panel on Climate Change (1991). *Climate Change: the IPCC Response Strategies*. Washington, DC, Island Press.

Joint Publications Research Service (1963). Review of research and development in cloud physics and weather modification. *Soviet-Block Research in Geophysics, Astronomy, and Space* **83**.

Keith, D. W. and H. Dowlatabadi (1992). Taking geoengineering seriously. *Eos, Transactions American Geophysical Union* **73**: 289–293.

Kheshgi, H. S. (1995). Sequestering atmospheric carbon-dioxide by increasing ocean alkalinity. *Energy* **20**: 915–922.

Kitzinger, U. and E. G. Frankel, eds. (1998). *Macro-engineering and the Earth: World Projects for the Year 2000 and Beyond*. Chichester, UK, Horwood Publishing.

Lackner, K., C. H. Wendt, D. P. Butt, E. L. Joyce, and D. H. Sharp (1995). Carbon dioxide disposal in carbonate minerals. *Energy* **20**: 1153–1170.

Lashof, D. A. and D. A. Tirpak (1990). *Policy Options for Stabilizing Global Climate*. New York, NY, Hemisphere.

MacCracken, M. C. (1991). *Geoengineering the Climate*. Livermore CA, Lawrence Livermore National Laboratory: 13.

MacDonald, G. J. F. (1968). How to wreck the environment. In *Unless Peace Comes; a Scientific Forecast of New Weapons*. N. Calder (ed.), Viking Press: 181–205.

Marchetti, C. (1977). On geoengineering and the CO_2 problem. *Climate Change* **1**: 59–68.

Martin, J. H., K. H. Coale, and K. S. Johnson (1994). Testing the iron hypothesis in ecosystems of the equatorial Pacific-Ocean. *Nature* **371**: 123–129.

McKay, C. P., O. B. Toon, and J. F. Kasting (1991). Making Mars habitable. *Nature* **352**: 489–496.

Michaelson, J. (1998). Geoengineering: a climate change Manhattan project. *Stanford Environmental Law Journal* **17**: 73–140.

Monastersky, R. (1995). Iron versus the greenhouse. *Science News* **148**: 220–222.

National Research Council (1979). *Carbon Dioxide and Climate: A Scientific Assessment*. Washington, DC, National Academy Press.

Norton, B. G. (1987). *Why Preserve Natural Variety*. Princeton, NJ, Princeton University Press.

Office of Technology Assessment (1991). *Changing by Degrees: Steps to Reduce Greenhouse Gases.* Washington, DC, Office of Technology Assessment.

Orville, H. T. (1957). *Final Report of the Advisory Committee on Weather Control.* Washington, DC, US Gov. Print Office.

Panel on Policy Implications of Greenhouse Warming (1992). *Policy Implications of Greenhouse Warming: Mitigation, Adaptation, and the Science Base.* Washington, D.C., National Academy Press.

Parson, E. A. and D. W. Keith (1998). Climate change – fossil fuels without CO_2 emissions. *Science* **282**: 1053–1054.

Peng, T. H. and W. S. Broecker (1991). Dynamic limitations on the Antarctic iron fertilization strategy. *Nature* **349**: 227–229.

President's Science Advisory Committee (1965). *Restoring the Quality of Our Environment.* Washington, DC, Executive office of the president.

Regan, T. (1983). *The Case for Animal Rights.* Berkeley, CA, UC press.

Reichle, D., J. Houghton, B. Kane, and J. Eckmann (1999). *Carbon Sequestration Science and Development.* Washington, DC, Department of Energy.

Rosenberg, N. J., R. C. Izaurralde, and E. L. Malone (1998). *Carbon Sequestration in Soils: Science, Monitoring, and Beyond.* Columbus, OH, Battelle.

Rusin, N. and L. Flit (1960). *Man Versus Climate.* Moscow, Peace Publishers.

Sagan, C. (1961). The planet Venus. *Science* **133**: 849–858.

Sagan, C. (1973). Planetary engineering on Mars. *Icarus* **20**: 513–514.

Schaefer, V. J. (1946). The production of ice crystals in a cloud of supercooled water droplets. *Science* **104**: 457–459.

Schelling, T. C. (1996). The economic diplomacy of geoengineering. *Climatic Change* **33**: 303–307.

Schneider, S. H. (1996a). Engineering change in global climate. *Forum App. Res. Publ. Pol.* **11**: 92–96.

Schneider, S. H. (1996b). Geoengineering: could or should we do it? *Climatic Change* **33**: 291–302.

Seidel, S. and D. L. Keyes (1983). *Can we Delay a Greenhouse Warming?* Washington, D.C., Environmental Protection Agency.

Seifritz, W. (1989). Mirrors to halt global warming? *Nature* **340**: 603.

Simon, H. A. (1996). *The Sciences of the Artificial.* Cambridge, MA., MIT Press.

Singer, P. (1990). *Animal Liberation.* New York, Avon.

Smil, V. (1985). *Carbon Nitrogen Sulfur: Human Interference in Grand Biospheric Cycles.* New York, NY, Plenum Press.

Study of Critical Environmental Problems (1970). *Man's Impact on the Global Environment: Assessment and Recommendations for Action.* Cambridge, MA, MIT Press.

Taubenfeld, H. J., ed. (1970). *Controlling the Weather: a Study of Law and Regulatory Procedures.* New York, NY, Dunellen.

Taylor, P. W. (1986). *Respect for Nature: a Theory of Environmental Ethics.* Princeton, NJ, Princeton University Press.

Teller, E., K. Caldeira, and G. Canavan (1999). *Long-Range Weather Prediction and Prevention of Climate Catastrophes: a Status Report.* Livermore, CA, Lawrence Livermore National Laboratory: 41.

Teller, E., L. Wood, and R. Hyde (1997). *Global Warming and Ice Ages: I. Prospects for Physics Based Modulation of Global Change.* Livermore, CA, Lawrence Livermore National Laboratory: 20.

Tribe, L. H. (1973). Technology assessment and the fourth discontinuity: the limits of instrumental rationality. *So. Cal. Law. Rev.* **46**: 617–660.

Tribe, L. (1974). Ways not to think about plastic trees: new foundations for enviromental law. *Yale Law J.* **83**: 1325–1327.

United Nations (1976). Convention on the prohibition of military or any other hostile use of environmental modification techniques. *United Nations, Treaty Series* **1108**: 151.

Watson, A. J. (1997). Volcanic iron, CO_2, ocean productivity and climate. *Nature* **385**: 587–588.

Watson, R. T., M. C. Zinyowera, and R. H. Moss, eds. (1996). *Climate Change, 1995: Impacts, Adaptations, and Mitigation of Climate Change: Scientific-Technical Analyses*. Cambridge, UK, Cambridge University Press.

Wexler, H. (1958). Modifying weather on a large scale. *Science* **128**: 1059–1063.

Yudin, M. I. (1966). The possibilities for influencing large scale atmospheric movements. *Modern Problems of Climatology*. M. Budyko. Wright-Patterson AFB, Foreign Technology Division.

Zikeev, N. T. and G. A. Doumani (1967). *Weather Modification in the Soviet Union, 1946-1966: a Selected Annotated Bibliography*. Washington, DC, Library of Congress, Science and Technology Division.

Index